Biomedical Physics

Biomedical Physics

Edited by **Thomas Jackson**

SYRAWOOD
PUBLISHING HOUSE
New York

Published by Syrawood Publishing House,
750 Third Avenue, 9th Floor,
New York, NY 10017, USA
www.syrawoodpublishinghouse.com

Biomedical Physics
Edited by Thomas Jackson

International Standard Book Number: 978-1-68286-292-6 (Hardback)

Printed in China.

Contents

Preface

The theories and applications of various concepts in physics are widely used in contemporary healthcare and medical studies across the world. Biomedical physics as a specialized field is aimed at perceiving the underlying principles and concepts of physics in bio-molecular structure, nano-medicine, nuclear medicine, etc. It plays a key role in understanding techniques and principles used in molecular imaging (cells and tissues) as well as medical imaging (pathology and anatomy through CT, MRI, etc.). This book elucidates the significant issues and studies performed by renowned experts and practitioners from various countries. It also elucidates diverse applications of theories from physics in analyzing biological systems and their interactions, cancer and oncology, neuroscience and brain imaging, computational research and imaging, etc. This text is a very useful reference tool for postgraduate students, medical practitioners, scientists, etc. who are looking for an updated and comprehensive resource guide on biomedical physics.

After months of intensive research and writing, this book is the end result of all who devoted their time and efforts in the initiation and progress of this book. It will surely be a source of reference in enhancing the required knowledge of the new developments in the area. During the course of developing this book, certain measures such as accuracy, authenticity and research focused analytical studies were given preference in order to produce a comprehensive book in the area of study.

This book would not have been possible without the efforts of the authors and the publisher. I extend my sincere thanks to them. Secondly, I express my gratitude to my family and well-wishers. And most importantly, I thank my students for constantly expressing their willingness and curiosity in enhancing their knowledge in the field, which encourages me to take up further research projects for the advancement of the area.

Editor

1

Modeling domain formation of MARCKS and protein kinase C at cellular membranes

Sergio Alonso* and Markus Bär

*Correspondence:
sergio.alonso@ptb.de
Physikalisch-Technische Bundesanstalt, Abbestrasse 2-12, 10587 Berlin, Germany

Abstract

Background: Phosphorylation and dephosphorylation of proteins are mechanisms of activation and deactivation which regulate many cellular processes. Both mechanisms have been usually described in well mixed environments. MARCKS is a protein which binds to the membrane by electrostatic interaction. It is translocated from the membrane and phosphorylated by Protein Kinase C. Back in the cytoplasm the translocated MARCKS proteins are dephosphorylated by the enzyme phosphatase and can reattach to the membrane. These three processes (membrane binding, translocation, dephosphorylation) give rise to a cyclic dynamics known as the myristoyl-electrostatic switch.

Methods: We employ a reaction-diffusion model for the concentrations of MARCKS and PKC in the cell in a circular domain.

Results: Herein, we start from a reaction-diffusion model taking into account mass conservation of the MARCKS proteins. Then, we extend the model by including the dynamics of binding and unbinding of PKC enzymes, which are in turn activated by spikes of calcium. Furthermore, we show that the model fits previous experimental results well and predicts, in addition, the formation of domains with high concentration of MARCKS proteins at the membrane.

Conclusions: We have developed a simple model of binding, phosphorylation and desphosphorylation for MARCKS protein. The main prediction emerging from numerical simulations of the model is the spontaneous appearance of domains of high concentration of membrane proteins.

Keywords: Reaction-diffusion models; Nonlinear dynamics; Phosphorylation; Membrane domains; Membrane proteins; Pattern formation

Background

Living cells process external perturbations by biochemical pathways which are regulated by complex metabolic networks. If a receptor at the membrane is activated by an external signal, a cascade of biochemical reactions and interactions inside the cell is triggered. Proteins, enzymes and small molecules interact by mutual activation and inhibition. Multiple molecules participate in metabolic networks, where phosphatases and kinases are particularly active in cascades that involve phosphorylation and dephosphorylation of proteins involved in the signaling [1].

The number of components of such networks is usually very large. To model the relations among the components, the interactions are simplified and the pathways are described by ordinary differential equations of the component concentrations in the cell. However, some molecules and reactions are restricted to particular compartments of the cell [2]. As a result, the spatio-temporal distribution of proteins in the cell is important [3]. The communication between steps of the pathways restricted to different compartments is typically achieved by diffusion and active transport of the involved molecules. For sufficiently high molecular concentrations, the spatial distribution can be modelled by reaction-diffusion equations. These equations permit also the implementation of spatially inhomogeneous conditions, which reflect the differences among separate compartments inside the cell.

The myristoylated alanine-rich C kinase substrate (MARCKS) is a charged and unfolded protein which appears in living cells at high concentrations of around $10\mu M$ [4]. This protein is active in many processes occurring at the cell membrane demanding the regulation of cytosketetal dynamics, e.g. cell movement [5], phagocytosis [6], exocytosis [7], and chemotaxis [8]. MARCKS is a hydrophilic protein which can bind to the membrane by electrostatic interactions with acidic lipids like phosphatidylinositol 4,5-biphosphate (PIP_2) or phosphatidylserine (PS) [4]. The functions of MARCKS are correlated with its ability to sequester PIP_2 at the membranes and to avoid thereby its hydrolysis by phospholipase C (PLC). Thus, MARCKS prohibits the generation of two important second messengers: inositol 1,4,5-triphosphate (IP_3) and diacylglycerol (DAG), which participate in the regulation of calcium (Ca^{2+}) in the cell.

MARCKS proteins lose their affinity to membranes when they are phosphorylated by protein kinase C (PKC). The phosphates reduce the positive charge of the protein and cause the unbinding from the membrane to the cytoplasm. In the cytoplasm, phosphatases remove the phosphates again from the protein. Consequently, MARCKS can bind again at the membrane. This cyclic dynamics of MARCKS binding and unbinding is known as the myristoyl-electrostatic switch [9]. It is controlled by Ca^{2+} which activates and promotes the binding of PKC at the membrane, where it phosphorylates MARCKS proteins. Therefore, phosphorylation and dephosphorylation of MARCKS appear in different regions of the cell, in contrast to the earlier Goldbeter-Koshland model of protein phosphorylation and dephosphorylation [10,11], where these processes are assumed to occur in a well-mixed environment inside the cell. The localization of both processes imposes gradients inside the cell [3] and, as we discuss below, may produce the polarization of the cell.

The experimental study of the interactions among MARCKS proteins, phospholipids and PKC has followed three different strategies: in vitro experiments wherein MARCKS interacts with vesicles composed by neutral and acidic phospholipids [12-17], in vitro experiments wherein MARCKS and PKC interacts with a Langmuir monolayer formed by a mixture of acidic and neutral phospholipids [18-21], and in vivo experiments with living cells subject to external perturbations [22-25]. In vitro experiments allow to control the phospholipid distributions on vesicles and in Langmuir monolayers and the concentrations of the proteins. They permit the characterization of the protein response to different concentrations of acidic phospholipids and the realization of dynamical experiments by the introduction of activated PKC [15,21]. A recent study for the interaction of MARCKS and PKC with a Langmuir monolayer [21] shows oscillations based on the nonlinear nature of protein-phopholipid interactions. In vivo experiments are more realistic, however, additional components in the cell may change the dynamics. The external introduction of a chemical signal produces variations of the Ca^{2+} concentration, and can activate PKC. Experiments with living cells show the co-distribution of domains of high MARCKS concentration at the membrane with domains of acidic lipids [26] or domains of calmodulin [27]. The interaction of lipids and MARCKS proteins can lead to formation of acidic lipids covered by electrostatically bound proteins [28].

Experiments with vesicles as well as in-vivo experiments have motivated the derivation of different reaction-diffusion models of the interaction of MARCKS with membranes [29,30]. These models conserve the total number of MARCKS proteins, i.e., neither synthesis nor degradation of proteins are considered, because the time scales studied here is small in comparison with those of protein synthesis. Such reaction-diffusion models with mass conservation have been previously employed in the study of other systems, for example to explain how cell polarity emerges [31-34] to describe the protein system that controls cellular division in the bacterium E. Coli [35-37].

In this paper, we add the dynamics of calcium and PKC to a mass-conserved reaction-diffusion model for binding, phosphorylation and dephosphorylation of MARCKS proteins in living cells. The enzyme PKC is activated by Ca^{2+}. The activated PKC binds at the membrane, where phosphorylates MARCKS proteins. In contrast with previous models [29,30], here, the parameter values were adapted to quantitatively reproduce the experimental results by Mogami et al. [23]. We compare the outcome of experiments in living cells after an increase of Ca^{2+} concentration with the numerical results obtained with our model following an equivalent increase of calcium. Then, we perform numerical simulations of the spatio-temporal dynamics keeping a constant Ca^{2+} concentration and the parameter values obtained by the comparison with experiments. Such calculations show the spontaneous formation of domains with high concentration of MARCKS proteins at the membrane. It suggests that the formation of domains of MARCKS proteins are likely to occur in living cells.

The manuscript is organized as follows. In the next section we describe the model for the PKC, MARCKS and Ca^{2+} spatio-temporal dynamics and the numerical methods employed for the integration of the resulting model. Then, we obtain the parameters of the model from the literature and by comparing the resulting numerical simulations with the experimental results obtained from [23]. Once the parameter values are chosen, extended numerical simulations show the formation of domains of MARCKS proteins at the cellular membranes by two mechanisms. We analyze both mechanisms and calculate the phase diagram. Finally, we discuss the main results of the manuscript and, in particular, their biological implication.

Methods

We employ a reaction-diffusion model for the concentrations of MARCKS and PKC in the cell. First only the temporal dynamics of the concentrations described by ordinary differential equations is considered, then the complete model is derived taking into account transport processes and the spatial aspects of the dynamics.

Temporal reaction dynamics

The binding and unbinding processes of MARCKS and PKC proteins, together with the phosphorylation and dephosphorylation reactions of MARCKS proteins are studied.

MARCKS dynamics

Phosphorylation and dephosphorylation of proteins are usually modeled by the Goldbeter-Koshland mechanism [10]. Using a quasi-steady approximation of the intermediate complexes, the original dynamics can be simplified to a couple of equations for the free and phosphorylated MARCKS (respectively $[MRK]_f$, and $[MRK]_p$) using Michaelis-Menten rates [11]:

$$\begin{aligned}\frac{d[MRK]_f}{dt} &= -k_{pkc}^{mrk}\frac{[MRK]_f}{k_{mrk}+[MRK]_f}[PKC] + K_{ppa}^{mrk}\frac{[MRK]_p}{k_1+[MRK]_p}[PPA],\\ \frac{d[MRK]_p}{dt} &= k_{pkc}^{mrk}\frac{[MRK]_f}{k_{mrk}+[MRK]_f}[PKC] - K_{ppa}^{mrk}\frac{[MRK]_p}{k_1+[MRK]_p}[PPA];\end{aligned} \tag{1}$$

where [PKC] and [PPA] are respectively the concentrations of the kinase and phosphatase, see sketch in Figure 1a. However, the phosphorylation of MARCKS is controlled by a binding-unbinding process. After considering $k_1 \gg [MRK]_p$ and defining $k_{ppa}^{mrk} = K_{ppa}^{mrk}[PPA]/k_1$, we extend Eq. (1) to a more adequate model of the evolution of the membrane, cytosolic and phosphorylated MARCKS (respectively $[MRK]_m$, $[MRK]_c$ and $[MRK]_p$), see sketch in Figure 1b:

$$\begin{aligned}\frac{d[MRK]_m}{dt} &= \mathcal{S}_m k_b^{mrk}[MRK]_c - k_u^{mrk}[MRK]_m - k_{pkc}^{mrk}\left(\frac{[MRK]_m}{k_{mrk}+[MRK]_m}\right)[PKC]_m,\\ \frac{d[MRK]_c}{dt} &= -\mathcal{S}_m k_b^{mrk}[MRK]_c + k_u^{mrk}[MRK]_m + k_{ppa}^{mrk}[MRK]_p,\\ \frac{d[MRK]_p}{dt} &= k_{pkc}^{mrk}\left(\frac{[MRK]_m}{k_{mrk}+[MRK]_m}\right)[PKC]_m - k_{ppa}^{mrk}[MRK]_p;\end{aligned} \tag{2}$$

where the total concentration of MARCKS is preserved. The terms k_b^{mrk} and k_u^{mrk} are the binding and unbinding rates of MARCKS. The rate of phosphorylation at the membrane and dephosphorylation in the cytoplasm are determined by the parameters k_{pkc}^{mrk} and k_{ppa}^{mrk}. The phosphorylation depends on the concentration of PKC at the membrane $[PKC]_m$ and, therefore, indirectly on Ca^{2+}. The dephosphorylation is assumed to be independent of $[Ca^{2+}]$ [38] because MARCKS proteins can be dephosphorylated by the catalytic subunit of protein phosphatase 2A (PP2Ac), and protein phosphatase 1 (PP1c), which activation is independent of Ca^{2+} [39]. A possible dependence of the phosphatase on Ca^{2+} would induce a reciprocal regulation of the enzymes producing more complex dynamics [40].

The binding of MARCKS depends on the surface term: $\mathcal{S}_m = (1-[MRK]_m/[MRK]_o)$, where the concentration $[MRK]_o$ is the saturation concentration, and corresponds to the theoretical concentration of MARCKS which completely covers the membrane. Assuming that $[MRK] \gg [PKC]$, $\mathcal{S}_m$ is the fraction of accessible space at the membrane.

PKC dynamics

We consider a simple binding and unbinding dynamics of the PKC enzymes at the membrane. The total concentration of enzymes is conserved giving rise to the next two coupled equations:

$$\begin{aligned}\frac{d[PKC]_m}{dt} &= \mathcal{S}_m k_b^{pkc}\left(\frac{[Ca^{2+}]}{k_{Ca}+[Ca^{2+}]}\right)[PKC]_c - k_u^{pkc}[PKC]_m,\\ \frac{d[PKC]_c}{dt} &= -\mathcal{S}_m k_b^{pkc}\left(\frac{[Ca^{2+}]}{k_{Ca}+[Ca^{2+}]}\right)[PKC]_c + k_u^{pkc}[PKC]_m;\end{aligned} \tag{3}$$

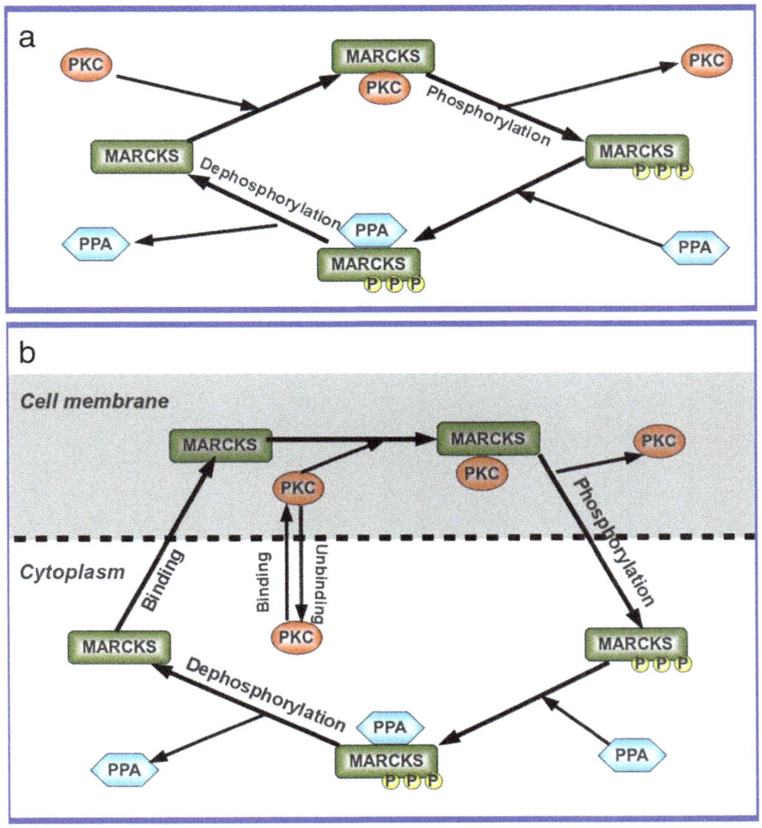

Figure 1 Sketches of the mechanisms. a) Sketch of the Goldbeter-Koshland mechanism of protein phosphorylation and dephosphorylation applied to MARCKS protein. **b)** Sketch of the myristoyl-electrostatic switch. The binding and unbinding of MARCKS and PKC proteins, and the phosphorylation and dephosphorylation of MARCKS proteins by kinase (PKC) and phosphatase (PPA) is shown.

where $[PKC]_m$ and $[PKC]_c$ are respectively the concentration of PKC at the membrane and in the cytoplasm. While we assume that the unbinding rate from the membrane k_u^{pkc} is constant, the binding rate k_b^{pkc} is modulated by the saturation factor $\mathcal{S}_m$.

The activity of PKC depends on the concentration of Ca^{2+}. The enzyme PKC is inactive under low concentration of Ca^{2+}. It becomes active for large concentrations and it binds at the membrane where it phosphorylates MARCKS proteins. The activation by Ca^{2+} follows a Michaelis-Menten dynamics.

Ca^{2+} dynamics

Usually, $[Ca^{2+}]$ is constant and stays at rest, however external signals or internal processes produce random spikes of large $[Ca^{2+}]$. In such case, the global dynamics of $[Ca^{2+}]$ is simply modeled by a relaxation dynamics to the rest state and random spike generation:

$$\frac{d[Ca^{2+}]}{dt} = -\frac{[Ca^{2+}] - [Ca^{2+}]_o}{\tau} + I(t); \tag{4}$$

where τ is the characteristic relaxation time to the rest state $[Ca^{2+}]_o$. This time is controlled by the concentration of buffers in the cell. The term $I(t)$ represents the random spikes. To incorporte the spikes we increase instanstaneously the calcium concentration by $[\Delta Ca^{2+}] \sim 0.03 - 0.1\ \mu M$.

Reaction-diffusion equations

MARCKS and PKC are hydrophilic proteins and they transport to the membrane by diffusion. Here, for simplicity, we neglect effects of molecular crowding [41] and consider standard diffusion, with diffusion coefficients D_m at the membrane and D_c in the cytoplasm. The diffusion of both macromolecules at the membrane is smaller than the diffusion in the cytoplasm $D_m \ll D_c$.

While the free macromolecules $[PKC]_c$, $[MRK]_c$ and $[MRK]_p$ can diffuse in the whole cell, the concentrations $[PKC]_m$ and $[MRK]_m$ are restricted to the membrane of the cell. To account for this we divide the system in two regions: one corresponding to the membrane and its immediate vicinity, where all the translocation processes between membrane and cytoplasm take place, and the other to the bulk of the cytoplasm. For more details see the subsection on numerical methods below. The set of reaction-diffusion equations for the concentrations at the translocation zone reads:

$$\begin{aligned}
\frac{\partial[MRK]_m}{\partial t} &= S_m k_b^{mrk}[MRK]_c - k_u^{mrk}[MRK]_m - k_{pkc}^{mrk}\frac{[MRK]_m}{k_{mrk}+[MRK]_m}[PKC]_m \\
&\quad + D_m\nabla^2[MRK]_m\,, \\
\frac{\partial[MRK]_c}{\partial t} &= -S_m k_b^{mrk}[MRK]_c + k_u^{mrk}[MRK]_m + k_{ppa}^{mrk}[MRK]_p + D_c\nabla^2[MRK]_c\,, \\
\frac{\partial[MRK]_p}{\partial t} &= k_{pkc}^{mrk}\frac{[MRK]_m}{k_{mrk}+[MRK]_m}[PKC]_m - k_{ppa}^{mrk}[MRK]_p + D_c\nabla^2[MRK]_p\,, \\
\frac{\partial[PKC]_m}{\partial t} &= S_m k_b^{pkc}\left(\frac{[Ca^{2+}]}{k_{Ca}+[Ca^{2+}]}\right)[PKC]_c - k_u^{pkc}[PKC]_m + D_m\nabla^2[PKC]_m\,, \\
\frac{\partial[PKC]_c}{\partial t} &= -S_m k_b^{pkc}\left(\frac{[Ca^{2+}]}{k_{Ca}+[Ca^{2+}]}\right)[PKC]_c + k_u^{pkc}[PKC]_m + D_c\nabla^2[PKC]_c\,;
\end{aligned} \tag{5}$$

The equations describing the dynamics in the bulk of the cytosol read:

$$\begin{aligned}
\frac{\partial[MRK]_c}{\partial t} &= k_{ppa}^{mrk}[MRK]_p + D_c\nabla^2[MRK]_c\,, \\
\frac{\partial[MRK]_p}{\partial t} &= -k_{ppa}^{mrk}[MRK]_p + D_c\nabla^2[MRK]_p\,, \\
\frac{\partial[PKC]_c}{\partial t} &= D_c\nabla^2[PKC]_c\,;
\end{aligned} \tag{6}$$

where the only chemical transformation is the dephosphorylation of MARCKS. We restrict our study to global signals of $[Ca^{2+}]$ given by Eq. (4). A full description of the spatio-temporal dynamics of $[Ca^{2+}]$ requires request a detailed description of the mechanisms of the intracellular $[Ca^{2+}]$ dynamics, which is beyond the purpose and scope of this manuscript.

Parameter values

Some of the parameters of the model can be obtained or estimated from the literature. Typical values of the diffusion coefficient of proteins in cells are around $D \approx 1 - 10$ $\mu m^2 s^{-1}$ [42,43]. The diffusion coefficient of a kinase has been explicitly measured [44], and we use this typical result as characteristic diffusion coefficient of PKC $D_c^{pkc} \approx 5$ $\mu m^2 s^{-1}$.

Although the molecular weight of MARCKS ($32Kda$) is smaller than the weight of PKC ($80Kda$), its apparent weight is similar ($80Kda$) [45]. Therefore we employ the same value for the diffusion coefficient of MARCKS $D_c \approx 5\ \mu m^2 s^{-1}$.

Membrane diffusion of proteins is smaller than the cytoplasmatic diffusion [46]. In particular, for mammalian cells, membrane diffusion coefficient of proteins has been estimated to be $D_m \approx 0.1\ \mu m^2 s^{-1}$ [47].

Calcium rest concentration has been previously estimated [48]. Complete dephosphorylation of MARCKS protein is obtained after less than 3 minutes in experiments with neutrophils [9], therefore, we infer a characteristic time of around 20 seconds. The total concentration of MARCKS $[MRK] = 10\mu M$ [4] is obtained from experiments with living cells. The coverage concentration of MARCKS is estimated for a typical cell in [29,30] and is of the same order than the total concentration.

Following [49,50] we take $[PKC]_0 \approx 0.5\ \mu M$ as typical value of PKC concentration. However, some groups have considered lower [51] or higher [52] concentration of PKC, in the range of $[PKC]_0 \approx 0.1 - 1\ \mu M$.

The rest of model parameters are estimated by the fitting of experimental curves shown in [23]. The list of parameter values is shown in Table 1.

Numerical methods

For a better comparison with the experiments shown in [23], we consider a circular domain representing a cell with radius $R = 10\ \mu m$. Note that the use of three-dimensional environments with two-dimensional membranes does not qualitatively change the results, for more details see [29].

Table 1 Parameter values of the model

Molecule	Parameter	Value	Meaning	Ref.
MARCKS				
	$[MRK]$	$7.5\ \mu M$	Concentration	[4]
	$[MRK]_0$	$15\ \mu M$	Coverage	[29,30]
	k_b^{mrk}	$4 \times 10^3\ s^{-1}$	Binding	[23]
	k_u^{mrk}	$0.1\ s^{-1}$	Unbinding	[23]
	k_{pkc}^{mrk}	$0.25\ \mu M^{-1}\ s^{-1}$	Phosphorylation	[23]
	k_{mrk}	$0.15\ \mu M$	Michaelis-Menten binding	[15]
	k_{ppa}^{mrk}	$0.03\ s^{-1}$	Dephosphorylation	[9]
	D_c	$5\ \mu m^2 s^{-1}$	Cytoplasmic Diffusion	[42,45]
	D_m	$0.1\ \mu m^2 s^{-1}$	Membrane Diffusion	[47]
PKC				
	$[PKC]$	$0.5\ \mu M$	Concentration	[49]
	k_b^{pkc}	$140\ s^{-1}$	Binding	[23]
	k_u^{pkc}	$1\ s^{-1}$	Unbinding	[23]
	k_{Ca}	$0.4\ \mu M$	Michaelis-Menten Ca^{2+}	[23]
	D_c	$5\ \mu m^2 s^{-1}$	Cytoplasmic Diffusion	[44]
	D_m	$0.1\ \mu m^2 s^{-1}$	Membrane Diffusion	[47]
Ca^{2+}				
	τ	$8\ s$	Relaxation time	[23]
	$[Ca^{2+}]_0$	$0.1\ \mu M$	Rest concentration	[48]

In the center we introduce a circular passive nucleus which does not interact with the dynamics, thus we consider non-flux boundary conditions in the internal radius ($R_i = 2.5$ μm) of the integration domain.

We discretize the resulting corona using polar coordinates. We employ finite differences, adapted to polar coordinates, for the calculation of the diffusion terms. While the radial discretization is constant and corresponds to $\Delta r = 0.125$ μm, the angular spatial discretization depends on the radial coordinate.

The exterior layer, corresponding to the external radius of the cell, corresponds to the thin layer where the binding and unbinding dynamics to the membrane during time Δt occurs. The discretization of the system implies that this external layer is 0.125 μm thick, much larger than the real thickness of the cellular membrane, however, the use of finite differences permits the definition of this interaction length, for more details see [29]. Within this discretization, we employ the membrane concentrations $[MRK]_m$ and $[PKC]_m$ as volume concentrations.

We employ an explicit Euler method with $\Delta t = 0.2$ ms for the numerical integration of the equations.

Results

First we employ experimental data obtained from [23] to build a model by fitting the parameter values. Then, we study by numerical simulations of the resulting model different types of symmetry breaking mechanisms.

Building a quantitative model

We consider experiments on MARCKS and PKC interaction in INS-1 cells, an insulin-secreting cell [23] to obtain the values of the parameters.

Parametrizing PKC and MARCKS binding

The enzyme PKC is activated by Ca^{2+} [38]. Once the enzyme is activated it can bind to the membrane. The PKC dependence on Ca^{2+} seems to follow a Michaelis-Menten dynamics [53,54], however, the explicit form is derived from the experiments on living cells [23]. In such experiments there is a sharp transition between the concentrations of low active PKC and high active PKC at $[Ca^{2+}] = 400$ nM. Low Ca^{2+} concentrations ($[Ca^{2+}] < 400$ nM) do not activate PKC and high Ca^{2+} concentrations ($[Ca^{2+}] > 400 nM$) activate PKC and produce the binding of PKC at the membrane [23]. The relaxation times (τ) observed in experiments are distributed between 5 and 15 seconds. For the numerical simulations shown here we use the value $\tau = 8$ s. A close estimation is obtained in a similar context in [55].

In Figure 2 the responses of both, PKC and MARCKS, after a spike of Ca^{2+} are shown simultaneously. Experimental data has been extracted form [23]. The ratio between the concentrations after and before the Ca^{2+} spike is plotted for the experimental and numerical cases. A spike of $[Ca^{2+}]$ decreases the concentration of PKC in the cytoplasm, see Figure 2. The peak corresponds to the continuous binding of PKC at the cellular membrane during the spike. The recovery shows the slow unbinding of the enzyme from the membrane during and after the decrease of free Ca^{2+} in the cell. From the characteristic times of the experimental binding and unbinding processes we estimate the values for the parameters k_b^{pkc} and k_u^{pkc}. Similar quantities have been estimated in experiments with

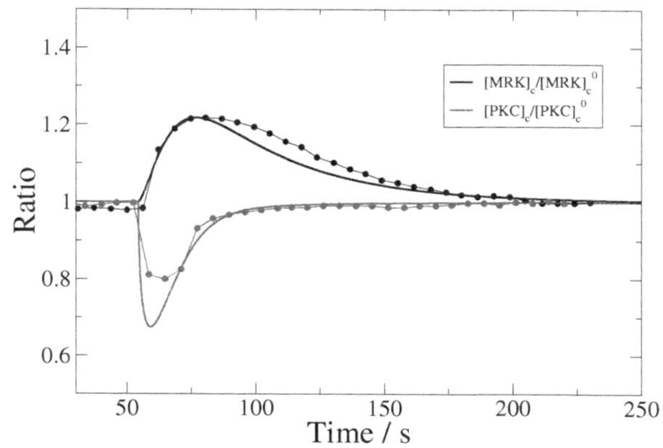

Figure 2 Comparison between numerical simulations and experiments. Evolution of PKC and MARCKS concentration on the cytoplasm after a Ca^{2+} spike. Points correspond to the experimental results, data extracted from [23] and solid lines to our numerical simulations. Total concentration of the proteins are: $[MRK] = 7.5\mu M$ and $[PKC] = 0.5\mu M$.

vesicles formed by acidic phospholipids [53] and high concentrations of Ca^{2+}. The results obtained from experiments with vesicles differ in one order of magnitude with the results estimated with living cells. A comparison between in vitro values and the results with living cells is shown in Table 2.

The binding of the PKC enzymes at the membranes causes the unbinding of MARCKS proteins. The processes of binding and phosphorylation of MARCKS are fitted from the experiments to give rise to the similar behavior shown in Figure 2. We employ the comparison between these curves to fit the parameters for the binding (k_b^{mrk}) and unbinding (k_u^{mrk}) of MARCKS, and the phosphorylation rate (k_{pkc}^{mrk}).

Dynamics of MARCKS and PKC

The absolute responses of MARCKS and PKC proteins to the Ca^{2+} spike are shown in Figure 3. The concentration of PKC at the membrane increases, see Figure 3b, and phosphorylates some membrane MARCKS proteins. It produces an increase of phosphorylated MARCKS, see Figure 3c, and the subsequent dephosphorylation in the cytoplasm. Note that the low dephosphorylation rate produces a delay in the increase of free MARCKS proteins on the cytoplasm with respect the phosphorylated MARCKS. It increases the recovery time of MARCKS to the initial state with respect to the PKC dynamics. Such increase is shown in Figure 2 and has been observed in different experimental realizations [23].

Table 2 Comparison between in vitro and in vivo parameter values

Molecule	Parameter	Meaning	Value in vitro	Fit on Figure 2
PKC				
	k_b^{pkc}	Binding	$40\ s^{-1}$ [53]	$400\ s^{-1}$
	k_u^{pkc}	Unbinding	$10\ s^{-1}$ [53]	$1\ s^{-1}$
MARCKS				
	k_b^{mrk}	Binding	$10 - 400\ s^{-1}$ [12]	$4 \times 10^3\ s^{-1}$
	k_u^{mrk}	Unbinding	$10\ s^{-1}$ [12]	$0.1\ s^{-1}$
	k_{pkc}^{mrk}	Phosphorylation	-	$0.25\ \mu M^{-1}\ s^{-1}$

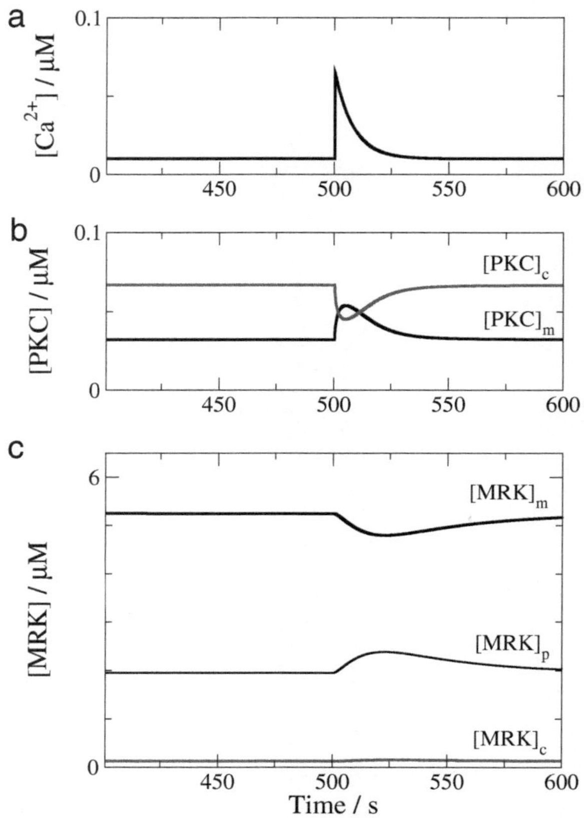

Figure 3 Effects of an spike of Ca^{2+} on the distribution of MARCKS and PKC. a) Evolution of Ca^{2+} concentration. **b)** Evolution of PKC at the membrane and in the cytoplasm. **c)** Evolution of MARCKS at the membrane and in the cytoplasm: free and phosphorylated.

Response to a set of [Ca^{2+}] spikes

The model can reproduce other experimental results. In Figure 4 we show the response of a different cell of the same type to a set of Ca^{2+} spikes. Keeping the same parameter values obtained from Figure 2 the model reproduces the PKC and MARCKS responses, where we have only modulated the amplitude of the spikes of calcium.

Whereas the response of MARCKS in the numerical simulation is comparable to the experiments, the comparison of the PKC response is not completely satisfactory. Tuning the value of some kinetic parameters or total concentrations may improve the results. However, we obtain a qualitative agreement without any further tunning.

Pattern formation in living cells

After the characterization of the model parameters from the above simulations, we perform large simulations for the whole cell, keeping the parameters constant and [Ca^{2+}] at rest.

Spontaneous domain formation

The evolution of the membrane concentration of MARCKS and PKC proteins are shown in Figure 5. The spatio-temporal plot, see Figure 5a, shows the generation of protein

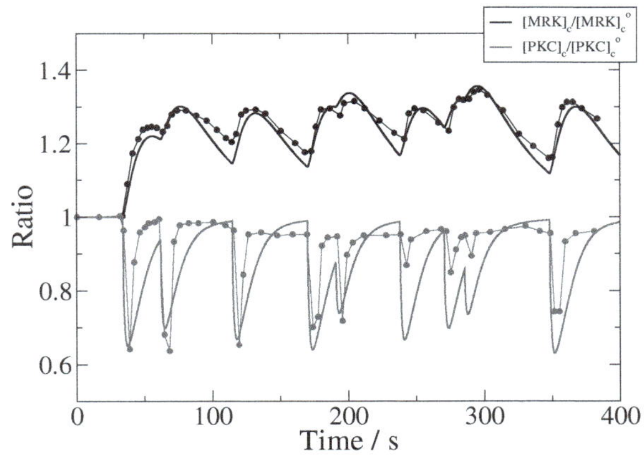

Figure 4 Evolution of PKC and MARCKS concentration in the cytoplasm after a set of Ca^{2+} spikes. Points correspond to the experimental results, data extracted from [23] and solid lines to our numerical simulations. Total concentration of the proteins are: $[MRK] = 7.5\ \mu M$ and $[PKC] = 0.5\ \mu M$.

domains at the membrane. First, the proteins are homogeneously distributed at the membrane, see Figure 5b. However, small spatial differences in the concentrations induced by small random perturbations lead to a long-wave instability. The spatial differences on the concentration increase with time and give rise to domains. [PKC] accumulates at the membrane outside the domains of MARCKS proteins.

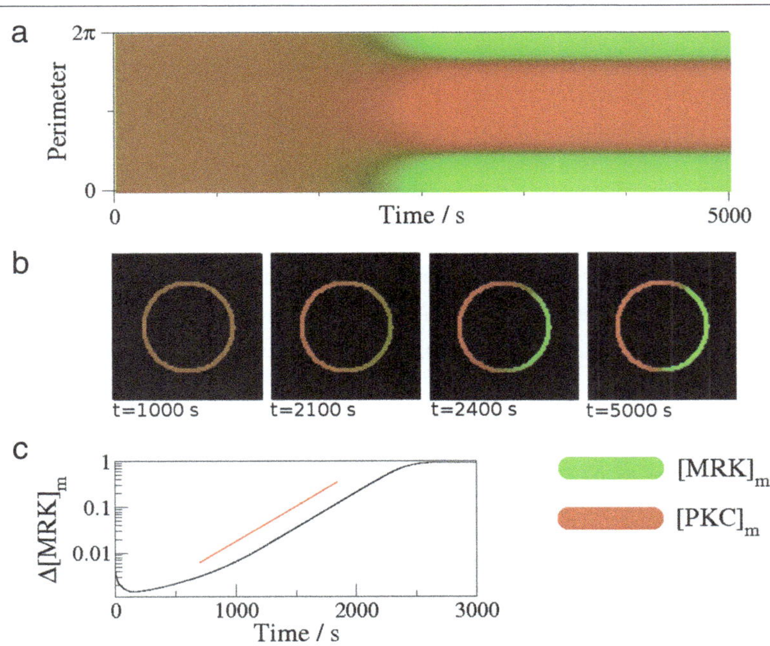

Figure 5 Two-dimensional numerical simulations of Eqs. (5-6) in a circular domain. From an initial homogeneous distribution of MARCKS and PKC with a weak random perturbation, domains of high protein concentration appear at the membrane. **a)** Spatio-temporal plot of the concentration of PKC and MARCKS at the membrane. **b)** Spatial distribution of MARCKS and PKC at the membrane at four different times. **c)** Evolution of the relative difference between the highest $[MRK]_m^+$ and the lowest $[MRK]_m^-$ value of MARCKS concentration at the membrane $\Delta[MRK]_m = ([MRK]_m^+ - [MRK]_m^-)/[MRK]_m^+$. The growing is exponential ($\Delta[MRK]_m \propto e^{-t/\tau_g}$) and the thin line shows a exponent of $\tau_g = 285$ s. Total concentration of the proteins are: $[MRK] = 7.5\ \mu M$ and $[PKC] = 0.5\ \mu M$.

Cells with larger number of transitory domains can be observed upon change of the values of some parameters or the cell size. These domains may be relatively stable, however, coarsening of domains is observed (results not shown) and, finally, a single large domain survives at the membrane.

To characterize the growing rate of the domains, the relative difference ($\Delta[MRK]_m$) between the local maximum and the minimum of $[MRK]_m$ is plotted in Figure 5c. This difference corresponds to $([MRK]_m^+ - [MRK]_m^-)/[MRK]_m^+$ and saturates after an exponential growth. This growth defines a characteristic temporal scale for the growing rate ($\tau_g = 285\ s$).

Bistability-induced domain formation

A second mechanism of pattern formation is associated with the bistability of two different homogeneous steady states. An inhomogeneous initial perturbation produces the simultaneous appearance of two stable solutions, which compete and finally develop a stationary profile.

A numerical simulation with higher MARCKS and PKC concentrations is shown in Figure 6a,b. The initial condition distributes the protein concentrations inhomogeneously along the cell giving rise to two different regions with high and low protein concentrations. It produces a polar orientation of the cell.

Small perturbations of the homogeneous initial state as in the previous section do not produce the formation of the domain, see Figure 6c,d. The distribution of MARCKS and PKC at the membrane arrives to two completely different states depending on the initial condition.

Change on MARCKS and PKC concentrations

In order to check the robustness of the pattern formation mechanisms, we have performed a systematic numerical study of the conditions of appearance of the membrane domains. We have chosen the total concentrations of MARCKS and PKC proteins as control parameters because they may be controlled in experiments [51]. The two types of pattern formation appearing in the model are shown in Figure 7. Simulations done with an small random perturbation on the initial condition can produce the formation of domains, light gray region in Figure 7. A single realization is shown in Figure 5. On the other hand, dark gray region corresponds to simulations which homogeneous solution is stable under small perturbations, but unstable if the perturbation is strong enough, see Figure 6.

Large concentrations of MARCKS saturate the membrane, above gray regions in Figure 7, whereas low concentrations do not allow the formation of domains, below gray regions in Figure 7. The absence of PKC precludes domain formation, in contrast, at fixed MARCKS concentration, large concentrations of PKC phosphorylate the membrane too fast and also suppress the domain formation.

The dashed line in Figure 7 corresponds to the cover concentration of MARCKS $[MRK]_o$. This quantity is a threshold for MARCKS concentration at the membrane.

Change on the diffusion coefficients

The two mechanisms responsible of domain formation are controlled by the values of the diffusion coefficients. In Figure 8 we study the dependence of the pattern formation

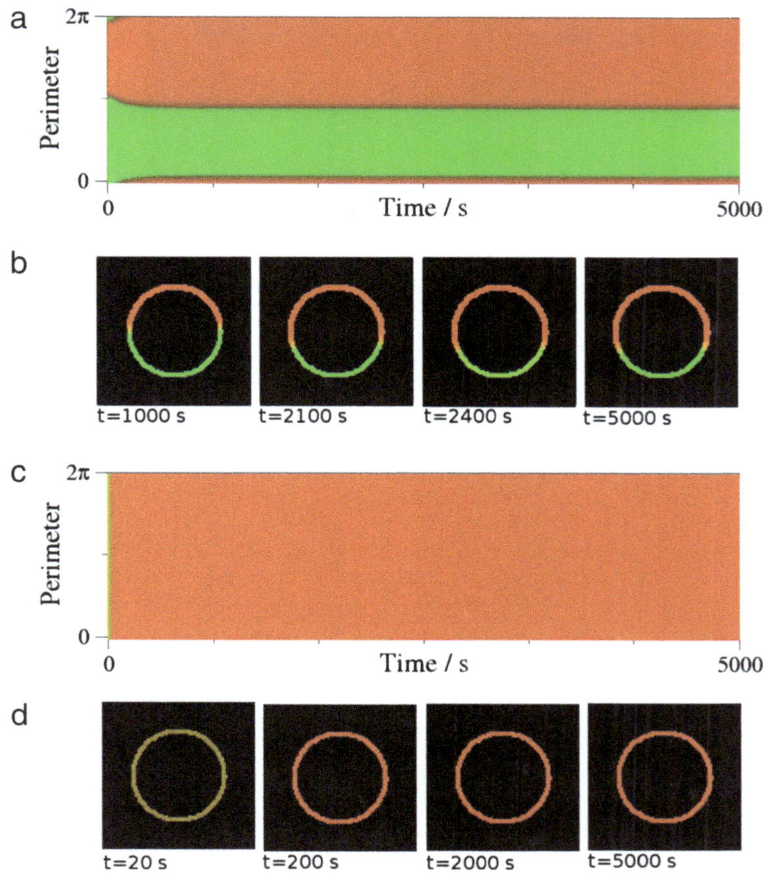

Figure 6 Two-dimensional numerical simulations of Eqs. (5-6) in a circular domain. Whereas from an initial asymmetric inhomogeneous distribution of MARCKS and PKC, domains of high protein concentration appear at the membrane (a-b), from an initial homogeneous distribution of MARCKS and PKC with a weak random perturbation, no pattern formation is obtained **(c-d)**. **a)** and **c)** Spatio-temporal plot of the concentration of PKC and MARCKS at the membrane. **b)** and **d)** Spatial distribution of MARCKS and PKC at the membrane at four different times. Total concentration of the proteins are: $[MRK] = 10.5\ \mu M$ and $[PKC] = 2.5\ \mu M$.

process on the values of D_m and D_c for spontaneous domain formation in Figure 8a and for bistability-induced domain formation in Figure 8b.

For a given value of D_c there is a critical value of D_m. Above this value there is no pattern formation and a homogeneous distribution is stable. Below this critical value protein domains are observed. Two different regimes are observed: for small D_c the critical membrane diffusivity follows a power law with D_c, and for large D_c the critical membrane diffusion saturates and is constant.

Cells are not identical and, there is a large variety of properties and sizes. Thus, we have also studied the effect of the size even within the population of one type of cells. The critical value of D_m depends on the size of the cell, see simulation results with different cell radius in Figure 8. Small cells with large D_c permit domain formation only for small membrane diffusion. Increasing the size of the cells facilitates pattern formation.

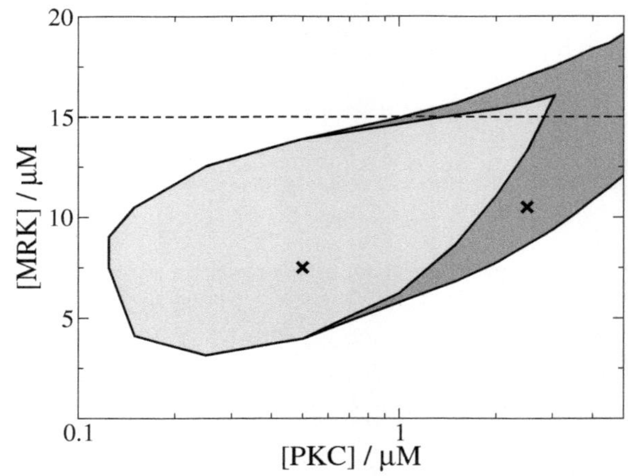

Figure 7 Phase diagram of the model for different values of [*MRK*] and [*PKC*]. Light gray region corresponds to protein domain formation at the membrane under small random initial perturbations. Dark gray region corresponds to bistability-induced protein domain formation under large perturbations. White region corresponds to homogeneous protein distribution at the membrane. Dashed line shows to the saturation concentration ($[MRK]_o$). Crosses mark the parameter values employed in Figures 5 and 6.

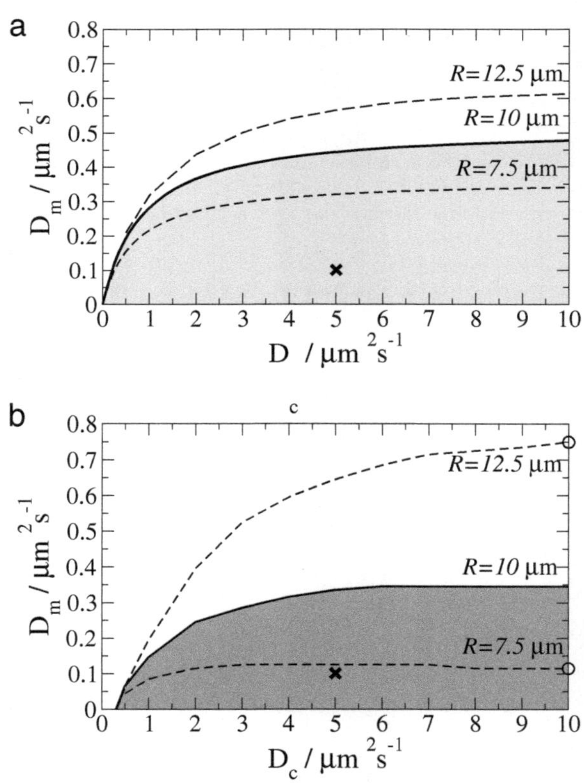

Figure 8 Phase diagram of the model for different values of the diffusion at the membrane D_m and in the cytosol D_c. Solid lines corresponds to simulations of a cell with radius $R = 10\ \mu m$. Dashed lines correspond to simulations of cells with radius $R = 7.5\ \mu m$ and $R = 12.5\ \mu m$. Above the line the homogeneous protein distribution is stable. Below the line, protein domain formation at the membrane under small random initial perturbations **(a)** and bistability-induced protein domain formation under large perturbations **(b)** are observed. Crosses mark the parameter values employed in Figures 5 and 6.

Discussion

Formation of MARCKS domains has been observed in experiments in hippocampal neuron growth cones [26]. They are correlated with domains of acidic phospholipids as PIP_2 or PS. Experiments with vesicles show that phospholipid domains do not form in absence of MARCKS. The introduction of MARCKS proteins induces the formation of domains [17]. Our results are obtained without considering the dynamics of the phospholipids. The electrostatic interaction of the phospholipids with MARCKS may induce the formation of phospholipid domains following the protein domains. Domains of MARCKS proteins at the membrane of smooth muscle cells co-distributed with calmodulin have been observed in experiments [27].

We propose two different mechanisms of pattern formation for MARCKS protein at membranes: A long wave instability and a bistability-related mechanism. The long-wave instability is responsible for the pattern formation for the parameter values fitted from [23]. Increasing the concentration of MARCKS and PKC produces the appearance of domains induced by the second mechanism.

The formation of domains is robust under change of parameter values. Figure 7 shows the window of MARCKS and PKC concentrations which permit pattern formation. The relation among the values of the two diffusion coefficients (D_m and D_c) which produce domain formation is shown in Figure 8. For a given value of D_m pattern formation happens above a critical size of the cell.

The predicted domains may have a regulatory function by making the protection of the acidic phospholipids through membrane-bound proteins more efficient. In this way a local accumulation of proteins may prevent the effect of PKC inside the domain, which produces the phosphorylation and membrane-unbinding of MARCKS and the subsequent hydrolysis of the unprotected PIP_2, which in turn induces a release of Ca^{2+} [56].

The relation between MARCKS proteins and the induction of Ca^{2+} spikes is beyond the scope of the present work. Here we have assumed that Ca^{2+} follows a simple relaxation dynamics, see Eq. (4). However, it is known that Ca^{2+} has a spatio-temporal dynamics, and traveling waves are generated by a calcium-induced calcium release mechanism [11]. The use of an adequate model for the Ca^{2+} dynamics combined with the generation of Ca^{2+} produced by the hydrolysis of the PIP_2 may provide a feedback loop between the processes at the membrane and in the interior of the cell, giving rise to a suitable mechanism of cell signaling control.

Possible stochastic effects due to a low number of proteins have been neglected here. The large concentration of MARCKS proteins in cells permits the use of deterministic dynamics. The concentration of PKC is smaller and stochastic effects may become relevant. The use of a stochastic model may enhance domain formation because the initially small perturbations necessary for pattern formation will naturally appear in a stochastic framework.

Living cells are three-dimensional. The model studied here is, however, a two-dimensional approximation. The two-dimensional view is, however, sufficient to calculate proteins concentrations at the membrane. The realization of three-dimensional simulations would increase the computational time and the complexity of the numerical methods substantially.

The modelling results obtained in this work were compared to the dynamics of MARCKS and PKC in experiments with living insulin-secreting cells. However, methods and results are equally relevant for other systems involving membrane proteins. For example, the growth-associated protein 43 (GAP 43) and the cortical cytoskeleton associated protein (CAP 23) are proteins with similar properties as MARCKS [4]. They form also domains at the membrane of living cells [26] based on the interaction with acidic phospholipids and PKC [4]. On the other hand, the myelin basic protein (MBP), one of the most abundant proteins in the nervous system, has a structure analogous to MARCKS [57], electrostatically interacts with the PIP_2 of the membranes and responds to elevated Ca^{2+} signals [58]. Furthermore, phosphorylation by PKC and other kinases regulate multiple processes in living cells, e.g. the formation of polarity of cells induced by PAR proteins [59], the cyclic dynamics of Rho GTPases [31] or the regulation of the cell division of E. Coli controlled by the Min proteins [37]. Hence, similar extensions of the Goldbeter-Koshland mechanism may be applicable to a large variety of biological systems.

Conclusions

In this paper, we have formulated a simple model for the binding and unbinding of MARCKS and PKC to the cellular membrane. Phosphorylation of MARCKS by PKC occurs at the membrane, while dephosphorylation of MARCKS in the cytosol. The parameter values of the model have been fitted from experiments shown in [23], wherein the binding dynamics of PKC and MARCKS are investigated. The resulting model with fitted parameters predicts the formation of protein domains at the membrane of the cell.

We employ an extended version of the classical Goldbeter-Koshland mechanism of protein phosphorylation and dephosphorylation. The separation of the two processes in two different compartments of the cell (phosphorylation at the membrane and dephosphorylation in the cytoplasm) combined with the difference of the diffusion coefficients at the membrane and in the cytoplasm ($D_m \ll D_c$) produces the formation of domains. Another important factor in the model is the conservation law included in the system.

In summary, we have developed a simple model of binding, phosphorylation and desphosphorylation for MARCKS protein. The main prediction of our numerical simulations is the spontaneous appearance of domains of high concentration of membrane proteins. The interaction of this protein domains with phospholipids at the membrane, in particular PIP_2 may have important consequences in the process of formation of IP_3, and, therefore, in the control of the signaling mechanisms of Ca^{2+}.

Competing interests

The authors declare that they have no competing interests.

Authors' contributions

SA has performed the numerical simulations, SA and MB have written the paper and all authors have given final approval for its publication. Both authors read and approved the final manuscript.

Acknowledgements

We thank Hideo Mogami for stimulating discussions. We acknowledge financial support from the German Science Foundation DFG within the framework of SFB 910 "Control of self-organizing nonlinear systems".

References

1. Alberts B, Johnson A, Lewis J, Raff M, Roberts K, Walter P: *Molecular Biology of the Cell*. New York: Garland Science; 2002.
2. Kholodenko BN, Hancock JF, Kolch W: **Signalling ballet in space and time.** *Nat Rev Mol Cell Biol* 2010, **11:**414–426.
3. Kholodenko BN: **Cell-signalling dynamics in time and space.** *Nat Rev Mol Cell Biol* 2006, **7:**165–176.
4. McLaughlin S, Wang J, Gambhir A, Murray D: **The electrostatic properties of membranes.** *Annu Rev Biophys Biomol Struct* 2002, **31:**151–175.

5. Kalwa H, Michel T: **The MARCKS protein plays a critical role in Phosphatidylinositol 4,5-Bisphosphate metabolism and directed cell movement in vascular endothelial cells.** *J Biol Chem* 2011, **286:**2320–2330.
6. Allen LH, Aderem A: **A role for MARCKS , the isozyme of protein kinase C and myosin I in zymosan phagocytosis by macrophages.** *J Exp Med* 1995, **182:**829–840.
7. Salli U, Saito N, Stormshak F: **Spatiotemporal interactions of Myristoylated alanine-rich C kinase substrate (MARCKS) protein with the actin cytoskeleton and exocytosis of oxytocin upon prostaglandin F2a stimulation of bovine luteal cells.** *Biol Repro* 2003, **69:**2053–2058.
8. Miller JD, Lankford SM, Adler KB, Brody AR: **Mesenchymal stem cells require MARCKS protein for directed chemotaxis in vitro.** *Am J Respir Cell Mol Biol* 2010, **43:**253–258.
9. Thelen M, Rosen A, Nairn AC, Anderem A: **Regulation by phosphorylation of reversible association of a myristoylated protein kinase C substrate with the plasma membrane.** *Nature (London)* 1991, **351:**320–322.
10. Goldbeter A, Koshland DE: **An amplified sensitivity arising from covalent modification in biological systems.** *Proc Nati Acad Sci USA* 1981, **78:**6840–6844.
11. Keener J, Sneyd J: *Mathematical Physiology*. New York: Springer; 2009.
12. Arbuzova A, Wang J, Murray D, Jacob J, Cafiso DS, McLaughlin S: **Kinetics of interaction of the Myristoylated alanine-rich C kinase substrate, membranes, and calmodulin.** *J Biol Chem* 1997, **272:**27167–21177.
13. Wang J, Gambhir A, Mihalyne G, Murray D, Golebiewska U, McLaughlin S: **Lateral sequestration of phosphatidylinositol 4,5-bisphosphate by the basic effector domain of Myristoylated alanine-rich C kinase substrate is due to nonspecific electrostatic interactions.** *J Biol Chem* 2002, **277:**34401–34412.
14. Rusu L, Gambhir A, McLaughlin S, Rädler J: **Fluorescence correlation spectroscopy studies of peptide and protein binding to phospholipid vesicles.** *Biophys J* 2004, **87:**1044–1053.
15. Verghese GM, Johnson JD, Vasulka C, Haupt DM, Stumpo DJ, Blackshear PJ: **Protein kinase C-mediated phosphorylation and calmodulin binding of recombinant Myristoylated alanine-rich C kinase substrate (MARCKS) and MARCKS-related protein.** *J Biol Chem* 1994, **269:**9361–9367.
16. Blackshear PJ, Verghese GM, Johnson JD, Haupt DM, Stumpo DJ: **Characteristics of the F52 protein, a MARCKS homologue.** *J Biol Chem* 1992, **267:**13540–13546.
17. Glaser M, Wanaski S, Buser CA, Boguslavsky V, Rashidzada W, Morris A, Rebecchi M, Scarlata SF, Runnels LW, Prestwich GD, Chen J, Aderemi A, Ahni J, McLaughlin S: **Myristoylated alanine-rich C kinase substrate (MARCKS) produces reversible inhibition of phospholipase C by sequestering phosphatidylinositol 4,5-Bisphosphate in lateral domains.** *J Biol Chem* 1996, **271:**26187–26193.
18. Rebecchi M, Boguslavsky V, Boguslavsky L, McLaughlin S: **Phosphoinositide-specific phospholipase C-δ1: effect of monolayer surface pressure and electrostatic surface potentials on activity.** *Biochemistry* 1992, **31:**12748–12753.
19. Wang J, Arbuzova A, Hangyás-Mihályné G, McLaughlin S: **The effector domain of myristoylated alanine-rich C kinase substrate binds strongly to phosphatidylinositol 4,5-Bisphosphate.** *J Biol Chem* 2001, **276:**5012–5019.
20. Dietrich U, Krüger P, Gutberlet T, Käs JA: **Interaction of the MARCKS peptide with PIP2 in phospholipid monolayers.** *Biochim Biophys Acta* 2009, **557:**1474–1481.
21. Alonso S, Dietrich U, Händel C, Käs JA, Bär M: **Oscillations in the lateral pressure of lipid monolayers induced by nonlinear chemical dynamics of the second messengers MARCKS and protein kinase C.** *Biophysical J* 2011, **100:**939–947.
22. Ohmori S, Sakai N, Shirai Y, Yamamoto H, Miyamoto E, Shimizu N, Saito N: **Importance of protein kinase C targeting for the phosphorylation of its substrate, myristoylated alanine-rich C-kinase substrate.** *J Biol Chem* 2000, **275:**26449–26457.
23. Mogami H, Zhang H, Suzuki Y, Urano T, Saito N, Kojima I, Petersen OH: **Decoding of short-lived Ca2 influx signals into long term substrate phosphorylation through activation of two distinct classes of protein kinase C.** *J Biol Chem* 2003, **278:**9896–9904.
24. Uchino M, Sakai N, Kashiwagi K, Shirai Y, Shinohara Y, Hirose K, Iino M, Yamamura T, Saito N: **Isoform-specific phosphorylation of metabotropic glutamate receptor 5 by Protein Kinase C (PKC) blocks Ca2 oscillation and oscillatory translocation of Ca2-dependent PKC.** *J Biol Chem* 2004, **279:**2254–2261.
25. Sawano A, Hama H, Saito N, Miyawaki A: **Multicolor imaging of Ca2 and protein kinase C signals using novel epifluorescence microscopy.** *Biophys J* 2002, **82:**1076–1085.
26. Laux T, Fukami K, Thelen M, Golub T, Frey D, Caroni P: **GAP43, MARCKS , and CAP23 Modulate PI(4,5)P2 at plasmalemmal rafts, and regulate cell cortex actin dynamics through a common mechanism.** *J Cell Biol* 2000, **149:**1455–1472.
27. Gallant C, You JY, Sasaki Y, Grabarek Z, Morgans KG: **MARCKS is a major PKC-dependent regulator of calmodulin targeting in smooth muscle.** *J Cell Sci* 2005, **118:**3595–3605.
28. John K, Bär M: **Alternative mechanisms of structuring biomembranes: self-assembly versus self-organization.** *Phys Rev Lett* 2005, **95:**198101.
29. Alonso S, Bär M: **Phase separation and bistability in a three-dimensional model for protein domain formation at biomembranes.** *Phys Biol* 2010, **7:**046012.
30. John K, Bär M: **Travelling lipid domains in a dynamic model for protein-induced pattern formation in biomembranes.** *Phys Biol* 2005, **2:**123–132.
31. Otsuji M, Ishihara S, Co C, Kaibuchi K, Mochizuki A, Kuroda S: **A mass conserved reaction-diffusion system captures properties of cell polarity.** *PLoS Comput Biol* 2007, **3:**1040–54.
32. Mori Y, Jilkine A, Edelstein-Keshet L: **Wave-pinning and cell polarity from a bistable reaction-diffusion system.** *Biophys J* 2008, **94:**3684–3697.
33. Jilkine A, Edelstein-Keshet L: **A comparison of mathematical models for polarization of single eukaryotic cells in response to guided cuess.** *PLoS Comput Biol* 2011, **7:**e1001121.
34. Goryachev AB, Pokhilko AV: **Dynamics of Cdc42 network embodies a turing-type mechanism of yeast cell polarity.** *FEBS Lett* 2008, **582:**1437–1443.

35. Meinhardt H, de Boer PAJ: **Pattern formation in escherichia coli: a model for the pole-to-pole oscillations of Min proteins and the localization of the division site.** *Proc Natl Acad Sci USA* 2001, **98:**14202.
36. Howard M, Rutenberg AD, de Vet S: **Dynamic compartmentalization of bacteria: accurate division in.** *E. Coli. Phys Rev Lett* 2001, **87:**278102.
37. Howard M, Kruse K: **Cellular organization by self-organization: mechanisms and models for Min protein dynamics.** *J Cell Biol* 2005, **168:**533–36.
38. Newton AC, Protein kinase C: **structural and spatial regulation by phosphorylation, cofactors, and macromolecular interactions.** *Chem Rev* 2001, **101:**2353–2364.
39. Clarke PR, Siddhanti SR, Cohen P, Blackshear PJ: **Okadaic acid-sensitive protein phosphatases dephosphorylate MARCKS, a major protein kinase C substrate.** *FEBS Lett* 1993, **336:**37–42.
40. Straube R, Conradi C: **Reciprocal enzyme regulation as a source of bistability in covalent modification cycles.** *J Theor Biol* 2013, **330:**56–74.
41. Weiss M, Elsner M, Kartberg F, Nilsson T: **Anomalous subdiffusion is a measure for cytoplasmic crowding in living cells.** *Biophys J* 2004, **87:**3518–3524.
42. Brown GC, Kholodenko BN: **Spatial gradients of cellular phospho-proteins.** *FEBS Lett* 1999, **457:**452–454.
43. Luby-Phelps K: **Cytoarchitecture and physical properties of cytoplasm: volume, viscosity, diffusion, intracellular surface area.** *Int Rev Cytol* 1997, **192:**189–221.
44. Arrio-Dupont M: **Mobility of creatine phosphokinase and β-Enolase in cultured muscle cells.** *Biophys J* 1997, **73:**2667–2673.
45. Bhat NR: **Phosphorylation of MARCKS (8O-kDa) protein, a major substrate for protein kinase C in oligodendroglial progenitors.** *J Neurosci Res* 1991, **308:**447–454.
46. Edidin M, Zagyansky Y, Lardner TJ: **Measurement of membrane protein lateral diffusion in single cells.** *Science* 1976, **191:**466–468.
47. Valdez-taubas J, Pelham RB: **Slow diffusion of proteins in the yeast plasma membrane allows polarity to be maintained by endocytic cycling.** *Curr Biol* 2003, **13:**1636–1640.
48. Rüdiger S, Shuai JW, Huisinga W, Nagaiah C, Warnecke G, Parker I, Falcke M: **Hybrid stochastic and deterministic simulations of calcium blips.** *Biophys J* 2007, **93:**1847–1857.
49. Bhalla US, Iyengar R: **Emergent properties of networks of biological signaling pathways.** *Science* 1999, **283:**381–387.
50. Bhalla US: **Signaling in small subcellular volumes. I. stochastic and diffusion effects on individual pathways.** *Biophys J* 2004, **87:**733–744.
51. Ozawa K, Szallasie Z, Kazanietzs MG, Blumberge PM, Mischakll H, Mushinskill JF, Beaven MA: **Ca2+-dependent and Ca2+-independent isozymes of protein kinase C mediate exocytosis in antigen-stimulated rat basophilic RBL-2H3 cells.** *J Biol Chem* 1993, **268:**1749–1756.
52. Kang M, Othmer HG: **The variety of cytosolic calcium responses and possibles roles of PLC and PKC.** *Phys Biol* 2007, **4:**325–343.
53. Nalefski EA, Newton AC: **Membrane binding kinetics of protein kinase C II mediated by the C2 domain.** *Biochemistry* 2001, **40:**13216–13229.
54. Mosior M, Golini ES, Epand RM: **Chemical specificity and physical properties of the lipid bilayer in the regulation of protein kinase C by anionic phospholipids: Evidence for the lack of a specific binding site for phosphatidylserine.** *Proc Natl Acad Sci USA* 1996, **93:**1907–1912.
55. Newton AC: **Lipid activation of protein kinases.** *J Lipid Res* 2009, **50:**S266–S271.
56. Allbritton NL, Meyer T, Stryer L: **Range of messenger action of calcium ion and inositol 1,4,5-trisphosphate.** *Science* 1992, **258:**1812–1815.
57. Harauz G, Ishiyama N, Bates IR: **Analogous structural motifs in myelin basic protein and in MARCKS.** *Mol Cell Biochem* 2000, **209:**155–63.
58. Nawaz S, Kippert A, Saab AS, Werner HB, Lang T, Nave K-A, Simons M: **Phosphatidylinositol 4,5-bisphosphate-dependent interaction of myelin basic protein with the plasma membrane in oligodendroglial cells and its rapid perturbation by elevated calcium.** *J Neurosci* 2009, **29:**4794–4807.
59. Tostevin F, Howard M: **Modeling the establishment of PAR protein polarity in the one-cell C. elegans embryo.** *Biophys J* 2008, **95:**4512–4522.

2

Travelling waves in models of neural tissue: from localised structures to periodic waves

Hil GE Meijer[1*] and Stephen Coombes[2]

*Correspondence: h.g.e.meijer@utwente.nl
[1] Department of Applied Mathematics, MIRA Institute of Biomedical Technology and Technical Medicine, University of Twente, Postbus 217, 7500 AE, Enschede, The Netherlands
Full list of author information is available at the end of the article

Abstract

We consider travelling waves (fronts, pulses and periodics) in spatially extended one dimensional neural field models. We demonstrate for an excitatory field with linear adaptation that, in addition to an expected stable pulse solution, a stable anti-pulse can exist. Varying the adaptation strength we unravel the organizing centers of the bifurcation diagram for fronts and pulses, with a mixture of exact analysis for a Heaviside firing rate function and novel numerical schemes otherwise. These schemes, for non-local models with space-dependent delays, further allow for the construction and continuation of periodic waves. We use them to construct the dispersion curve – wave speed as a function of period – and find that they can be oscillatory and multi-valued, suggesting bistability of periodic waves. A kinematic theory predicts the onset of wave instabilities at stationary points in the dispersion curve, leading to period doubling behaviour, and is confirmed with direct numerical simulations. We end with a discussion of how the construction of dispersion curves may allow a useful classification scheme of neural field models for epileptic waves.

PACS codes: Primary 87.19.lj; 87.19.le; 87.19.lq; 87.19.lf

Keywords: Neural field theory; Brain wave equation; Numerical continuation; Anti-pulse; Dispersion curve

Background

The analysis of waves in models arising from the study of the nervous system has a long tradition. The seminal example is the development and numerical analysis of the model for action potential propagation in an excitable axonal fibre by Hodgkin and Huxley [1], and see [2] for an excellent review. This has been followed by more rigorous mathematical analysis using tools from geometric singular perturbation theory for the existence of pulses (homoclinics) and wave trains (periodics) [3], as well as the development of a stability theorem [4]. However, the detailed properties of waves in detailed biophysical models is often best pursued with a mixture of both mathematical and numerical analysis. This is nicely exemplified by the work of Miller and Rinzel [5], who numerically construct the dispersion relation (between speed and period) for steadily propagating periodic wave trains, and then use a mathematical kinematic theory, which applies the dispersion relation, instantaneously, to individual pulses to predict how interspike time intervals can be distorted during propagation. Going beyond the single neuron scenario traveling waves in neurobiology are often studied at the tissue level, using neuroimaging methods such as electroencephalography (EEG). For example these waves can take the form of spindle waves seen at the onset of sleep [6], the propagation of synchronous discharge during an epileptic seizure [7] and waves of excitation associated with sensory processing [8]. Such waves are a consequence of synaptic interactions and the intrinsic behaviour of local neuronal circuitry. The propagation speed of these waves is of the order $\mathrm{cm\,s^{-1}}$, an order of magnitude slower than that of action potential propagation along axons. The class of computational models that are believed to support synaptic waves differ radically from classic models of waves in excitable systems. Most importantly, synaptic interactions are non-local (in space), involve communication (space-dependent) delays (arising from the finite propagation velocity of an action potential) and distributed delays (arising from neurotransmitter release and dendritic processing). In contrast models of axonal fibres are typically based on local partial differential equation models (that track the opening and closing of ion channels and model voltage transport along a fibre with a diffusive process). A common way to model the spatiotemporal evolution of coarse grained variables such as synaptic or firing rate activity in populations of neurons is through the use of a *neural field model* (in which space is continuous). Despite a growing body of analysis on synaptic waves in neural field models, focusing mainly on fronts and pulses (and see [9,10] for further discussion) surprisingly relatively little analysis has been performed on periodic waves. Rather in most neural field models of EEG only spatially homogeneous solutions and their instabilities to pattern states have been analysed e.g. [11-13]. Interestingly in some of these models such instabilities have been shown to give rise to periodic travelling waves, e.g. in the work of Curtu and Ermentrout on neural field models with adaptation [14] and that of Venkov *et al.* [15] in models with short-range inhibition, long range excitation and axonal delays, though not subsequently analysed in the full nonlinear regime (far from bifurcation). This begs the question as to whether, like many models for action potential wave trains in axons, it is useful to analyse and classify neural field models in terms of their dispersion curves and the large period limit where one recovers the description of a travelling pulse. This is precisely the question we address in this paper, by considering a minimal one dimensional single population neural field model. Importantly the techniques we develop are readily applicable to more general neural field models, including those with multiple populations in higher dimensions and with various forms of feedback, such as spike frequency adaptation.

In Section "Neural tissue models" we discuss some simple neural field models with a focus on those with purely excitatory connectivity and linear adaptation as these are perhaps the simplest ones that support travelling waves. Direct simulations are used to show that not only do they manifest travelling pulses and periodic wave trains, but also a localised wave of decreased activity that we term an anti-pulse. The existence and stability of both pulses and anti-pulses is determined in Section "Travelling fronts and pulses" for the case of an idealised Heaviside firing rate function. Novel numerical techniques for handling more general firing rate functions are developed in Section "Numerical techniques for analysing neural fields with sigmoidal firing rates". Not only do these allow us to show that the analytical results obtained for Heaviside firing rates are similar to those for steep sigmoidal functions, they also open the door for the numerical study of periodic waves and their dispersion curves. These are computed in Section "Dispersion curves of wavetrains" and discussed in the context of a kinematic theory that has previously been used so effectively for excitable models of axons to predict and organise irregular patterns of wave propagation. Similarly we find that, when combined with a kinematic theory, dispersion curves for neural field models are a powerful tool for understanding exotic patterns of travelling wave activity (beyond periodic waves). Finally in Section "Discussion and conclusions" we review the results obtained and discuss natural extensions of the work in this paper.

Neural tissue models

Neural field models describe the coarse-grained activity of neuronal populations, and make no attempt to model the detailed behaviour of single neurons. Rather they model the statistical properties of tissue activity that is more directly relevant to EEG. Due to the long-range axonal connectivities in cortex neural field models are typically formulated as non-local integral or integro-differential equations. For a recent review see [16]. Compared to the analysis of local models (of ordinary or partial differential equation type) their analysis (mathematical and numerical) is not as thoroughly developed. Fortunately, in certain physiologically realistic regimes equivalent partial differential equation (PDE) models can be formulated. In particular this then allows the use of powerful techniques from nonlinear PDE theory to be brought to bear, especially the numerical analysis of travelling waves. To illustrate this possibility we now formulate a simple continuum neural field model in one spatial dimension and present some simulations of travelling wave behaviour.

Neural field models of cortex typically assume a density of neurons at a point with inputs that arise from the delayed and weighted contribution of activity at other points in the tissue. Let us introduce the synaptic activity at a point x in the tissue at time t and call this $u(x, t)$. The output from this activity will be taken to be a population firing rate of the form $f(u)$, for some sigmoidal shape f

$$f(u) = \frac{1}{1 + \exp(-\beta(u - \theta))}, \tag{1}$$

with steepness parameter $\beta > 0$ and threshold θ. A simple neural field model then takes the symbolic form

$$Qu = \psi, \tag{2}$$

where ψ is the *drive* to the synapse and captures information about both anatomical connectivity patterns and the distribution of axonal delays, while the temporal differential operator Q describes synaptic filtering. For example to capture a post synaptic potential with a bi-exponential response of the form $e^{-\alpha_1 t} - e^{-\alpha_2 t}$ for $t > 0$ (up to a normalisation factor $\alpha_1\alpha_2/(\alpha_2 - \alpha_1)$) we would use:

$$Q = \left(1 + \frac{1}{\alpha_1}\frac{\partial}{\partial t}\right)\left(1 + \frac{1}{\alpha_2}\frac{\partial}{\partial t}\right). \tag{3}$$

To generate a normalised α-function response, $\alpha^2 t e^{-\alpha t}$, we would simply set $\alpha_1 = \alpha = \alpha_2$, and to model a synapse with a simple exponential response $\alpha e^{-\alpha t}$ we would let $\alpha_2 \to \infty$ and set $\alpha_1 = \alpha$. The drive to the synapse takes the integral form

$$\psi(x, t) = \int_\Omega dy w(|y|) f(u(x - y, t - |y|/v)), \tag{4}$$

where Ω sets the domain of the tissue, w describes the anatomical connectivity and v represents the velocity of an action potential. Models of the type (2) have a long history in the field of neural tissue modelling and have been particularly useful in understanding the generation of EEG rhythms, as in the work of Nunez [17], Jirsa and Haken [18], and Liley *et al.* [11]. Moreover, they are naturally extended to include adaptive processes, such as *spike frequency adaptation* [19] (i.e. the addition of a current that activates in the presence of high activity). These are known to favour the generation of travelling pulses over travelling fronts. In this case we would modify the drive according to $\psi \to \psi - a$, and choose a simple dynamic model for a such as

$$\left(1 + \tau \frac{\partial}{\partial t}\right) a = \kappa u. \tag{5}$$

Here τ and κ are constants describing the timescale separation between activity and adaptation and how strong the (linear) adaptation becomes, respectively. More sophisticated models of metabolic processes whose combined effect is to modulate neuronal response could also be considered, though for the purposes of this paper we shall work with (5). Moreover, for illustration we shall work with the explicit choice $w(x) = \exp(-|x|)/2$ and $Q = 1 + \partial/\partial t$, i.e. $\alpha_1 = 1, \alpha_2 \to \infty$, namely an exponentially decaying synaptic footprint and a synapse with an exponentially decaying response.

Figure 1 shows simulations of model (2) for two different values of the adaptation strength κ. We choose $1/v = 0, \theta = 1/4, \tau = 7$ and a steep sigmoidal activation function (1) with $\beta = 42$. We use an equidistant spatial discretisation with periodic boundary conditions and compute ψ using fast Fourier transforms (FFT) (as for $1/v = 0$ it has a spatial convolution structure). With the same initial conditions quite different behaviour emerges. For $\kappa = 1.1$ a travelling pulse appears, while for $\kappa = 0.9$, we observe a moving localised area with lower activity and high background activity. We will refer to this solution as a travelling anti-pulse. A solution of this type has previously been observed in [20], though not exhaustively analysed. In the next section we will do so, for a Heaviside firing rate function ($\beta \to \infty$), and show that $2\theta(\kappa + 1) = 1$ determines a critical value where the dynamics change from pulse to anti-pulse.

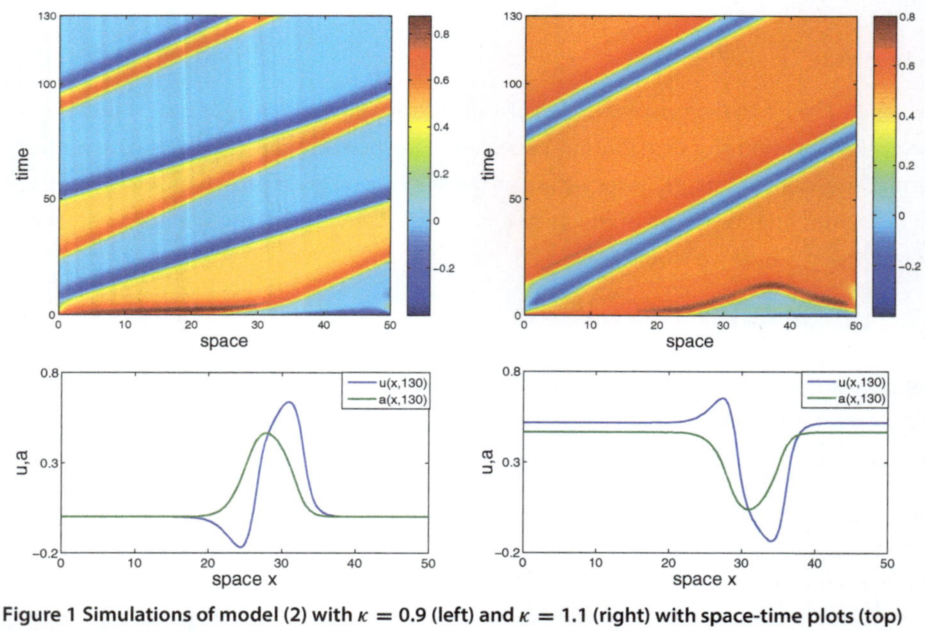

Figure 1 Simulations of model (2) with $\kappa = 0.9$ (left) and $\kappa = 1.1$ (right) with space-time plots (top) and the final profile of u, a (bottom).

Travelling fronts and pulses

The simulation in Figure 1 indicates that a stable travelling anti-pulse solution exists. We will consider the existence and stability for a Heaviside firing rate function, making use of an Evans function approach. This is a powerful tool for the stability analysis of nonlinear waves on unbounded domains. It was originally formulated by Evans [4] in the context of a stability theorem about excitable nerve axon equations of Hodgkin-Huxley type. The extension to integral models is far more recent, see [21] for a discussion. For neural field models with axonal delays it has previously been noted these do not typically induce any change of wave stability [22,23] (though it will affect the shape and speed of a wave). Since our simulations indicate similar effects for the anti-pulse, we will initially consider the case $1/v = 0$ for simplicity. The discussion below is adapted from that presented in [22,23], to which we refer the reader for further details.

Existence of the anti-pulse

To show existence and stability it is convenient to first rewrite the integro-differential equation (2) in integral form. We define $\eta(t) = \mathrm{e}^{-t}$ and $\eta_a = \mathrm{e}^{-t/\tau}/\tau$ with $\eta(t) = \eta_a(t) = 0$ if $t < 0$. The model with adaptation can then be written in the integral form

$$u = \eta * \psi - \kappa\eta * (\eta_a * u), \tag{6}$$

where $*$ denotes temporal convolution (e.g. $[\eta * \psi](x,t) = \int_0^\infty \eta(s)\psi(x, t-s)\mathrm{d}s$). The construction of travelling waves, e.g. fronts and pulses, proceeds by introducing the co-moving frame $\xi = x - ct$ with $c > 0$. Travelling waves are then stationary solutions $u(x,t) = u(\xi)$ where

$$u(\xi) = \int_0^\infty \mathrm{d}s\eta(s)\left(\psi(\xi + cs) - \kappa\int_0^\infty \mathrm{d}s'\eta_a(s'-s)u(s')\right), \tag{7}$$

with

$$\psi(\xi) = \int_{-\infty}^\infty \mathrm{d}y w(y) f(u(\xi - y)). \tag{8}$$

Introducing the Fourier transform,

$$\widehat{g}(k) = \int_{-\infty}^\infty g(\xi)e^{-ik\xi}\mathrm{d}\xi, \tag{9}$$

equation (6) can be arranged to the following form

$$\widehat{u}(k) = \widehat{\eta}_c(k)\widehat{\psi}(k), \qquad \widehat{\eta}_c(k) = \frac{\widehat{\eta}(k)}{1 + \kappa\widehat{\eta}(k)\widehat{\eta}_a(k)}. \tag{10}$$

Since $\widehat{\eta}(k) = (1+ik)^{-1}$ and $\widehat{\eta}_a(k) = (1+ik\tau)^{-1}$ we find

$$\widehat{\eta}_c(k) = \frac{-(1+ik\tau)}{\tau(k - i\lambda_+)(k - i\lambda_-)}, \quad \lambda_\pm = \frac{1}{2\tau}\left(1 + \tau \pm \sqrt{(1-\tau)^2 - 4\tau\kappa}\right). \tag{11}$$

Exploiting the pole structure of (11), an inverse Fourier transform yields

$$\eta_c(t) = \frac{1}{\tau(\lambda_- - \lambda_+)}\left((1 - \tau\lambda_+)\mathrm{e}^{-\lambda_+ t} - (1 - \tau\lambda_-)\mathrm{e}^{-\lambda_- t}\right), \quad t \geq 0. \tag{12}$$

A travelling wave solution can then be written succinctly as

$$u(\xi) = \int_0^\infty \mathrm{d}s\eta_c(s)\psi(\xi + cs). \tag{13}$$

For $f(u) = H(u - \theta)$, a Heaviside step function, the pulse solution can be found by specifying where the solution is above the threshold θ. An anti-pulse has a profile $u(\xi)$ above threshold everywhere except for $\xi = x - ct \in [-\Delta, 0]$. So the profile is formally given by

$$u(\xi) = \int_0^\infty \mathrm{d}s \eta_c(s) \left(1 - \int_{\xi+cs}^{\xi+cs+\Delta} w(y)\mathrm{d}y\right), \tag{14}$$

where we note that the width Δ and speed c are yet unknown. These can be found by imposing the crossing conditions $u(-\Delta) = u(0) = \theta$. This yields the following two equations

$$2\theta = \frac{2\tau c^2 + (2 + \tau - \tau\kappa)c + (1 + \kappa) + (1 + \kappa)(c\tau + 1)\mathrm{e}^{-\Delta}}{(\tau c^2 + (1 + \tau)c + 1 + \kappa)(1 + \kappa)}, \tag{15}$$

$$\begin{aligned}\tau(\lambda_- - \lambda_+) 2\theta = {} & \frac{1 - \lambda_+\tau}{\lambda_+ (c^2 - \lambda_+^2)} \left(2c^2\mathrm{e}^{-\lambda_+\Delta/c} + \lambda_+ c\left(1 - \mathrm{e}^{-\Delta}\right) + \lambda_+^2 \left(1 + \mathrm{e}^{-\Delta}\right)\right) \\ & - \frac{1 - \lambda_-\tau}{\lambda_- (c^2 - \lambda_-^2)} \left(2c^2\mathrm{e}^{-\lambda_-\Delta/c} + \lambda_- c\left(1 - \mathrm{e}^{-\Delta}\right) + \lambda_-^2 \left(1 + \mathrm{e}^{-\Delta}\right)\right).\end{aligned} \tag{16}$$

When we take the limit $\Delta \to \infty$, we see that (15) and (16) with $\varepsilon = \tau^{-1}$ reduce to

$$2\theta - \frac{2c^2 + c(1 - \kappa + 2\varepsilon) + \varepsilon(1 + \kappa)}{(c^2 + c(1 + \varepsilon) + \varepsilon(1 + \kappa))(1 + \kappa)} = 0, \tag{17}$$

which recovers a result in [24] for right-moving inactivating fronts (left-moving waves in their case). For completeness we note that the travelling pulse satisfies

$$2\theta = \frac{(1 + c\tau)\left(1 - \mathrm{e}^{-\Delta}\right)}{(1 + c)(1 + c\tau) + \kappa}, \tag{18}$$

$$\begin{aligned}2(\lambda_- - \lambda_+)\tau\theta = {} & \frac{(1 - \lambda_-\tau)}{\left(c^2 - \lambda_-^2\right)\lambda_-} \left(2c^2 \left(\mathrm{e}^{-\lambda_-\Delta/c} - 1\right) + \lambda_- (\lambda_- + c)\left(1 - \mathrm{e}^{-\Delta}\right)\right) \\ & - \frac{(1 - \lambda_+\tau)}{\left(c^2 - \lambda_+^2\right)\lambda_+} \left(2c^2 \left(\mathrm{e}^{-\lambda_+\Delta/c} - 1\right) + \lambda_+ (\lambda_+ + c)\left(1 - \mathrm{e}^{-\Delta}\right)\right).\end{aligned} \tag{19}$$

These are determined analogously as for the anti-pulse but with the requirement that the profile is above threshold only for $\xi \in (-\Delta, 0)$ [24]. The speed of activating fronts (moving to the right) can be obtained from (18) in the limit $\Delta \to \infty$. The above equations apply to right-moving waves, i.e. $c > 0$.

Stability of the anti-pulse

We will now construct the Evans function. First, we linearise (6) around the travelling solution $\bar{u}(\xi)$ by writing $u(\xi, t) = \bar{u}(\xi) + z(\xi, t)$. In particular, we will look for exponentially decaying/growing separable solutions of the form $z(\xi, t) = z(\xi)e^{\lambda t}$. Collecting $O(z)$ terms yields the eigenvalue equation

$$z(\xi) = \frac{1}{c}\int_{-\infty}^{\infty} \mathrm{d}y w(y) \int_{\xi-y}^{\infty} \mathrm{d}r \eta_c((r + y - \xi)/c)\mathrm{e}^{\lambda(\xi-r-y)/c} f'(\bar{u}(r))z(r). \tag{20}$$

For the Heaviside firing rate function we have $f'(u(r)) = \delta(u(r) - \theta)/|u'(r)|$, which appears inside the integral only and hence poses no difficulties. Since we have crossing points at $\xi = 0$ and $\xi = -\Delta$, we get $z(\xi) = A(\lambda, \xi)z(0) + B(\lambda, \xi)z(-\Delta)$, where

$$A(\lambda,\xi) = \frac{1}{c|u'(0)|}\int_{\xi}^{\infty} \mathrm{d}y\, w(y)\eta_c\left(\frac{y-\xi}{c}\right) \mathrm{e}^{\lambda(\xi-y)/c}, \tag{21}$$

$$B(\lambda,\xi) = \frac{1}{c|u'(-\Delta)|}\int_{\xi+\Delta}^{\infty} \mathrm{d}y\, w(y)\eta_c\left(\frac{y-\xi-\Delta}{c}\right) \mathrm{e}^{\lambda(\xi+\Delta-y)/c}. \tag{22}$$

The eigenvalue problem should be self-consistent at $\xi = 0$ and $\xi = -\Delta$ giving the system of equations

$$\begin{bmatrix} z(0) \\ z(-\Delta) \end{bmatrix} = \mathcal{A}(\lambda) \begin{bmatrix} z(0) \\ z(-\Delta) \end{bmatrix}, \qquad \mathcal{A}(\lambda) = \begin{bmatrix} A(\lambda,0) & B(\lambda,0) \\ A(\lambda,-\Delta) & B(\lambda,-\Delta) \end{bmatrix}. \tag{23}$$

A nontrivial solution for z exists if $\mathcal{E}(\lambda) := \det(\mathcal{A}(\lambda) - I) = 0$. The function $\mathcal{E}(\lambda)$ is the Evans function for the travelling anti-pulse. Note that the same construction applies to the travelling pulse. Solving (15) and (16) we find that $c(\kappa)$ has a turning point, so that we have a fast and a slow branch, denoted by c_+ and c_- respectively. Figure 2 shows the roots of the Evans function along the pulse and anti-pulse branches. There is a trivial eigenvalue $\lambda = 0$ due to translation invariance. The nontrivial eigenvalue is positive for the slow (anti-) pulse c_-, and negative for the fast branch c_+. Hence, the latter is stable.

A combined view on fronts and pulses

From equations (15) and (16), it is straightforward to determine the speed as a function of the adaptation strength κ, see Figure 3. For completeness, we also plot the speeds of the (right-moving) travelling anti-front, front and pulse. What is apparent is that all curves meet at the point C given by $(\kappa_c, c) = (1/(2\theta) - 1, 1/(2\theta) - 1 - 1/\tau)$. Here we have a codimension 2 heteroclinic cycle bifurcation. We have checked using simulations that varying κ for Gaussian connectivity induces a similar transition between pulse and anti-pulse behaviour. The unfolding of this bifurcation involves two heteroclinic and two homoclinic bifurcation curves. It implies that the anti-pulse exists generically.

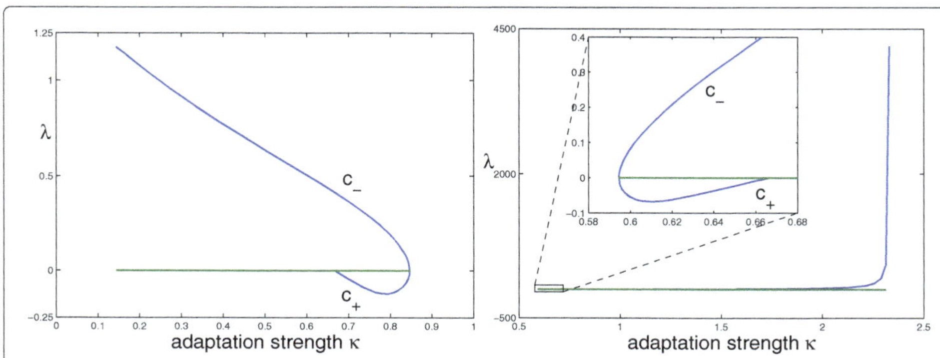

Figure 2 Eigenvalues for the travelling pulse (left) and anti-pulse (right). Parameters are $\theta = 0.3$ and $\tau = 7$.

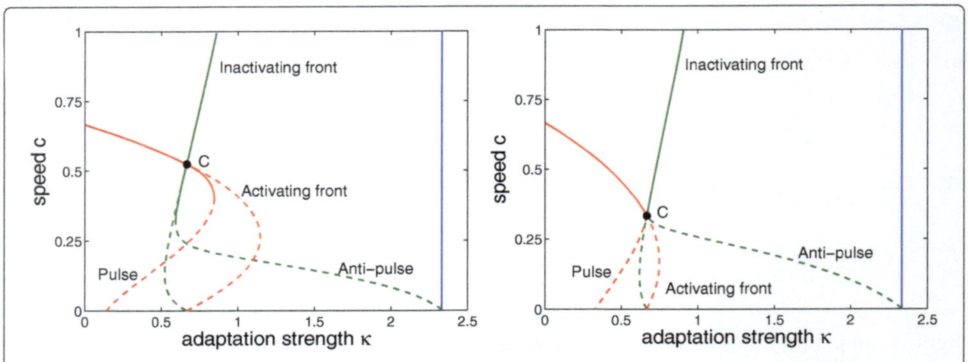

Figure 3 Speed c of fronts and waves as a function of adaptation strength κ for $\tau = 7$ (left) and $\tau = 3$ (right) with $\theta = 0.3$ fixed. Colour indicates activating fronts and travelling pulse (red) and inactivating fronts and travelling anti-pulses (green) and stable/unstable parts are solid/dashed. All curves come together at the heteroclinic cycle at ($\kappa_C = 2/3$ and $c = 11/21$ (left) and $c = 1/3$ (right). The vertical solid blue line indicates $(1 + \kappa)\theta = 1$ beyond which the nontrivial steady state does not exist.

We have also plotted the profiles for $\kappa = 0.65$ and $\kappa = 0.75$ with $\tau = 7$ and $\theta = 0.3$, see Figure 4. This shows the difference between the slower and faster solutions. The after-overshoot in activity is due to diminished adaptation during the anti-pulse. When τ is decreased, two things happen. First, the profile develops a slow and strongly damped oscillation as $\lambda_\pm$ become complex. Second, the fast branch of stable pulse solutions disappears, see Figure 3 (right) as the adaptation is fast and pulls down all activity quickly. This happens when the pulse and anti-pulse branch have become tangent to the inactivating front and front branches for $\Delta \to \infty$, respectively, i.e. they have interchanged their position in the (c, κ) plane.

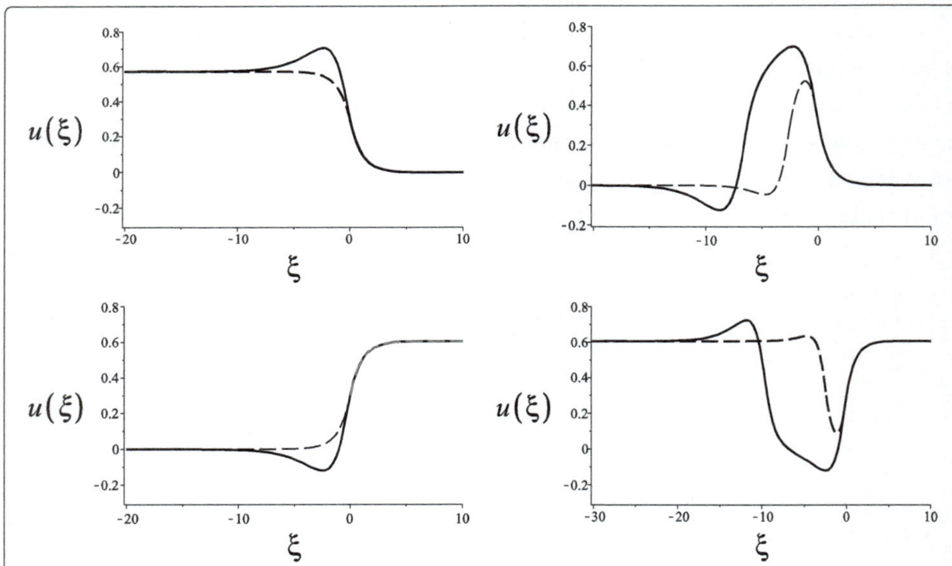

Figure 4 The profile of the front and pulse solutions all travelling to the right with $\theta = 0.3$, $\tau = 7$ and Heaviside activation function. The dashed solutions are slower and unstable. (Top) Travelling activating front (Left) and pulse (Right) for $\kappa = 0.75$. (Bottom) Travelling inactivating front (Left) and anti-pulse (Right) for $\kappa = 0.65$. The solid curve ($\Delta = 9.346, c = 0.4858$) corresponds to a stable anti-pulse and the dashed ($\Delta = 2.394, c = 0.243$) to an unstable solution.

Numerical techniques for analysing neural fields with sigmoidal firing rates

Despite the elegant analysis that can be done for a Heaviside firing rate function this is a special case and the techniques developed above for existence and stability do not generalise. A natural alternative approach is to develop numerical schemes for investigating the persistence of fronts and pulses for more general sigmoidal firing rate functions, such as given by equation (1), as well as the inclusion of axonal delays where $1/v$ is finite. In this section we develop two complementary numerical schemes, both of which are easily embedded within a continuation framework to track travelling wave solution branches in parameter space. The first exploits shapes of synaptic connectivity for which an equivalent PDE model can be obtained, as in the Jirsa-Haken-Nunez brain wave equation [18], and the second is a more general approach that has no restriction on the choice of kernel shape. In both cases the challenge is to compute the synaptic drive (for a travelling wave) given by:

$$\psi(\xi) = \int_{-\infty}^{\infty} w(|y|) f(u(\xi - y + c|y|/v))\mathrm{d}y, \tag{24}$$

in a computationally efficient and accurate fashion.

Numerical continuation I: Equivalent PDE

Although neural field models are inherently nonlocal, for special choices of connectivity an equivalent PDE model can be formulated, as described in [9]. This has also been used for simulation and continuation in, for instance, [11,25-28]. In this case travelling wave and pulse solutions are readily found with standard continuation techniques. In illustration of this point consider an exponentially decaying connectivity $w(x) = \exp(-|x|)/2$. Then on an infinite domain, it can be shown [29] that $\psi(x,t)$ given by (4) satisfies the PDE

$$\left[\mathcal{A}^2 - \partial_{xx}\right]\psi = \mathcal{A} f(u), \tag{25}$$

where

$$\mathcal{A} = \left(1 + \frac{1}{v}\partial_t\right). \tag{26}$$

Equation (25) is often referred to as the Jirsa-Haken-Nunez brain wave equation. We then look for waves $u(x,t) = u(\xi)$ with $\xi = x - ct$ of the PDE system given by (2), (5) and (25). Under the replacement $\partial_x \to \mathrm{d}_\xi$ and $\partial_t \to -c\mathrm{d}_\xi$, we obtain, for our explicit choices, the 4th order ordinary differential equation system:

$$\begin{aligned}
\frac{\mathrm{d}u}{\mathrm{d}\xi} &= (u - \psi + a)/c, \\
\frac{\mathrm{d}\psi}{\mathrm{d}\xi} &= \phi, \\
\frac{\mathrm{d}\phi}{\mathrm{d}\xi} &= \frac{f(u) + f'(u)(-u + \psi - a)/v + 2c\phi/v - \psi}{c^2/v^2 - 1}, \\
\frac{\mathrm{d}a}{\mathrm{d}\xi} &= (a - \kappa u)/(c\tau).
\end{aligned} \tag{27}$$

It is easily checked that the model has either one or three equilibria with $(u, \psi, \phi, a) = (\overline{u}, f(\overline{u}), 0, \kappa\overline{u})$, where $\overline{u}$ is a solution of $\overline{u} = f(\overline{u})/(1+\kappa)$. Orbits connecting one or two equilibria are homoclinic and heteroclinic solutions corresponding to travelling pulses and fronts, respectively. These are readily analysed using a numerical toolbox such as

MATCONT [30]. However, the equivalent PDE approach is limited to certain special choices of synaptic connectivity possessing a rational Fourier transform structure [25].

Numerical continuation II: Integral model

Many modelling studies employ Gaussian connectivity rather than a decaying exponential for which an equivalent (finite order) PDE cannot be derived. In such cases, we spatially discretise the neural field and do continuation of waves in the moving frame $\xi = x - ct$ directly as in [31] (for a neural field model with refractoriness) and in [32] for stationary patterns (in a model without axonal delays). Here we will work with the choice $w(x) = \sigma \exp(-(\sigma x/2)^2)/2\sqrt{\pi}$, with $\sigma > 0$. We discretize u and a on an equidistant mesh and compute derivatives u_ξ, a_ξ with central finite differences. Typically, we use $2^N, N = 11$ mesh points which gave stable numerical results. Next we use periodic boundary conditions and an integral phase condition, i.e. $\int (u - u_0)\dot{u} d\xi = 0$ where u_0 is some reference solution, e.g. the previously computed orbit. Next we compute the spatial convolution using FFT. When we consider axonal delays, i.e. $0 < v < \infty$, the convolution structure is not obvious. The crucial point is that the drive (for stationary solutions in the traveling wave frame) is a sum of convolutions that can each be computed efficiently using FFTs. To see this we break the integration of $\psi(\xi)$ into two parts, one over the left-half of the real line and the other over the right. Introducing $a_\pm = c/v \pm 1$ allows us to write (24) as

$$\begin{aligned} \psi(\xi) &= \int_0^\infty w(y) \left[f(u(\xi + y + cy/v)) + f(u(\xi - y + cy/v)) \right] dy \\ &= \frac{1}{a_-} \int_0^\infty w(z/a_-) f(u(\xi + z)) dz + \frac{1}{a_+} \int_0^\infty w(z/a_+) f(u(\xi + z)) dz, \end{aligned} \tag{28}$$

and in the last line we see that each term separately has a convolution structure. We conclude that $\psi(\xi)$ (with finite v) can be cast into a suitable form by an asymmetric scaling of the connectivity function. In the limit when the wavespeed equals the transmission speed, the scaled connectivity function $w(z/a_-)$ approaches a delta distribution. Initial data is taken from a stable wave found with simulations. We then compute periodic solutions of the integral model with standard pseudo-arclength continuation with free parameters the speed c and the size of the spatial domain T.

Numerical results

The results from the analysis with a Heaviside activation function (for fronts and pulses) are recovered for smooth, but steep sigmoidal activation functions. For illustration we choose $\beta = 42$ and keep $\tau = 7$ and $\theta = 0.30$, see Figure 5. The diagram with the heteroclinic cycle persists including the values of κ where the fronts and pulse emanate from $c = 0$. The curve for the speed of the anti-pulse also exhibits the same limiting behaviour for parameters near the heteroclinic cycle bifurcation. However, the homoclinic bifurcation curve corresponding to the anti-pulse now emanates from a Zero-Hopf (ZH) bifurcation point. Its unfolding involves homoclinic and heteroclinic bifurcation curves that oscillate towards the ZH-point generically. The corresponding waves have small amplitude and are unstable. Note also that the critical value of κ corresponding to a saddle-node bifurcation in which the higher steady state disappears, is lower compared to the case of Heaviside activation function. Additionally, there is a Hopf bifurcation curve

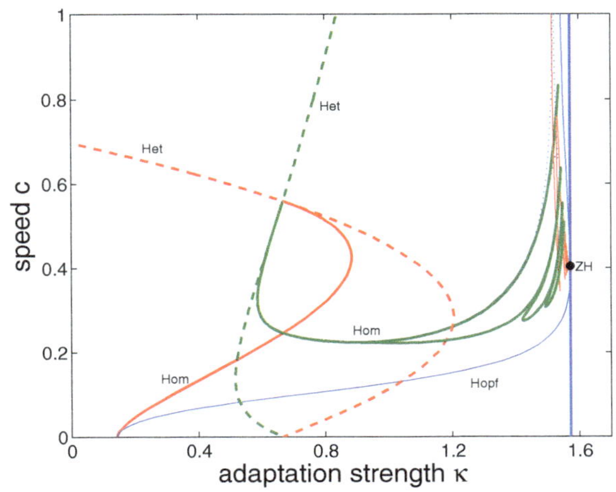

Figure 5 The wavespeed c as a function of κ for waves, pulses and fronts. Other parameters fixed at $\theta = 0.3, \beta = 42, \tau = 7$. Blue: saddle-node of equilibria (solid) and limit point of cycles (dotted); black: Hopf; solid red/green curves: homoclinic orbit; dashed red/green curves: heteroclinic.

starting from $(\kappa, c) \approx (1/\tau, 0)$ where the homoclinic curve corresponding to pulse solutions emanates as well. Fixing system parameters, we can follow a periodic orbit from the Hopf curve by continuation in c and the period T. Continuing further, the periodic orbits accumulate on a homoclinic curve corresponding to the anti-pulse or pulse, which ever occurs first. There is also a homoclinic orbit to the middle steady state (left purple curve in Figure 5), but this steady state is unstable itself and consequently the travelling wave as well. We do not consider this branch further.

Decreasing the slope parameter to $\beta = 9.5$ we obtain the bifurcation diagram shown in Figure 6 that is much more complex than for higher values of β. In essence, we find that the range of κ where three steady states coexist, shrinks. Consequently, also the range over which fronts exist shrinks, but slightly more. The travelling pulse, on the other hand,

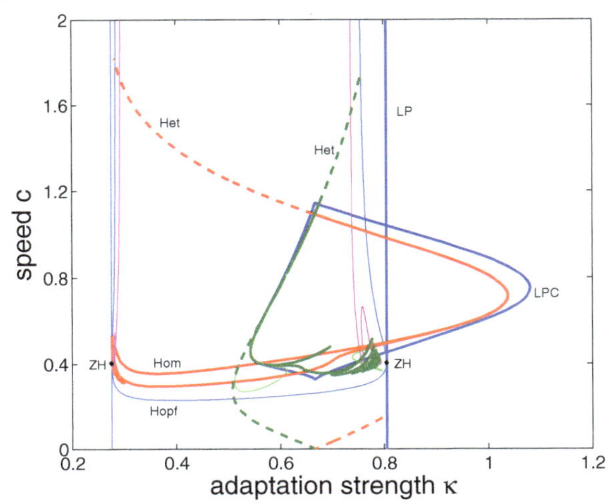

Figure 6 The wavespeed c as a function of κ for waves, pulses and fronts. Other parameters fixed at $\theta = 0.3, \beta = 9.5, \tau = 7$. Blue curves: fold and Hopf of equilibria (thin) and limit point of cycles (thick); solid red/green curves: homoclinic orbit; dashed red/green curves: heteroclinic.

exists over a much wider range of adaptation strengths, even where only the lower steady state exists. Next travelling waves have become complex as the appearance of the LPC curve indicates and as discussed in the Section "Dispersion curves of wavetrains".

The diagram contains an additional saddle-node curve and the Hopf curve no longer starts from $c = 0$, but turns. First we note that the heteroclinic cycle still persists. Also another Zero-Hopf point at the left saddle-node bifurcation curve is present together with a similar structure of homoclinic curves. One has a high speed and is associated to an unstable wave as explained above. The other homoclinic curve has several turns until at $\kappa \approx 0.5425$ it approaches the heteroclinic cycle tangentially to a heteroclinic curve. These are the anti-pulse and inactivating front, respectively. The heteroclinic curve is very similar to the one computed for $\beta = 42$, but the branch terminates when it meets the Hopf curve. The other heteroclinic curve representing the activating front is even split into two branches. Emanating from $c = 0$ it meets the right saddle-node curve tangentially and then exists for $0.2848 < \kappa < 0.7647$ between two Hopf curves for higher values of c. The upper branch is tangent to the homoclinic curve representing the travelling pulse. This homoclinic curve exists to the right of the saddle-node bifurcation where it turns and then undulates towards the left Zero-Hopf point. There is another homoclinic curve emanating from the left Zero-Hopf point with high speeds but similar to the other purple branches, it corresponds to unstable solutions. The stability of the front and pulse solutions is similar as for higher values of β. It is remarkable that many of the features obtained from the Heaviside analysis persist for not so steep sigmoidal firing rate functions (though the overall bifurcation structure can be more complex).

Travelling waves can be found by starting from a Hopf bifurcation and then to continue periodic orbits in two parameters, e.g. the period T plus the speed c or the strength κ. When the period goes to infinity, the wave approaches one of the homoclinic curves discussed above. In between, in the (c, T)-diagram the curve of periodic orbits makes several turns. These limit point of cycles (LPC) mark the increasing complexity of travelling wave solutions, as we will discuss below. Here the primary LPC's are indicated by a thick blue solid line.

Dispersion curves of wavetrains

As briefly discussed in the Background, dispersion curves for periodic wavetrains in axonal models have proved very useful for understanding the behaviour of more irregular wavetrains using a kinematic theory. Thus it is first useful to set the scene for dispersion curves in neural field models by reviewing this approach for a simple excitable fibre model with FitzHugh-Nagumo (FHN) dynamics.

Dispersion curves for a FHN model

We write the FHN model as

$$u_t = u_{xx} + u(u-1)(u-a) + v, \qquad v_t = bu, \tag{29}$$

where $(u, v) = (u(x, t), v(x, t))$, $0 < a < 1$, $b > 0$, $x \in \mathbb{R}$ and $t > 0$. Despite its simplicity this model has a rich dynamical structure with a variety of travelling waves and pulses, as discussed and analysed in [33-36]. The dispersion curve for periodic orbits (stationary in a co-moving frame) relates the speed of a wave to its period, giving $c = c(T)$. In the limit of a large period one recovers the homoclinic orbit describing a solitary travelling pulse.

For the FHN model the dispersion curve has a slow and fast branch, and a minimum period largely determined by the refractory time-scale. If the shape of the periodic wave (over one period) is pulse-like then one may invoke a kinematic theory to determine the stability of solution branches, as well as describe the evolution of inter-pulse intervals in more exotic wavetrain patterns (that can bifurcate from periodic orbits). For a review of this approach we refer the reader to [37]. Importantly periodic waves are predicted to be stable if $c'(T) > 0$, with wave bifurcations (found to be period-doubling bifurcations for the FHN model) occurring at stationary points in the dispersion curve where $c'(T) = 0$.

It is well known in the context of travelling waves in PDEs that the eigenvalues of the steady state corresponding to the homoclinic solution can strongly affect the shape of the dispersion curve [33]. The stable and unstable eigenvalues closest to the imaginary axis are called leading. When the leading eigenvalues are real, the dispersion curve approaches the asymptotic wavespeed (namely that of the solitary pulse) monotonically. When some are complex, the dispersion curve displays oscillations if the complex eigenvalues are the closest to the imaginary axis. Moreover, in the FHN model the saddle-focus boundary can be determined analytically. One can determine the dispersion curve for parameter values on both sides of the saddle-focus boundary, as in [35,38], see Figure 7. Fixing b, the origin has three real eigenvalues for larger values of a, while for lower values of a the origin is a saddle-focus. This is reflected in the dispersion curve, where for large values of a the dispersion curve is monotone and the origin is a "simple" saddle as the eigenvalues show. For smaller values of a the origin is a saddle-focus and the dispersion curve is oscillatory. Note that the monotonicity and oscillations are also visible in the activity profile.

Dispersion curves for neural field models

We turn now to the construction of dispersion curves in neural field models for pulse-like periodic waves. Although we may do this explicitly for the case of a Heaviside firing rate function (extending the technique for fronts and pulses used in Section "Travelling fronts and pulses") we prefer instead to focus on the numerical construction of dispersion curves using the approaches developed in Section "Numerical techniques for analysing neural

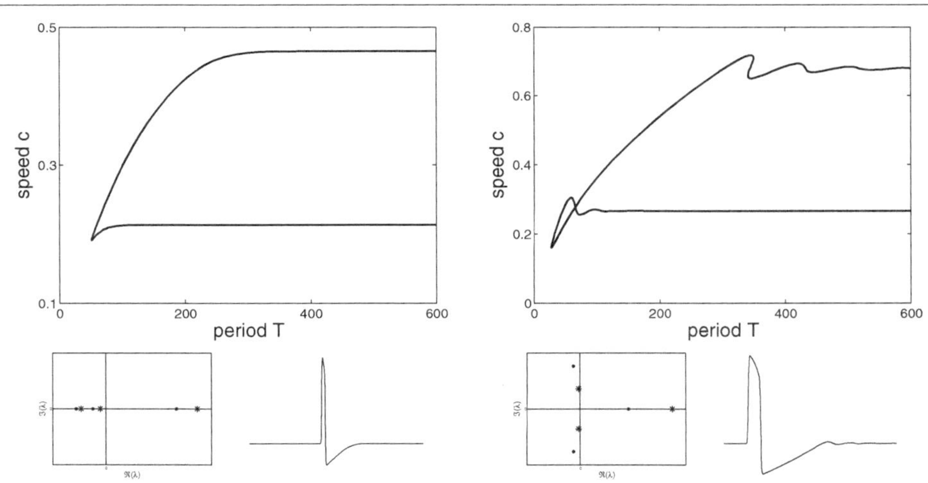

Figure 7 Dispersion curves for (left) $a = 0.15$ and (right) $a = 0.01$ and $b = 0.0025$. Stars/dots indicate eigenvalues of the origin for higher/lower asymptotic wavespeeds, respectively. The dispersion curve and the profile of u have an oscillatory tail for $a = .01$.

fields with sigmoidal firing rates". Moreover, as for the FHN model, we pinpoint the saddle-focus boundary as it predicts where one finds complex dispersion curves. Based on this we show several bifurcation diagrams where we collect the primary backbone of waves and pulses. From these diagrams it is straightforward to understand the shape of dispersion curves for various values of the system parameters.

We start with high values of β keeping $1/\nu = 0$ for now. We choose the adaptation strength κ such that there is both a fast and a slow pulse and compute the dispersion curve by solving the travelling wave equation for $u(\xi)$ for fixed speed c with the periodicity constraint that $u(\xi) = u(\xi + T)$, to determine the dispersion curve $c = c(T)$, see Figure 8. In this case the dispersion curve is monotone, which reflects that the leading eigenvalues of the steady state are real. Also the profile approaches the steady state without oscillating. The faster branch is stable according to the kinematic theory, which agrees with direct simulations.

For lower values of β, the dispersion curve is qualitatively different. As the wavenumber grows, both the fast and slow branch oscillate toward the asymptotic value, see Figure 9. The oscillations are due to complex eigenvalues of the steady state. Also the profile spirals towards the steady state. The stability of the wavetrains is much more delicate as with every other turn in the dispersion curve, there is a change of stability. We have found near every extremum a period-doubled branch $c_2(T)$ which also oscillates as $T \to \infty$ leading to period-doubled branches $c_4(T)$. This structure repeats itself, but we show only the main additional branches. Hence there exist complex stable wavetrains where one or more pulses are grouped together, see Figure 9.

The transition from saddle to (wild) saddle-focus occurs for decreasing values of β or τ or increasing κ and involves several codimension 2 homoclinic bifurcations. At the transition two scenarios are possible. First, two real leading eigenvalues can collide and become complex as shown in Figure 10. For $\tau = 7$ we found that the complex eigenvalues were already closer to the imaginary axis for $\beta < 15.78$. This is one of the two Belyakov bifurcations [39] also underlying the transition in the FHN model [35,38]. For most parameter choices, especially high τ this situation is typical. In the second scenario, for lower values of τ the transition from monotone to oscillatory dispersion curves occurs as the complex leading unstable eigenvalues are closer to the imaginary axis than the real stable one as illustrated in Figure 10 at $\beta = 22.73$ for fixed $\tau = 4.4$. For values of β below this critical value we then find oscillatory dispersion curves. We remark that at $\beta = 26.35$ the unstable eigenspace of the equilibrium of the homoclinic orbit is three-dimensional, which, to the best of our knowledge, is the first observation of this codimension 2 homoclinic

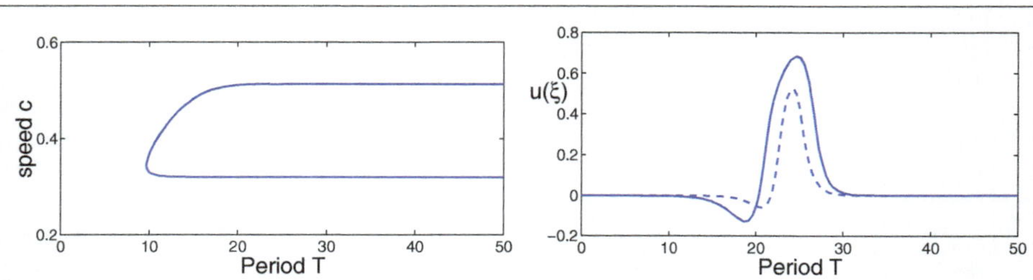

Figure 8 Left: Monotone dispersion curve for steep activation function. Parameters are $\tau = 7, \beta = 42, \theta = 0.3, \kappa = 0.8$. Right: There is a stable (solid) and an unstable wave (dashed). Wavetrains consist of this primary T-periodic profile.

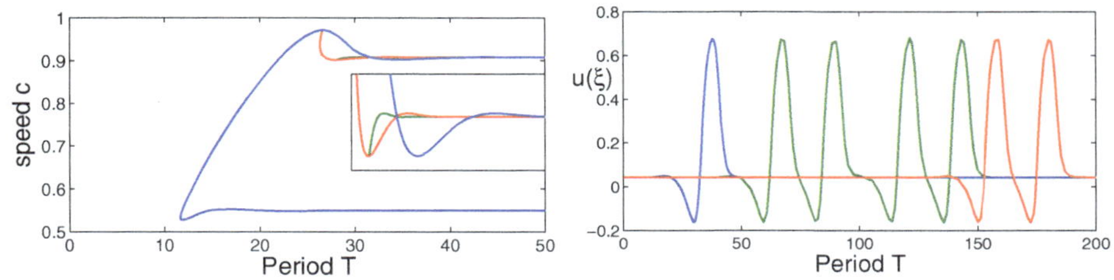

Figure 9 Left: Dependence of the wavespeed on the wavenumber with $\beta = 9.5, \kappa = 0.9, \theta = 0.3$, $\tau = 7$ and branches: 1 (blue), 2 (red), 4 (green). The spatial period has been divided by the wavenumber to illustrate that the additional branches start near the extrema. Inset shows that the green branch also emerges from an extremum. Right: The spatial arrangement of travelling pulses can be much more complex with one or more pulses grouped together. Colors correspond to the branches of the dispersion curve.

bifurcation. A detailed investigation is outside the scope of this paper, but we found no bifurcations emerging from this point. Note that when β is decreased further the complex pair of eigenvalues approaches the imaginary axis suggesting a Shilnikov-Hopf scenario, similarly as [35].

Moving parameters further, increasing κ and/or decreasing β, the dispersion curve oscillates more and more. At a certain moment, the fast branch is multi-valued. Here for given T the dispersion curve has two points with $c'(T) > 0$ on the faster branch suggesting bistability of waves. Indeed, simulations show that there are two stable waves with a different profile and velocity for the same wavelength, see Figure 11 (left). Changing parameters even more, the dispersion curve breaks up into two disconnected sets creating gaps, see Figure 11 (right). Here the primary T-periodic wave does not exist for certain values of T marking the birth of ever more complex wavetrains. Eventually, for lower values of β the part with the homoclinics disappears in a saddle-node bifurcation. Here the system is oscillatory and the dispersion relation behaves as $c = \gamma T$ for some constant γ.

Let us finally also mention the effects of model choices on dispersion curves for neural field models. Using the new numerical continuation scheme we have developed, we have computed the dispersion curves for a few model variants. We do this for parameters such that waves in all variants exist. First, and most importantly, we have not observed any qualitative changes when we change to Gaussian connectivity or alpha synapses or

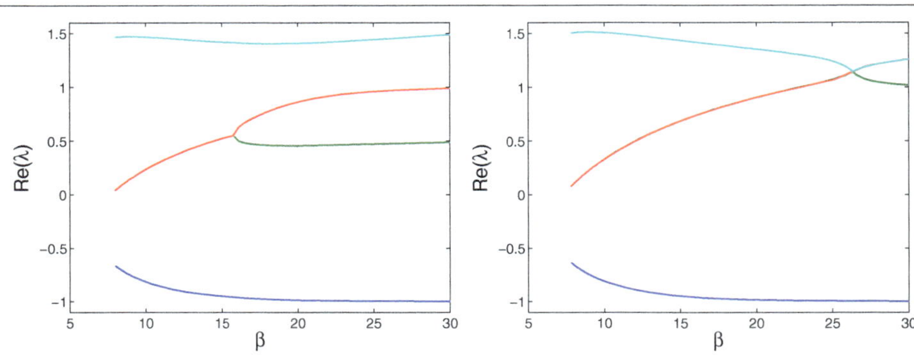

Figure 10 Eigenvalues of the homoclinic orbit related to the travelling pulse for $\kappa = 0.7, \theta = 0.3$ as β and c are varied for $\tau = 7$ (left) and $\tau = 4.4$ (right).

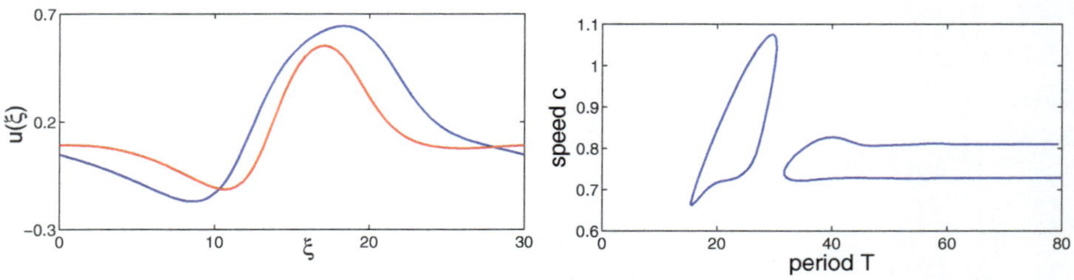

Figure 11 (left) A multi-valued (upper branch of) the dispersion curve leads to bistability of waves for $T = 30, \beta = 8, \kappa = 0.96, \tau = 7, \theta = 0.3$. The blue profile has $c = 1.215$ and the red profile has $c = 0.812$. Note that the profile of the slower wave has an additional oscillation near the steady state. (right) Creation of isolas and dispersion gaps. There are no T-periodic waves for $30.34 < T < 31.7$ with $\beta = 8, \kappa = 1.15, \tau = 10, \theta = 0.3$.

include delays. The quantitative aspects are still considerable though. The use of Gaussian connectivity can increase the speed, while transmission delays slow them down, see Figure 12 (left). Their combined effect shows that Gaussian connectivity still increases the speed. In addition, we have observed that as κ is increased, the waves first disappear for exponential connectivity and then still exist for Gaussian connectivity. For an alpha synapse the wavespeed decreases roughly twofold. Also the wavelength at which the first extremum in the dispersion curve occurs, is halved, see Figure 12 (right).

Discussion and conclusions

In this paper we have developed the numerical tools to explore and continue travelling wave solutions for non-local neural field models with space-dependent axonal delays. Moreover, we have validated our approach against the analytically tractable case of a Heaviside firing rate and shown how bifurcation diagrams of this special case are modified as one moves toward more physiologically realistic shallower sigmoidal firing rate shapes. Interestingly we have shown that as well as pulses and fronts expected for excitatory networks with inhibitory feedback, also anti-pulses are a robust travelling wave solution. Moreover, the bifurcation diagram for travelling localised states, i.e. fronts, pulses, and anti-pulses, is organised around a co-dimension 2 heteroclinic cycle bifurcation. Our main result, however, has been the numerical construction of dispersion curves. We have shown that they offer similar insight into the behaviour and instability of periodic travelling waves as originally found in their application to excitable reaction diffusion systems. Namely, that the eigen-spectrum of fixed points for a homoclinic orbit corresponding to a

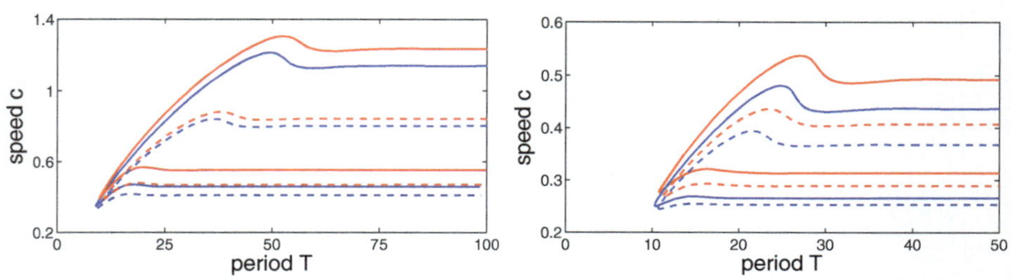

Figure 12 Dispersion curves using exponential (left) and alpha (right) synapses. Red curves for Gaussian connectivity and blue for exponential connectivity. Solid lines indicate no delay ($1/v = 0$), dashed lines indicate $v = 4$. Other parameters fixed at $\sigma = 1, \beta = 9, \kappa = 0.75, \tau = 10, \theta = 0.3$.

pulse can strongly affect the shape of the dispersion curve by complex leading eigenvalues arising through a saddle to saddle-focus transition leading to oscillations in the dispersion curve), as the LPC curve in Figure 6 indicates. Not only can this lead to wave instabilities at stationary points which predicted by a kinematic theory, are period doubling bifurcations, it also leads to multi-stability and waves of super-normal speed, as seen in Figures 5, 6 and 11. Such super-normal speeds are greater than that of the isolated pulse and can be read off from the dispersion curve for large values of the period. Furthermore it is possible to see gaps in dispersion curve for the case when β is small.

Importantly our work opens up a novel way to classify the behaviours (and the consequences for spatio-temporal wave propagation) for a broad spectrum of neural field models that have been used in the modelling of EEG, such as the Liley model [11], and in particular those for epilepsy, such as in the work of Marten *et al.* [40] and Goodfellow *et al.* [13]. Namely, we expect the main similarities, or differences, between these models to be captured through a comparison of their dispersion curves. We conjecture that this may clarify why coherent oscillations, as opposed to travelling waves, are found to be favoured in the model presented in [40]. This, and related work on how to sculpt the shape of dispersion curves through the detailed form of the neural field model using multiple populations, higher-order synapse models, distributions of axonal speeds and so on, will be reported upon elsewhere.

Competing interests
The authors declare that they have no competing interests.

Authors' contributions
HM and SC have written the paper. Both authors have given approval for its publication after reading the final paper.

Author details
[1]Department of Applied Mathematics, MIRA Institute of Biomedical Technology and Technical Medicine, University of Twente, Postbus 217, 7500 AE, Enschede, The Netherlands. [2]Center for Mathematical Medicine & Biology, School of Mathematical Sciences, University of Nottingham, NG7 2RD, Nottingham, UK.

References
1. Hodgkin AL, Huxley AF: **A quantitative description of membrane and its application to conduction and excitation in nerve.** *J Physiol* 1952, **117:**500–544.
2. Rinzel J: **Electrical excitability of cells, theory and experiment: review of the Hodgkin-Huxley foundation and an update.** *Bull Math Biol* 1990, **52:**3–23.
3. Carpenter G: **Traveling wave solutions of nerve impulse equations.** *PhD thesis.* University of Wisconsin, Madison; 1974.
4. Evans J: **Nerve axon equations: IV The stable and unstable impulse.** *Indiana University Math J* 1975, **24:**1169–1190.
5. Miller RN, Rinzel J: **The dependence of impulse propagation speed on firing frequency, dispersion, for the Hodgkin-Huxley model.** *Biophys J* 1981, **34:**227–259.
6. Steriade M, Jones EG, Línas RR: *Thalamic Oscillations and Signalling.* New York: Wiley; 1990.
7. Connors BW, Amitai Y: **Generation of epileptiform discharges by local circuits in neocortex.** In *Epilepsy: Models, Mechanisms and Concepts.* Edited by Schwartzkroin PA. Cambridge: Cambridge University Press; 1993:388–424.
8. Ermentrout GB, Kleinfeld D: **Traveling electrical waves in cortex: Insights from phase dynamics and speculation on a computational role.** *Neuron* 2001, **29:**33–44.
9. Coombes S: **Waves, bumps, and patterns in neural field theories.** *Biol Cybernetics* 2005, **93:**91–108.
10. Bressloff PC: **Spatiotemporal dynamics of continuum neural fields.** *J Phys A* 2012, **45:**033001.
11. Liley DTJ, Cadusch PJ, Dafilis MP: **A spatially continuous mean field theory of electrocortical activity.** *Network: Comput Neural Syst* 2002, **13**(1):67–113.
12. Breakspear M, Roberts JA, Terry JR, Rodrigues S, Mahant N, Robinson PA: **A unifying explanation of primary generalized seizures through nonlinear brain modeling and bifurcation analysis.** *Cerebral Cortex* 2006, **16:**1296–1313.
13. Goodfellow M, Schindler K, Baier G: **Self-organised transients in a neural mass model of epileptogenic tissue dynamics.** *NeuroImage* 2011, **55:**920–932.
14. Curtu R, Ermentrout B: **Pattern formation in a network of excitatory and inhibitory cells with adaptation.** *SIAM J Appl Dynamical Syst* 2004, **3:**191–231.

15. Venkov NA, Coombes S, Matthews PC: **Dynamic instabilities in scalar neural field equations with space-dependent delays.** *Physica D* 2007, **232:**1–15.
16. Coombes S: **Large-scale neural dynamics: simple and complex.** *NeuroImage* 2010, **52:**731–739.
17. Nunez PL: **The brain wave equation: a model for the EEG.** *Math Biosci* 1974, **21:**279–297.
18. Jirsa V K Haken, H: **Field theory of electromagnetic brain activity.** *Phys Rev Lett* 1996, **77:**960–963.
19. Benda J, Herz AVM: **A universal model for spike-frequency adaptation.** *Neural Comput* 2003, **15:**2523–2564.
20. Coombes S, Laing CR: **Delays in activity based neural networks.** *Philos Trans R Soc A* 2009, **367:**1117–1129.
21. Sandstede B: **Evans functions and nonlinear stability of travelling waves in neuronal network models.** *Int J Bifurcation and Chaos* 2007, **17:**2693–2704.
22. Coombes S, Lord GJ, Owen MR: **Waves and bumps in neuronal networks with axo-dendritic synaptic interactions.** *Physica D* 2003, **178:**219–241.
23. Coombes S, Owen MR: **Evans functions for integral neural field equations with Heaviside firing rate function.** *SIAM J Appl Dynamical Syst* 2004, **3:**574–600.
24. Bressloff PC, Folias SE: **Front bifurcations in an excitatory neural network.** *SIAM J Appl Dynamical Syst* 2004, **65:**131–151.
25. Laing CR, Troy WC: **PDE methods for nonlocal models.** *SIAM J Appl Dyn Syst* 2003, **2:**487–516.
26. Laing C: **Spiral waves in nonlocal equations.** *SIAM J Appl Dynamical Syst* 2005, **4**(3):588–606.
27. Shusterman V, Troy WC: **From baseline to epileptiform activity: a path to synchronized rhythmicity in large-scale neural networks.** *Phys Rev E* 2008, **77:**061911.
28. Steyn-Ross ML, Steyn-Ross DA, Sleigh JW: **Interacting Turing-Hopf instabilities drive symmetry-breaking transitions in a mean-field model of the cortex: a mechanism for the slow oscillation.** *Phys Rev X* 2013, **3:**021005.
29. Jirsa VK, Haken H: **A derivation of a macroscopic field theory of the brain from the quasi-microscopic neural dynamics.** *Physica D* 1997, **99:**503–526.
30. Dhooge A, Govaerts W, Kuznetsov YA, Meijer HGE, Sautois B: **New features of the software MatCont for bifurcation analysis of dynamical systems.** *Math Comput Modell Dynamical Syst* 2008, **14**(2):147–175.
31. Meijer HGE, Coombes S: **Travelling waves in a neural field model with refractoriness.** *J Math Biol* 2014, **68**(5):1249–1268. online first, DOI:10.1007/s00285-013-0670-x.
32. Rankin J, Avitabile D, Baladron J, Faye G, Lloyd DJ: **Continuation of localised coherent structures in nonlocal neural field equations.** *SIAM J Sci Comput*, **36-1**(2014):B70-B93, arXiv:1304.7206.
33. Kuznetsov YA: **Impulses of a complicated form in models of nerve conduction.** *Selecta Mathematica (formerly Sovietica)* 1994, **13:**127–142.
34. Champneys AR, Kirk V, Knobloch E, Oldeman BE, Sneyd J: **When Shilnikov meets Hopf in excitable systems.** *SIAM J Appl Dynamical Syst* 2007, **6:**663–693.
35. Röder G, Bordyugov G, Engel H, Falcke M: **Wave trains in an excitable FitzHugh-Nagumo model: bistable dispersion relation and formation of isolas.** *Phys Rev E* 2007, **75:**036202.
36. Guckenheimer J, Kuehn C: **Homoclinic orbits of the Fitz Hugh-Nagumo equation: bifurcations in the full system.** *SIAM J Appl Dynamical Syst* 2010, **9**(1):138–153.
37. Keener J, Sneyd J: *Mathematical Physiology.* New York: Springer; 1998.
38. Kuznetsov YA: *Elements of Applied Bifurcation Theory, 3rd edition.* New York: Springer; 2004.
39. Homburg AJ, Sandstede B: **Homoclinic and heteroclinic bifurcations in vector fields.** In *Handbook of Dynamical Systems. Volume III,* Chap. 8. Edited by Broer H, Takens F, Hasselblatt B. Amsterdam: Elevier; 2010:379–524.
40. Marten F, Rodrigues S, Benjamin O, Richardson MP, Terry JR: **Onset of poly-spike complexes in a mean-field model of human EEG and its application to absence epilepsy.** *Philos Trans R Soc A* 2009, **367:**1145–1161.

3

Extracting novel information from neuroimaging data using neural fields

Dimitris A Pinotsis* and Karl J Friston

* Correspondence: d.pinotsis@ucl.ac.uk
The Wellcome Trust Centre for Neuroimaging, University College London, Queen Square, London WC1N 3BG, UK

Abstract

We showcase three case studies that illustrate how neural fields can be useful in the analysis of neuroimaging data. In particular, we argue that neural fields allow one to: (i) compare evidences for alternative hypotheses regarding neurobiological determinants of stimulus-specific response variability; (ii) make inferences about between subject variability in cortical function and microstructure using non-invasive data and (iii) estimate spatial parameters describing cortical sources, even without spatially resolved data.

Keywords: Neural field theory; Dynamic causal modelling; Attention; Connectivity; Gamma oscillations; V1; Electrocorticography; Visual cortex; Electrophysiology

Introduction

This paper reviews some recent applications of neural field models in the analysis of neuroimaging data. Neural fields model current fluxes as continuous processes on the cortical sheet, using partial differential equations (PDEs). The key advance that neural field models offer, over other population models (like neural masses), is that they embody spatial parameters (like the density and extent of lateral connections). This allows one to model responses not just in time but also over space. Conversely, these models are particularly useful for explaining observed cortical responses over different spatial scales; for example, with high-density recordings, at the epidural or intracortical level. However, the impact of spatially extensive dynamics is not restricted to expression over space but can also have profound effects on temporal (e.g., spectral) responses at one point (or averaged locally over the cortical surface) [1]. This means that neural field models may also play a key role in the modelling of non-invasive electrophysiological data that does not resolve spatial activity directly. In what follows, we attempt to shed light on these different uses of neural fields and put forward three reasons why these models can be useful in the analysis of neuroimaging data. Each of these motivations is demonstrated by analysing a particular dataset obtained using three different modalities: electrocorticography (ECoG), magnetoencephalography (MEG) and local field potential recordings (LFPs). We argue that neural fields allow one to: (i) compare evidences for alternative hypotheses regarding the important neurobiological determinants of stimulus-specific response variability; (ii) make inferences about between subject variability in cortical function and microstructure using non-invasive data and (iii) obtain estimates of spatial parameters describing cortical sources in the absence of spatially resolved data. Our analyses exploit dynamic causal modelling [2] and include model space explorations that embody different hypotheses about the generation of observed responses in relation to model evidence - obtained using Variational Bayes [3]. This model comparison uses a variational free-energy bound to furnish optimized models in a manner similar to fitting empirical spectra with AR and ARMA models, see e.g. [4]. The advantage this approach has over other optimization criteria is that it provides an optimal balance between model fit and complexity; yielding models that are both parsimonious and accurate. The analyses presented here showcase particular instances where neural field models serve as a mathematical microscope, allowing us to extract information that is hidden in electrophysiological data.

Review

In the following, we present analyses of three different datasets obtained with ECoG, MEG and LFP recordings respectively; these illustrate how neural fields combined with dynamic causal modelling can disclose important neurobiological properties of cortical microcircuitry to which we do not have direct access through conventional analysis techniques.

Comparing alternative hypotheses about the determinants of stimulus-specific gamma peak variability

In our first study [5], we exploited high density spatially resolved data from multi-electrode electrocorticographic (ECoG) arrays to study the effect of varying stimulus contrast on cortical excitability and gamma peak frequency. These data were obtained at the lab of Pascal Fries at the Ernst Strungmann Institute for Neuroscience in collaboration with the Max Planck Society in Frankfurt. Details of the task and the surgical procedures are described elsewhere [6,7]. Neuronal signals were recorded during a task where a monkey first fixated its gaze (in a window with 1° radius) and later released a lever when it detected a colour change at fixation (see Figure 1). During the fixation

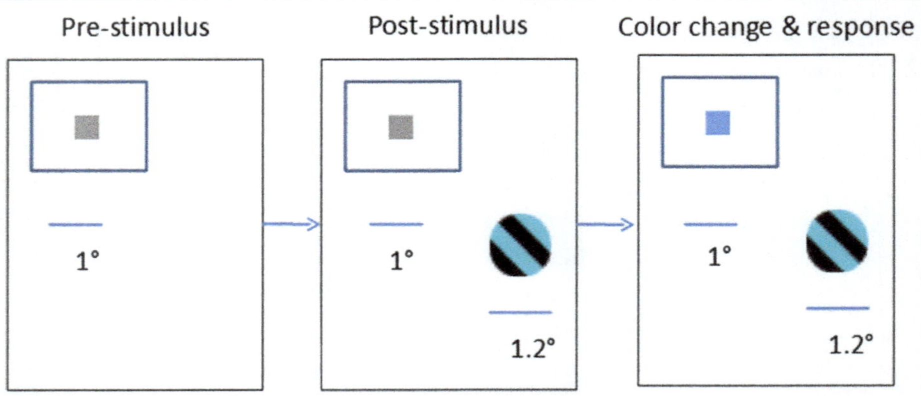

Figure 1 ***Experimental Task.*** The monkey released a lever when it detected a colour change at fixation. The colour change could occur at any moment in time following fixation onset, starting from 200 ms post-fixation onset to 3 seconds post fixation onset. While the monkey fixated, a stimulus located at 4° of eccentricity was displayed (a physically isoluminant sinusoidal grating; diameter: 1.2°; spatial frequency: 0.4-0.8 cycles/deg; drift velocity: 0.6 deg/s drifting within a circular aperture). Stimulus contrast varied between zero and an 82% contrast value.

period, a stimulus was presented at 4° eccentricity. Crucially, the contrast of the stimulus varied between trials.

We analysed the activity before the fixation colour change and the subsequent behavioural response. We obtained model predictions using a canonical microcircuit field model introduced in [8], see also [9] and Figure 2. Table 1 describes the priors over model parameters. During model optimisation, these prior means are multiplied by scale factors with lognormal priors.

The canonical microcircuit model we used conforms to known neuroanatomy and produces observed differences in gamma responses that reflect varying stimulus conditions. It provides a probabilistic mapping from hidden neuronal states to observed spectra, which are modelled as a mixture of induced activity, channel noise and sample noise. The schematic in Figure 2 summarizes the equations of motion or state equations that specify a canonical microcircuit neural mass model of a single source. This model contains four populations, each loosely associated with a specific cortical layer. The second-order differential equations describe changes in hidden states (e.g., voltage) that subtend observed local field potentials or EEG signals. These differential equations effectively mediate a linear convolution of presynaptic activity to produce postsynaptic depolarization. Average firing rates within each sub-population are then transformed through a nonlinear (sigmoid) voltage-firing rate function $\sigma(\cdot)$ to provide inputs to other

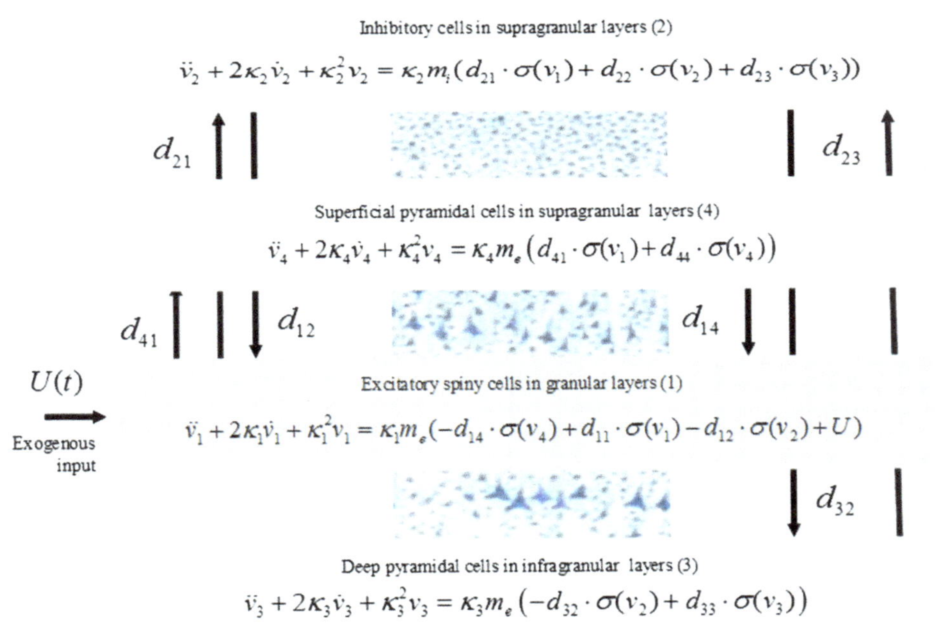

Figure 2 ***The Canonical Microcircuit (CMC) neural mass model.*** This figure shows the evolution equations that specify a CMC field model of a single source. This model contains four populations occupying different cortical layers: the pyramidal cell population of the Jansen and Rit (JR) model is here split into two subpopulations allowing a separation of the sources of forward and backward connections in cortical hierarchies. As with the JR model, second-order differential equations mediate a linear convolution of presynaptic activity to produce postsynaptic depolarization. This depolarization gives rise to firing rates within each sub-population that provide inputs to other populations.

Table 1 Prior expectations of model parameters

Parameter	Physiological interpretation	Prior mean
$\kappa_1, \kappa_2, \kappa_3, \kappa_4$	Postsynaptic rate constants	1/2, 1/35, 1/35, 1/2 (ms^{-1})
d_{11}, d_{14}, d_{12}		108,45,1.8
$d_{22}, d_{21}, d_{23}, d_{33}$	Amplitude of intrinsic connectivity kernels	9,162,18,45
d_{41}, d_{32}, d_{44}	($\times 10^3$)	36,18,9
m_e, m_i	Maximum postsynaptic depolarization	8,32 (mV)

populations. These inputs are weighted by connection strengths. In Figure 3, electrophysiological responses and model predictions are shown in dashed and solid lines. The real and imaginary parts are shown in the left and right panels respectively. In this figure, lines of different colours correspond to different contrast conditions, where the peak frequency ranges between 48 and 60 Hz and contrast levels vary between 0 and 82% of some maximum value. There are 9 contrast conditions; where in the first two there is no prominent gamma peak. It is known that gamma peak frequency is highly contrast-dependent, see e.g. [10,11]. This is in accordance to what we observe here, namely a monotonic relation between peak frequency and stimulus contrast.

Bipolar differences were extracted from ECoG sensors covering a large part of the primate brain. From the modeller's vantage point, the important issue here is that a single source in the visual cortex is sampled by several sensors. In our approach we have allowed for the optimization of both the parameters governing topographic aspects of cortical activity and the deployment of the lead fields (their dispersion and location). In brief, we endowed our observation model with sufficient degrees of freedom to characterize unknown aspects of sensor sensitivity.

In this study, we wanted to delineate important mechanisms underlying contrast effects. We tested these effects by allowing for trial-specific changes in model parameters. In

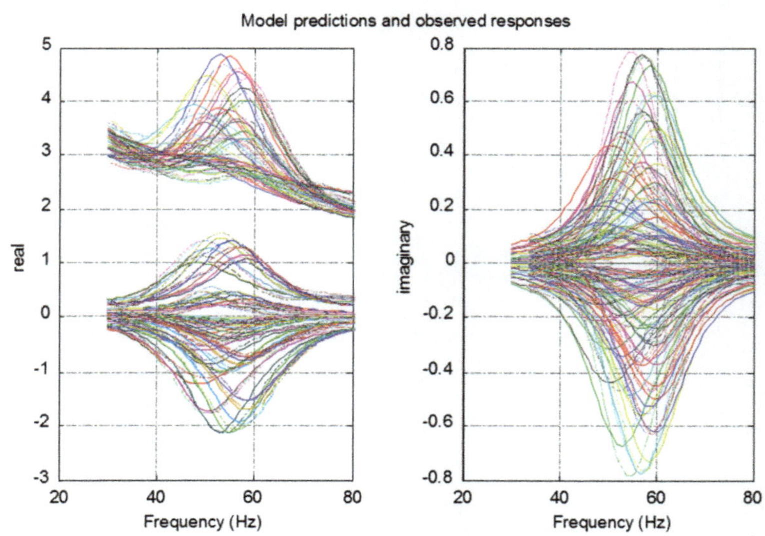

Figure 3 ***Model predictions and data for all conditions simultaneously.*** Empirical responses (cross spectral densities) and model fits are shown in dashed and solid lines. The real and imaginary parts of model fits and observed spectral responses are shown in the left and right panels respectively. The two sets of curves in the left panel correspond to auto- and cross-spectral densities from bipolar sites. Subsequent model comparisons rely on model evidence in data from all conditions.

particular, we asked which model can best explain spectral responses for different contrast conditions, see Figure 4. Our candidate DCMs allowed for contrast dependent changes in the strength of recurrent connections, the strength of intrinsic connections or the extent of intrinsic connections (or combinations thereof). The ability of each model to explain induced responses was evaluated in terms of their Bayesian model evidence; which provides a principled way to evaluate competing or complementary models or hypotheses.

The first set of parameters comprised the gains of neural populations that are thought to encode precision errors. These gain parameters correspond to the precision (inverse variance or reliability) of prediction errors in predictive coding. This fits comfortably with neurobiological theories of attention under predictive coding and the hypothesis that contrast manipulation amounts to changes in the precision of sensory input, as reviewed in [12]. These changes affect hierarchical processing in the brain and control the interaction between bottom up sensory information and top down modulation: here, we focus on the sensory level of the visual hierarchy.

The second set of parameters comprised intrinsic connection strengths among pyramidal and spiny stellate cells and inhibitory interneurons. This speaks to variations in cortical excitability, which modulates the contributions of different neuronal populations under varying contrast conditions – and a fixed dispersion of lateral connections. This hypothesis fits comfortably well with studies focusing on the activation of reciprocally connected networks of excitatory and inhibitory neurons [13-15].

The last set included the spatial extent of excitatory and inhibitory connections. From single cell recordings, it is known that as the boundary between classical and nonclassical receptive field depends crucially upon contrast [16]. As stimulus contrast is

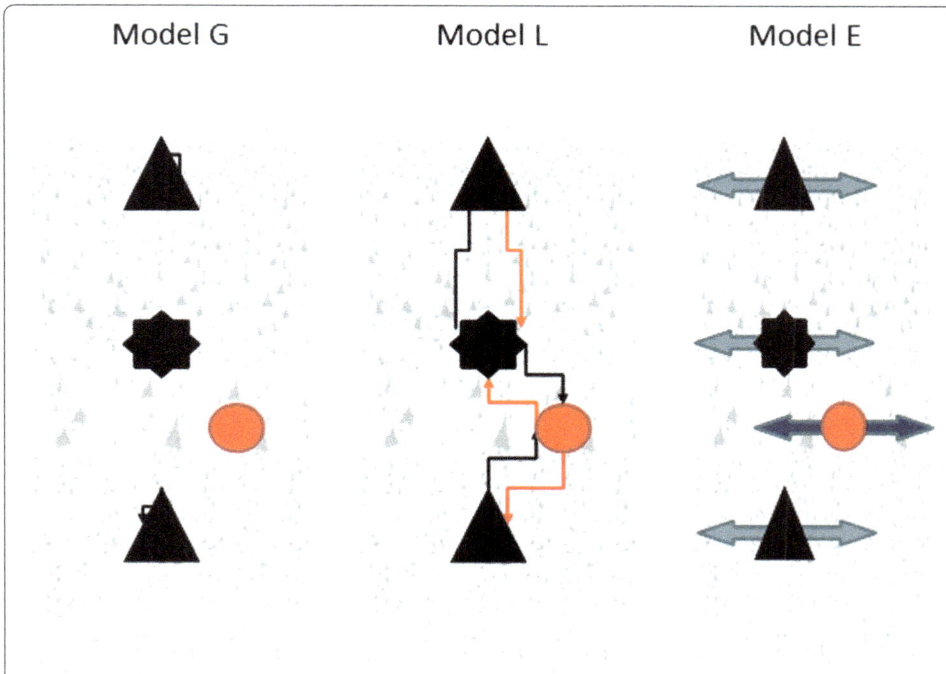

Figure 4 ***Candidate models.*** We consider three putative mechanisms of gain control that lead to models or hypotheses about trial-specific modulations of (G:) recurrent connections of neuronal populations, (L:) horizontal connections between distinct neuronal subpopulations and (E:) spatial dispersion of horizontal connections.

decreased, the excitatory region becomes larger and the inhibitory flanks become smaller. The hypothesis here rests on a modulation of the effective spatial extent of lateral connections; effective extent refers to the extent of neuronal populations that subserve stimulus integration - that are differentially engaged depending on stimulus properties (as opposed to the anatomical extent of connections).

In Figure 5, we see the results of Bayesian model comparison [17] between seven candidate models. These are the three models depicted in Figure 3 and their combinations.

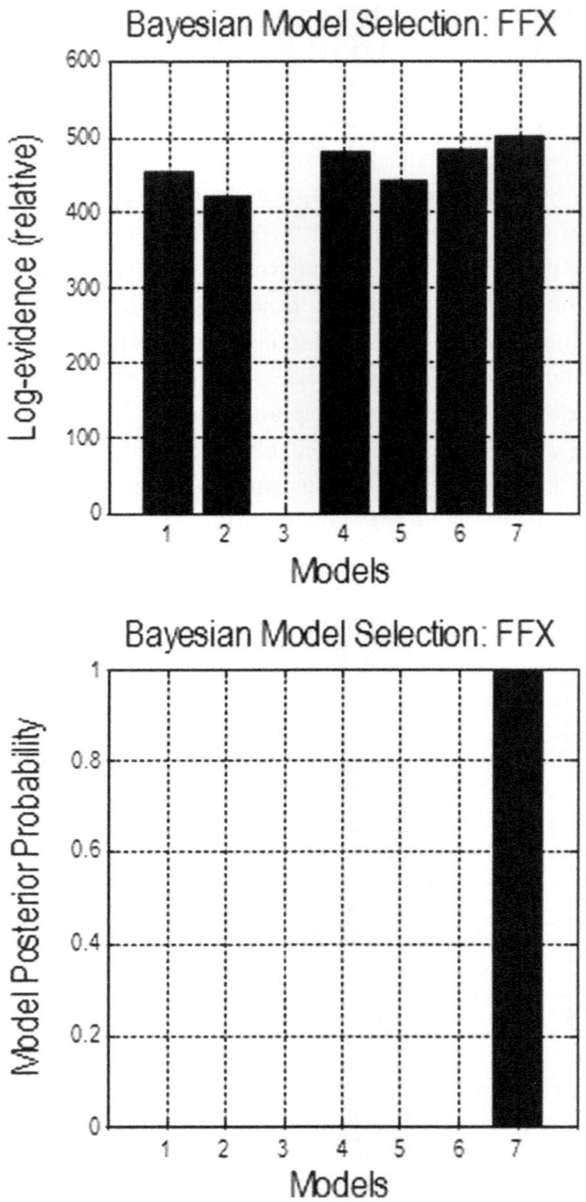

Figure 5 ***Results of Bayesian Model Comparison.*** The model involving modulations of all parameters (model 7) has the highest evidence with a relative log evidence of 17 with respect to the model that allows for modulations in all but the extent parameters (model 6). The first three models correspond to hypotheses (i), (ii) and (iii) while models 4 and 5 to combinations (i) + (ii) and (ii) + (iii). In the bottom panel we include model posterior probabilities. These results suggest that we can be almost certain that all three synaptic mechanisms contribute to the formation of cross spectral data features.

Model comparison pools the evidence for models in data from all conditions. By construction, the free energy $\mathcal{F}(G,q) \approx \ln p(G|m)$ approximates the log-evidence (marginal likelihood) of the data. In other words, the free energy reports the probability of obtaining the data under any given model. Variational free energy was introduced by Richard Feynman in the context of path integral formulations of quantum mechanics and has been used extensively in machine learning to finesse the difficult problem of exact Bayesian inference. The models we compared allowed only a subset of parameters to vary with contrast level, where each model corresponds to a hypothesis about contrast-specific effects on cortical responses. We found that the model involving modulations of all candidate parameters has the highest evidence (with a relative log evidence difference of 16 with respect to the model that allows for modulations of all but the extent parameters – model 6). This simply means that we can be confident that all three candidate mechanisms contribute to the modulation of spectral responses, which is consistent with a plethora of studies that have considered visual contrast effects in terms of adaptive (extraclassical) gain control mechanisms and predictive coding in the visual cortex.

Explaining inter-subject variability of peak gamma frequency of visually induced oscillations
Our second case study also involved spectral responses obtained from the visual cortex during a perception experiment [8]. This study was however based on non-invasive MEG data. These data were obtained during a task whose focus was on understanding how activity in the gamma band is related to visual perception [18]. In earlier work, the surface of the primary visual cortex was estimated via retinotopic mapping and was found to correlate with individual gamma peak frequencies. A similar visual MEG experiment found a correlation between gamma peak frequency and resting GABA concentration, as measured with MR spectroscopy [19].

Our focus here was on obtaining mechanistic explanations for this intersubject variability in peak gamma frequency, observed during visually induced oscillations. In particular, we wanted to disclose the origins of individual gamma peak variability and understand whether this variability can be attributed to cortical structure or function (the level of cortical inhibition as expressed by resting GABA concentration). The generative model we used in this study was similar to that of the study described above (but allowing for patchy horizontal connectivity, see [1,20]). This neural field DCM was particularly useful to discern the roles of cortical anatomy and function as it parameterises both structure and functional excitation-inhibition balance. Our model included parameters describing the dispersion of lateral or horizontal connections which we associate with columnar width and kinetic parameters describing the synaptic drive that various populations are exposed to. We focussed on estimates of the parameters describing columnar width and the excitatory drive to inhibitory interneurons. We then looked at the three-way relation between these estimates, the size of V1 and peak gamma frequency.

Our results show that both structure and function contribute to intersubject variability in gamma peaks. We found correlations between columnar width and gamma peak; this confirmed the crucial role of cortical anatomy (i.e., macroscopic V1 size) in explaining individual differences, see Figure 6 (top panel). We also found correlations with the parameter describing the drive to inhibitory interneurons, see Figure 7. This underlined the importance of cortical inhibition. This correlation remained significant

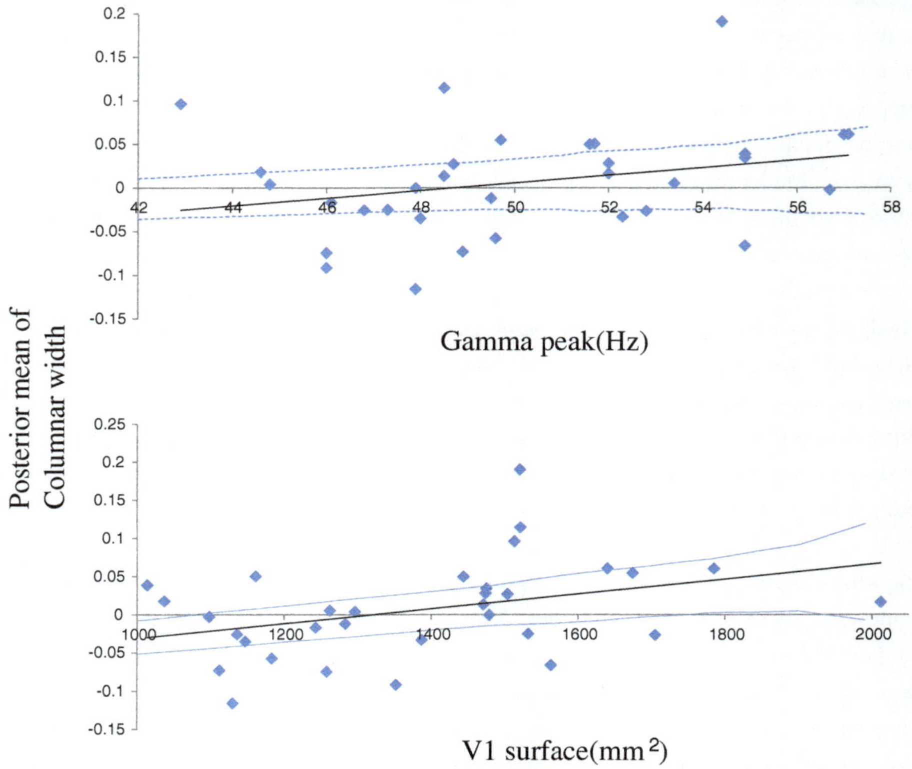

Figure 6 ***Correlations with columnar width.*** (Top panel): We found a correlation between the log scaling of the posterior estimate of columnar width and peak gamma frequency (Pearson $r = 0.271$, $p = 0.06$, 30 d.f., one-tailed test). This suggests that increases in peak gamma frequency across subjects can be attributed to a greater columnar width. (Bottom panel) Correlation between the log scaling of the posterior estimate of columnar width and V1 surface (Pearson $r = 0.36$, $p = 0.02$, 30 d.f., one-tailed test). This result suggests that a larger V1 is constituted by bigger macrocolumns. This finding, together with that reported in the top panel, confirms our earlier empirical finding that gamma peak and V1 size are correlated. Furthermore, our use of an underlying generative (mechanistic) model, offers insights into local cortical microstructure that would be hard (or impossible) to disclose otherwise. Confidence intervals are plotted as dotted lines.

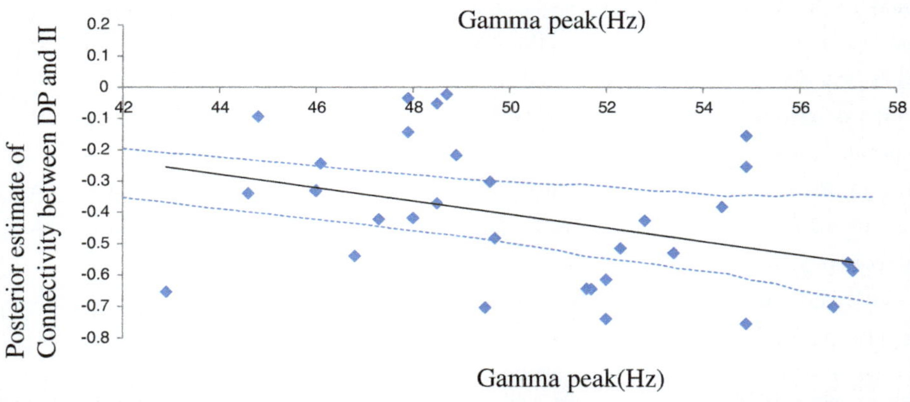

Figure 7 Correlation between the log scaling of the connectivity estimate between pyramidal cells and inhibitory interneurons and peak gamma frequency (Pearson $r = -0.37$, $p = 0.02$, 30 d.f.,two-tailed test). Confidence intervals are plotted as dotted lines.

when controlling for V1 size and columnar width, suggesting that this correlation cannot by accounted for by spatial effects at the microscopic or macroscopic level. The important thing to note here is that we have estimated microscopic properties of cortical sources (microanatomy and local inhibition) non-invasively. We also found a significant correlation between columnar width and V1 surface which implies that subject with a larger V1 might have bigger macrocolumns, see Figure 6 (bottom panel). This result is consistent with the previous empirical results and establishes the construct validity of DCM in this neural field setting.

Estimating the spatial properties of sources when there is no explicit spatial information

A common theme in the examples discussed so far has been the modelling of lateral interactions in sensory cortices. Both studies considered above exploited spatially resolved data (either invasive or non-invasive). However, it is possible to obtain estimates of parameters that describe the topographic properties of cortical sources like the extent of lateral connections, even when using spatially unresolved data, like data from a single LFP electrode. This is precisely what we did in this final case study [21]. This study considered LFP data obtained from the rat auditory cortex under anaesthesia. Local field potentials were recorded from primary (A1) auditory cortex in the Lister hooded rat, following the application of the anaesthetic agent Isoflurane; 1.4 mg (see [22] for details). Ten minutes of recordings were used to obtain frequency domain data-features $\mathbf{g}_Y(\omega)$ using a vector autoregression model [23].

In this analysis we first established the identifiability of our model using simulated data to ensure that Bayesian inversion scheme was able to recover veridical estimates. Almost universally, the fitting or inversion of Dynamic Causal Models optimizes a free energy bound on the log-evidence for a model m. This bound is optimized with respect to a variational density $q(\theta)$ over unknown model parameters. By construction, the free energy bound ensures that when the variational density maximizes free energy, it approximates the true posterior density over parameters, $q(\theta) \approx p(\theta|G, m)$. The (approximate) conditional density and (approximate) log-evidence are used for inference on parameters and models respectively (see also Figure 8).

Here, we examined the conditional mean of the transit time when the simulated log-scaling deviates from its prior expectation (of zero). This revealed how our model inversion operates in regions of parameter space that are remote from prior assumptions. Inferring axonal conduction speeds using single channel data illustrates our point that suitably informed models enable one to access the spatial attributes of neuronal infrastructures, even in the absence of spatially resolved data. We synthesized observed spectra by setting the values of the parameters equal to their prior expectations, with the exception of transit time, which was varied over a range of 36% to 272%. The synthetic data were inverted and the resulting conditional estimates are shown in terms of the conditional mean and 90% confidence intervals in Figure 9 (left panel). One can see that there is a remarkable agreement between the conditional mean and true values and that the (relatively precise) confidence intervals include the true value (with the exception of the most extreme deviations). This is an interesting result that establishes the identifiability of the model, at least in relation to conduction speed. Clearly this only establishes face validity (the inversion does what it is meant to) but does not speak to the physiological validity of the model, which is addressed using empirical data below. Having said this, this sort of validation speaks to the possibility

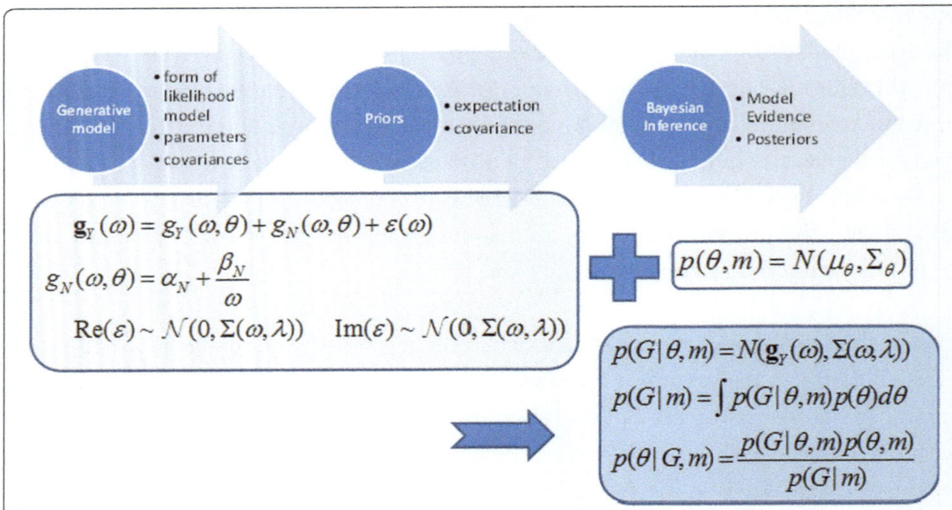

Figure 8 Summary of the approach. We assume the measured signal is a mixture of predicted spectra, channel noise (a mixture of white and pink noise with weights alpha and beta) and Gaussian observation noise epsilon. Bayesian inference uses a Laplace approximation to the posterior density over unknown model parameters to obtain approximate conditional density and log-evidence – that are then used for inference on parameters and models respectively.

of estimating small differences in conduction velocity that might be elicited experimentally through pharmacological or other manipulations.

Finally, we used the empirical LFP data described above to obtain conditional estimates of the (log-scaling of the) parameters. Figure 9 (right panel), presents these estimates in terms of their conditional means and 95% confidence intervals. Most confidence intervals include a log – scaling of zero, with the exception of inverse intrinsic conduction speed v and neuronal gain g, which increased by 160 and 120% respectively with respect to their

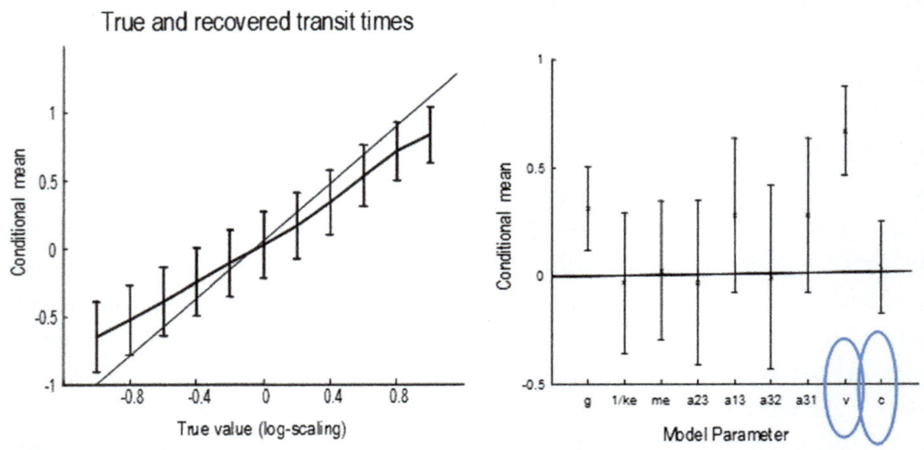

Figure 9 ***(Left panel) Sensitivity analyses of synthetic data.*** This figure shows the conditional estimates of the inverse conduction speed parameter, υ. These estimates were obtained from simulated data sets generated using υ as per Table 1, with a log-scaling from −1 to 1. We illustrate the agreement between the true parameter value (dotted line) and its conditional estimate. The bars represent 90% conditional confidence intervals. *(Right panel) Parameter estimates obtained using LFP data.* We show here the conditional estimates (90% confidence intervals) of the neural field model parameters using empirical data from a single trial recording of auditory responses in the rat. The estimates lie close to the prior values (a log-scaling of zero), which were motivated by animal physiological studies [24].

prior values. This is not surprising as the priors were selected to reproduce spectra that are typical of this experimental setup. There key thing to notice here is that that we were able to obtain fairly precise estimates of topographic parameters like the extent of intrinsic connections c despite the fact that we are only using data from a single electrode.

Conclusions

This paper reviews some recent results regarding the modelling of spatially extended neuronal dynamics using neural field models. We have focused on fitting these models to empirical data and entertained questions about cortical structure and function. Addressing these questions relies on exploiting neuroimaging data to invert neural field DCMs. Optimizing neural field models can be a hard task due to their intrinsic nonlinearities. We have here shown that using dynamic causal modelling it is possible to validate neural field models in relation to empirical data. Our approach comprises three important components: first, a careful selection of priors that conform to known neurobiology; where parameters about which we have little prior knowledge are given flat priors – and the parameters of the observation model are optimized concurrently. Second, an appropriate cost function: here a variational free energy bound on log model evidence. Third, we call on linearity and ergodicity assumptions that allow one to summarize cortical activity efficiently, in terms of frequency domain responses.

In our approach, neural fields serve as generative models that allow us to define a probabilistic mapping from free parameters to observed cross spectra. This assumes that the measured signal is a mixture of predicted spectra, channel and observation noise and can furnish predictions for conventional measures of linear systems; like coherence, phase delay or cross correlation functions, as detailed in [25]. In brief, there is a mapping between model parameters (effective connectivity) and spectral characterizations (functional connectivity) that provides a useful link between the generative modelling of biophysical time series and dynamical systems theory. Neural field models prescribe a likelihood function; this function taken together with the priors specify a dynamic causal model that can be inverted using standard variational procedures [3]. This Variational Laplace scheme approximates model evidence with a variational free energy. The (approximate) posterior density and (approximate) log-evidence are used for inference on parameters and models respectively. In other words, one can compare different models (e.g., neural field and mass models) using their log-evidence and also make inferences on parameters, under the model selected.

In brief, our approach is based on a combination of neural fields with appropriate observation models that are optimised in relation to observed data and are – crucially – compared in terms of their evidence. This provides a principled way to adjudicate among different models or hypotheses about functional brain architectures and the physiological correlates of neuronal computations.

Software note

The procedures described in this review can be accessed as part of the SPM academic freeware (in the DCM toolbox: http://www.fil.ion.ucl.ac.uk/spm/). This code has been written in a modular way that allows people to select from a suite of neural mass and field models to analyse their data – or indeed analyse simulated data that can be generated by the routines. It has also been written in a way that allows people to specify their own models – and connectivity architectures – in terms of Matlab routines (using an equation of motion and a nonlinear mapping from hidden states to observed measurements).

Competing interests

The authors declare that they have no competing interests.

Authors' contributions

DAP and KJF have written the paper. Both authors have given approval for its publication after reading the final paper.

References

1. Pinotsis DA, Friston KJ: **Neural fields, spectral responses and lateral connections.** *Neuroimage* 2011, **55:**39–48.
2. Friston KJ, Harrison L, Penny W: **Dynamic causal modelling.** *Neuroimage* 2003, **19:**1273–1302.
3. Friston K, Mattout J, Trujillo-Barreto N, Ashburner J, Penny W: **Variational free energy and the Laplace approximation.** *Neuroimage* 2007, **34:**220–234.
4. Hannan EJ: *Multiple Time Series.* New York: Wiley. com; 2009:38.
5. Pinotsis DA, Brunet N, Bastos A, Bosman CA, Litvak V, Fries P, Friston KJ: **Contrast gain-control and horizontal interactions in V1: a DCM study.** *Neuroimage* 2014, **92:**143–155. to appear, http://dx.doi.org/10.1016/j.neuroimage.2014.01.047.
6. Rubehn B, Bosman C, Oostenveld R, Fries P, Stieglitz T: **A MEMS-based flexible multichannel ECoG-electrode array.** *J Neural Eng* 2009, **6:**036003.
7. Bosman CA, Schoffelen J-M, Brunet N, Oostenveld R, Bastos AM, Womelsdorf T, Rubehn B, Stieglitz T, De Weerd P, Fries P: **Attentional stimulus selection through selective synchronization between monkey visual areas.** *Neuron* 2012, **75:**875–888.
8. Pinotsis DA, Schwarzkopf DS, Litvak V, Rees G, Barnes G, Friston KJ: **Dynamic causal modelling of lateral interactions in the visual cortex.** *Neuroimage* 2013, **66:**563–576.
9. Bastos AM, Usrey WM, Adams RA, Mangun GR, Fries P, Friston KJ: **Canonical microcircuits for predictive coding.** *Neuron* 2012, **76:**695–711.
10. Sceniak MP, Hawken MJ, Shapley R: **Visual spatial characterization of macaque V1 neurons.** *J Neurophysiol* 2001, **85:**1873–1887.
11. Sceniak MP, Chatterjee S, Callaway EM: **Visual spatial summation in macaque geniculocortical afferents.** *J Neurophysiol* 2006, **96:**3474–3484.
12. Feldman H, Friston KJ: **Attention, uncertainty, and free-energy.** *Front Hum Neurosci* 2010, **4:**215. doi:10.3389/fnhum.2010.00215.
13. Kang K, Shelley M, Henrie JA, Shapley R: **LFP spectral peaks in V1 cortex: network resonance and cortico-cortical feedback.** *J Comput Neurosci* 2010, **29:**495–507.
14. Brunel N, Wang X-J: **What determines the frequency of fast network oscillations with irregular neural discharges? I. Synaptic dynamics and excitation-inhibition balance.** *J Neurophysiol* 2003, **90:**415–430.
15. Traub RD, Jefferys JG, Whittington MA: **Simulation of gamma rhythms in networks of interneurons and pyramidal cells.** *J Comput Neurosci* 1997, **4:**141–150.
16. Kapadia MK, Westheimer G, Gilbert CD: **Dynamics of spatial summation in primary visual cortex of alert monkeys.** *Proc Natl Acad Sci* 1999, **96:**12073–12078.
17. Penny WD, Stephan KE, Daunizeau J, Rosa MJ, Friston KJ, Schofield TM, Leff AP: **Comparing families of dynamic causal models.** *Plos Comput Biol* 2010, **6:**e1000709.
18. Schwarzkopf DS, Robertson DJ, Song C, Barnes GR, Rees G: **The frequency of visually induced gamma-band oscillations depends on the size of early human visual cortex.** *J Neurosci* 2012, **32:**1507–1512.
19. Muthukumaraswamy SD, Edden RA, Jones DK, Swettenham JB, Singh KD: **Resting GABA concentration predicts peak gamma frequency and fMRI amplitude in response to visual stimulation in humans.** *Proc Natl Acad Sci* 2009, **106:**8356.
20. Grindrod P, Pinotsis DA: **On the spectra of certain integro-differential-delay problems with applications in neurodynamics.** *Physica D: Nonlinear Phenomena* 2011, **240:**13–20.
21. Pinotsis DA, Moran RJ, Friston KJ: **Dynamic causal modeling with neural fields.** *Neuroimage* 2012, **59:**1261–1274.
22. Moran RJ, Jung F, Kumagai T, Endepols H, Graf R, Dolan RJ, Friston KJ, Stephan KE, Tittgemeyer M: **Dynamic causal models and physiological inference: a validation study using isoflurane anaesthesia in rodents.** *PLoS One* 2011, **6:**e22790.
23. Roberts SJ, Penny WD: **Variational Bayes for generalized autoregressive models.** *Signal Processing, IEEE Trans* 2002, **50:**2245–2257. doi:101109/TSP2002801921.
24. Jansen BH, Rit VG: **Electroencephalogram and visual evoked potential generation in a mathematical model of coupled cortical columns.** *Biol Cybern* 1995, **73:**357–366.
25. Friston KJ, Bastos A, Litvak V, Stephan KE, Fries P, Moran RJ: **DCM for complex-valued data: cross-spectra, coherence and phase-delays.** *Neuroimage* 2012, **59:**439–455.

4

Associative learning and self-organization as basic principles for simulating speech acquisition, speech production, and speech perception

Bernd J Kröger[1,2*], Jim Kannampuzha[1] and Emily Kaufmann[3]

* Correspondence: bernd.kroeger@rwth-aachen.de
[1]Neurophonetics Group, Department of Phoniatrics, Pedaudiology, and Communication Disorders, Medical School, RWTH Aachen University, Aachen, Germany
[2]Cognitive Computation and Applications Laboratory, School of Computer Science and Technology, Tianjin University, Tianjin, China
Full list of author information is available at the end of the article

Abstract

Background: Quantitative neural models of speech acquisition and speech processing are rare.

Methods: In this paper, we describe a neural model for simulating speech acquisition, speech production, and speech perception. The model is based on two important neural features: associative learning and self-organization. The model describes an SOM-based approach to speech acquisition, i.e. how speech knowledge and speaking skills are learned and stored in the context of self-organizing maps (SOMs).

Results: The model elucidates that phonetic features, such as high-low, front-back in the case of vowels, place and manner or articulation in the case of consonants and stressed vs. unstressed for syllables, result from the ordering of syllabic states at the level of a supramodal phonetic self-organizing map. After learning, the speech production and speech perception of speech items results from the co-activation of neural states within different cognitive and sensorimotor neural maps.

Conclusion: This quantitative model gives an intuitive understanding of basic neurobiological principles from the viewpoint of speech acquisition and speech processing.

Keywords: Speech production; Speech perception; Speech acquisition; Babbling; Imitation; Associative learning; Self-organization; Neural maps; Self-organizing maps; Sensorimotor learning

Background

While a great deal of research has been carried out in order to investigate brain locations of different parts or modules which comprise the speech production and speech perception system (e.g. [1-3]), little is known about the *neural functioning* of these modules during speech acquisition, speech production, and speech perception. In order to fill this gap, quantitative functional neural models have been developed (e.g. [4-8]).

One model, the neuroanatomically grounded Hebbian-learning model [8], establishes highly specialized functional units called "Hebbian neuronal circuits" (HNCs, see also [9]). This model appears to be especially neurobiologically realistic since it learns to associate sensory and motor speech items in a similar way to the early phases of speech acquisition in children. In order to maintain balance between neurobiological realism and computational tractability, this approach does not model single neurons and spike chains, but rather uses "cells" or "nodes" as basic neuron-like elements which represent a local set of neurons, and thus it realizes a lumped-type or mean-field type model in which the primary objects of modeling are the average activity rate of the neuron set (or cell) and cell exhibitory and inhibitory connectivity.

Self-organizing map approaches (SOM, Kohonen) belong to the group of lumped element rate based approaches as well, but it should be noted that the degree of abstraction is much higher in SOM models than in models such as the neuroanatomically grounded Hebbian-learning model of [8]. On the other hand, SOM approaches – as well as more neuroantomically grounded approaches – are capable of representing the basic principles of neural systems, i.e. self-organization, associative learning, Hebbian learning, adaptation, and neural plasticity.

Quantitative neural models of speech processing (i.e. speech production and speech perception) and speech acquisition which include the generation and/or perceptual processing of articulatory and acoustic speech signals are rare. One of the most cited approaches in this direction is the DIVA model [7,10,11]. The DIVA approach mainly concentrates on modeling the relationship between sensory feedback and speech articulation. That model has been successfully applied e.g. to exemplifying motor adaptation in speech production [7,10,12]. The approach introduced in the present paper concentrates on the questions of how speech knowledge, including knowledge concerning speech motor skills, is learned and how this knowledge is stored. In contrast, no assumptions concerning knowledge or skill storage are given in the DIVA approach. Thus, the goal of the present paper is to introduce a comprehensive model of speech acquisition, speech production, and speech perception, based on SOM theory, which includes knowledge and skill storage.

Methods

Structure of the model

Biologically-based neural models that describes complex behavior or complex human functions, such as speaking, separate functional structure and knowledge [13]. The *functional structure* of such a system is basically composed of neural maps and neural mappings [7].

A *neural map* is an assembly of model neurons which represents a specific *neural state*, i.e. a phonemic, phonetic, motor plan, or sensory state in the case of our modeling approach. These maps are located in specific cortical regions. A neural map, e.g. neural map A, comprises N_i model neuron n_i ($i = 1, \ldots, N_i$). Each of these model neurons may be activated to a certain degree $a_i(t)$ at each time instant t. The whole activation pattern $a_i(t)$ ($i = 1, \ldots, N_i$) of a neural map represents a specific neural state, e.g. a motor plan, a sensory state, or a phonemic state at a certain time instant. The strength of activation of each model neuron varies between zero (0, no activation) and one (1, full activation).

All model neurons n_i ($i = 1, \ldots, N_i$) of the neural map A and all neurons n_j ($j = 1, \ldots, N_j$) of the neural map B can be connected with each other (Figure 1). The entirety of N_i x N_j *neural links* or *neural connections* between a neural map A and B is called a *neural mapping*. The strength (or connectivity) of each neural link is called *synaptic*

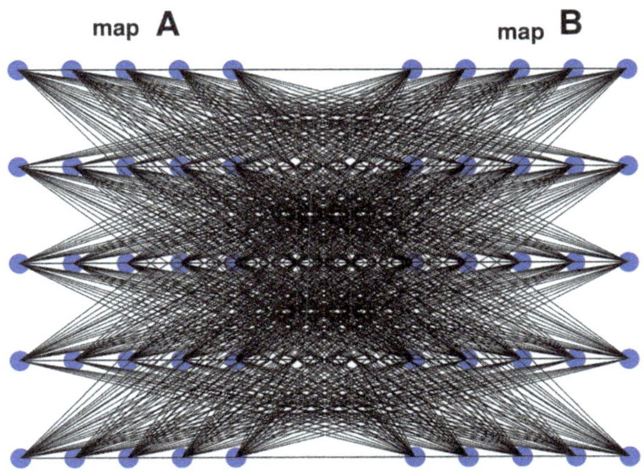

Figure 1 Neural map A, B and neural mapping between the two maps. Map **A** comprises 5 x 5 model neurons (N_i = 25); map **B** comprises the same number of model neurons (N_j = 25); neural mapping (black lines) comprises 25 × 25 = 625 synaptic connections.

link weight w_{ij}. Like neuron activity, each synaptic link weight is typically quantified on a scale between 0 and 1, where 0 represents no connection and 1 represents maximal exhibitory connection.

Additionally, neurons *within* map A or *within* map B can be interconnected. But in our modeling approach, this occurs only in the case of the self-organizing phonetic map and is included within the learning equation for the development of that map.

Only partial speech *knowledge* exists at birth or even prenatally [14]. Especially speech motor skills must be acquired step by step during the first years of life (ibid.). In our approach, speech knowledge is stored by *adjusting* (or changing) the *synaptic link weights* (w_{ij}). The way our neural model acquires speech knowledge is described in detail in Results Section.

In contrast to contemporary spiking neuron approaches for modeling brain functions (e.g. [15,16]), the *model neurons* defined in our approach represent an *ensemble* of natural cortical neurons which are spatially and functionally closely connected. Each model neuron may represent, for example, a cortical column. (This concept is used mainly in vision, e.g. [17,18]) Thus, the activation of a model neuron, and the forwarding of that activation by the axon of that neuron, (i) is the average activation of a bundle of natural neurons and (ii) the average over a specific time interval, i.e. the time interval of syllable processing in the case of our approach. This kind of averaging is widely used in approaches to modeling higher-level brain functions e.g. for *self-organizing maps* [19-21] and working memory [17]. These types of neural models can be summarized as *activation rate models*. In contrast to spiking neuron models, activation rate models do not focus on neurophysiological details such as neural spike trains of individual neurons. Rather, the simplification on the "microscopic" neural level allows us to model large-scale and higher-level brain functions and thus allows us to model "macroscopic" behavior (e.g. speech learning and speech processing) on the basis of neurofunctional principles.

The most important neurofunctional principle used in our modeling approach is *Hebbian learning*, i.e. a synaptic link between two model neurons is strengthened (w_{ij} increases over time) if both neurons are activated during the same time interval. A further neurofunctional principle, neural self-organization, results from Hebbian learning and is described in detail in Results Section.

Basic components of the model

In the present paper we will focus on the description of the higher-level aspects (cortical aspects) of speech processing, while the lower-level aspects (subcortical and peripheral processing) implemented in our model have already been discussed in other papers [22-24]. Figure 2 gives an overview concerning the cortical neural maps in our model. The blue colored boxes represent *state maps* (i.e. motor plan, phonemic, auditory, and somatosensory state map). These neural maps are assumed to be part of short-term memory because the neural states occurring in these maps – e.g. a *motor plan, a phonemic, or a sensory state* of a specific speech item – are activated only during a specific time interval, i.e. the time interval in which that speech item is processed by the model (e.g. up to a few hundred milliseconds for a syllable). Auditory and somatosensory states are summarized in this paper as sensory states. The green colored map (self-organizing phonetic map) and the neural mappings between the state maps and the phonetic map (green colored mappings) are assumed to be part of long-term memory. Here, the motor plan as well as sensory information is stored for each frequent syllable of the language being acquired (the target language).

The motor plan map and the sensory maps are also connected via the *sensorimotor processing loop* (Figure 2). The feedforward part of this loop (*execution feedforward pathway*) generates speech articulator movements at the level of the articulatory-acoustic vocal tract model on the basis of currently activated motor plan states for a syllable, word or utterance. (These are also referred to as "vocal tract action scores" in [24-26]). The vocal tract model generates geometrical data (i.e. a vocal tract shape and a set of articulator positions) for each time instant during the production of each

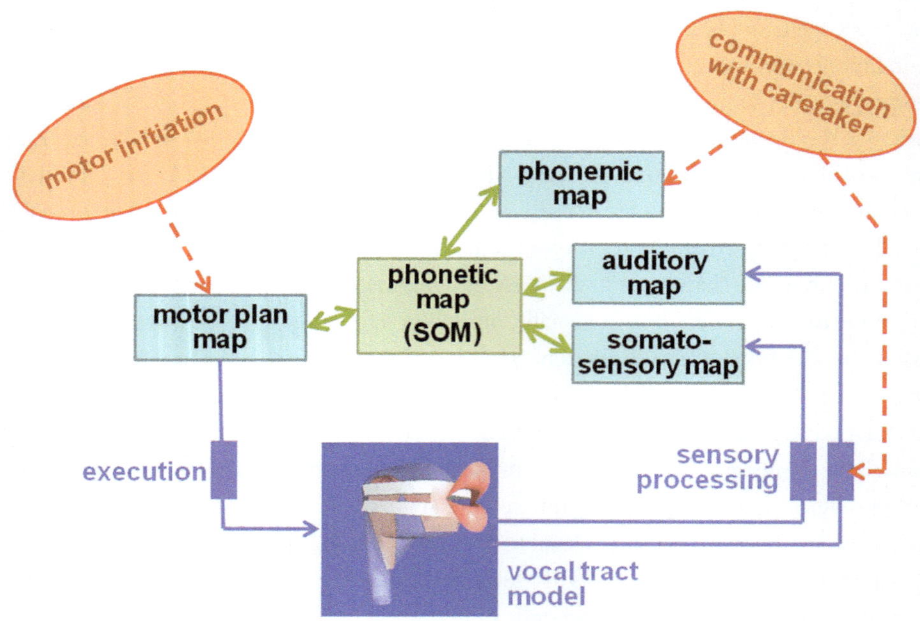

Figure 2 The structure of our neural model for simulating speech acquisition, speech production, and speech perception. Light blue and light green boxes represent cortical neural maps, while green double arrows indicate cortical neural mappings (see text). Dark blue arrows and boxes represent the cortical, subcortical and peripheral sensorimotor processing loops (see text). Red ovals represent babbling and imitation training sets which control the model during speech acquisition for implementing knowledge.

speech item as well as an acoustic speech signal for the speech item being produced [26,27]. Both signals are processed within the *sensory feedback pathway* or *sensory processing pathway* in order to generate an appropriate auditory and somatosensory neural state for the speech item being produced.

It should be noted that a mental lexicon (as defined in [28,29]) is not included in our model and lies beyond the scope of our modeling approach. The only symbolic linguistic representation used in our approach is the *phonemic representation* (see Figure 2). Here it is assumed that each frequent syllable occurring in the target language is represented by one model neuron within the phonemic state map.

Cortical state maps and neural representations

Motor plan, auditory, and somatosensory state maps are cortical neural maps which represent higher-level unimodal (i.e. motor or sensory) representations for a currently processed speech item (in most cases a syllable).

The existence of a *higher-level motor representation* (i.e. a *motor plan state*) is postulated in our approach on the basis of [2,30,31]. At this higher motor level, the overall temporal arrangement of the speech actions, which constitute a speech item, is specified, while the concrete muscle activation patterns are generated at lower cortical and subcortical levels. This (higher-level) motor plan state representation is part of short-term memory because this representation comprises at least the motor organization of a complete syllable, the duration of which may be up to several hundred milliseconds.

An example of a neural representation of a motor plan state is given in Figure 3. Eight vocal tract actions are needed in order to produce the syllable "ist" (Standard German "is"). Up to three tiers (i.e. horizontal rows) represent up to three onset consonant actions for each syllable. Up to two neuron rows represent the vocalic syllable nucleus; both vocalic rows are needed e.g. for representing a diphthong in German. Up to three neuron rows represent the coda consonant actions because up to three initial and final consonants may occur in Standard German. Two neuron rows represent velopharyngeal opening and closing actions, which are needed for the production of nasals and obstruents (plosives and fricatives). Two neuron rows represent glottal opening and glottal closing actions, which are needed for the production of voiceless and voiced

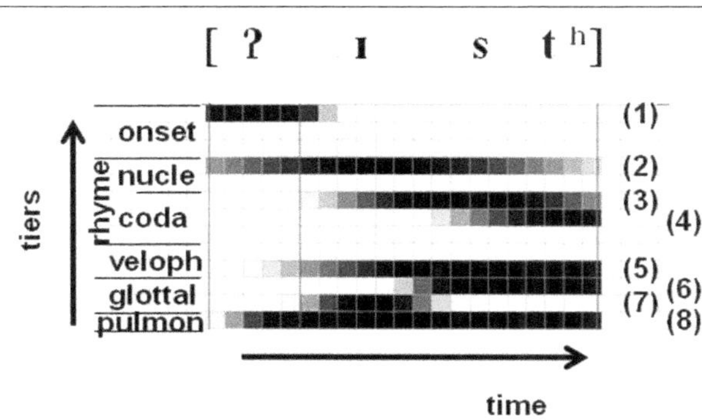

Figure 3 Neural representation of a motor plan state for the German word "ist" ("is"). Top line: phonetic transcription of the word. Below: a 21 x 13 model neuron map (motor plan state map) representing the motor plan of that word/syllable. Vertical columns: 12.5 ms time intervals; horizontal rows: temporal location and duration of each vocal tract action. The degree of activation (white = no activation; to black = full activation) of a model neuron represents the degree to which an articulator has approached the appropriate vocal tract action target.

speech sounds, respectively. The last neuron row represents the pulmonary pressure action which is needed as a basic power source for producing a speech item. The example syllable "ist" comprises eight vocal tract actions: (1) glottal tight-closing action for producing the initial glottal stop consonant [?]; (2) vocalic action for producing the German lax vowel [I]; (3) apical near-closing action for producing the fricative sound [s]; (4) apical full-closing action for producing the plosive sound [t]; (5) velopharyngeal tight-closing action in order to ensure a pressure build-up in the oral cavity during the production of the obstruent sounds [s] and [t]; (6) glottal opening action in order to ensure voicelessness during the production of [s] and [t]; (7) glottal closing action in order to ensure voicing during the production of the vowel [I]; (8) pulmonary pressure action in order to provide sufficient aerodynamic power for the production of the syllable. The neural representation for a further specification of each vocal tract action (e.g. articulator and target specification) is not shown here. A detailed description of our concept of vocal tract action scores (also called "gesture scores") is given in [25,32,33].

On the basis of this neural representation of a motor plan state (Figure 3), articulator movements controlling the vocal tract model can be calculated [25]. This represents the execution of a vocal tract action score by our vocal tract model. The neural representation of a motor plan state as introduced above is a *distributed neural representation,* because in principle all model neurons of the motor plan state map can contribute to the representation of a motor plan state. In contrast, the phonemic state map only comprises *local neural representations,* because here each syllable is represented by a specific neuron within the phonemic state map.

Evidence for a *higher-level auditory representation* is given by [34]. At this level, the sound impression of a whole syllable can be represented (i.e. how a speech item sounds). In our approach the higher-level auditory representation is assumed to be a "neural spectrogram" (see Figure 4). This auditory state representation is part of short-

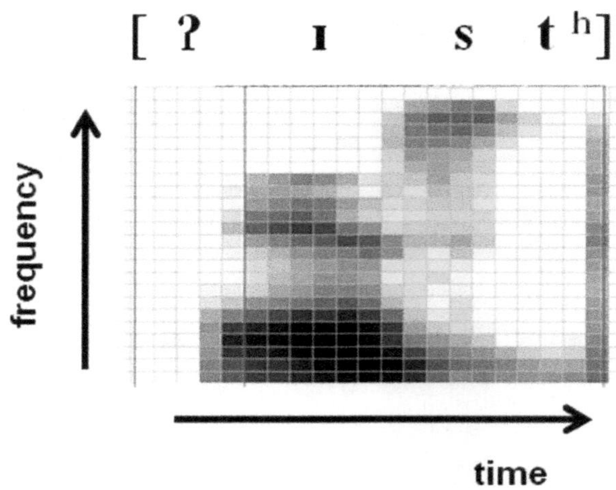

Figure 4 Neural representation of an auditory state within the auditory state map for the German word "ist" ("is"). Top line: phonetic transcription of the word. Below: a 21 x 24 neuron map (auditory state map) representing the auditory state of the word/syllable in the format of a Bark-scaled spectrogram. Horizontal rows: frequency in Bark groups from 1 to 24 (0 to 20000 Hz). Vertical columns: 12.5 ms time intervals. The degree of activation of a model neuron represents the acoustic intensity within that frequency-time-square. The blue vertical line represents the start time of the syllable nucleus, while the two red vertical lines represent the beginning and end of the syllable. (See also the motor plan state map, Figure 3.)

term memory because it comprises the auditory sound of a complete speech item (at least one syllable) and it must be activated over the whole time period of the production of at least one syllable in order to be able to check whether the syllable is produced correctly.

The calculation of the neural representation of an auditory state (i.e. auditory processing as a part of sensory processing, see Figure 2) is done by applying a spectral analysis (amplitude part of a Fourier analysis from 0 to 20000 Hz) to the acoustic speech signal (Figure 4). A 25 ms Hamming window is used. Thus, the spectral analysis of adjacent columns of model neurons overlaps in time. Each Bark group neural excitation is estimated by calculating the mean amplitude of all frequency bands occurring within that specific Bark group. The degree of neural activation (i.e. from 0 or white = no activation; to 1 or black = full activation) is proportional to the logarithm of spectral amplitude within that Bark group.

The existence of a *higher-level somatosensory representation* is postulated e.g. by [10,35]. Each somatosensory representation comprises a tactile and a proprioceptive aspect. A tactile higher-level representation in speech comprises at least the spatio-temporal pattern of contact between articulators (i.e. between the upper and lower lips and between the tongue and the palate). A higher-level proprioceptive representation comprises at least the spatio-temporal pattern of distance between articulators and intended targets for each vocal tract action and thus is comparable to a higher-level motor plan representation as defined in the present paper. Thus, for this preliminary version of our modeling approach we used the motor plan state representation also as a rough estimate for a higher-level somatosensory state representation. However, for a further refinement of our model, it would be possible to use geometrical data from the vocal tract model in order to estimate a detailed somatosensory representation (cf. [36,37]).

The existence of a *phonemic state representation* is postulated e.g. by [10], referred to as a "speech sound map" in that paper. In contrast to higher-level motor and sensory representations, which were implemented in our modeling approach as *distributed* neural representations, on the level of the *phonemic state representation*, each model neuron is assumed to represent exactly one phonemic state (i.e. a specific syllable or word; cf. [ibid.]). Thus, if a specific syllable is processed (produced or perceived), one model neuron becomes maximally activated at the level of the phonemic map. This is called a *local neural representation*.

Because speech processing is an ongoing flow of production and/or perception of syllables, it is assumed in our modeling approach that the activation of syllable-related model neurons within the phonemic map may overlap in time in order to allow the simulation of the succession of syllables over time.

Because motor plan and sensory state maps must be capable of representing all syllables (or at least all frequently occurring syllables) of a target language, the total number of model neurons n_i constituting the motor plan is set to $60 \times 13 = 780$ neurons, while the total number of model neurons composing the sensory state map is set to $60 \times 24 = 1440$ neurons. This allows for the modeling of a maximum syllable length of 750 ms, which occurs only if syllables are uttered in isolation. Typically, syllable length is much shorter, which – in our model representations – leads to many non-activated neurons at the temporal beginning and end of motor plan and sensory representations.

These non-activated neurons could be used for the representation of previous or subsequent syllables in a further version of our model. The beginning time of the syllable nucleus (i.e. the time of release of the last consonant within the syllable onset) is always set as a reference time and thus is always represented by the same neurons in the motor plan as well as in the sensory state maps (represented by the vertical blue line in Figure 3 and Figure 4). It should be noted that in addition to the required model neurons just described, 132 additional model neurons are needed at the level of the motor plan map for specifying each vocal tract action (for more detailed information see [24,25]). The phonemic map is of a comparably small size because the number of model neurons within this map directly reflects the number of syllables which occur in the target language at the phonemic level.

The phonetic map

The phonetic map in our modeling approach is a self-organizing map (SOM; for an introduction to SOMs, see e.g. [21]). The size of a self-organizing map increases during learning, which in our case means during speech acquisition. It results in an increasing amount of neural storage for speech knowledge within the mapping between the SOM and the state maps (see Growing-SOM approaches, e.g. [5,38]). In this preliminary version of modeling speech acquisition, we use phonetic maps of a fixed size, e.g. comprising 15 × 15, 20 × 20, or 25 × 25 model neurons.

It will be shown that phonetic maps – as well as self-organizing maps in general – exhibit *local* neural representations. We will see that a model neuron within a phonetic map represents a *specific phonetic realization* or *"exemplar"* of a phonemic state. This may be a specific phonetic realization of a syllable due to different contexts in which that syllable occurs. Thus, a model neuron within the phonetic map, as in the phonemic map, represents a syllable, but in the case of the phonetic map, it represents a concrete phonetic realization of a syllable. Thus, it is easy to see why a model neuron within the phonetic map must be directly connected with a motor plan and a sensory state (see Figure 2). This results from the fact that a fully activated model neuron at the level of the phonetic map (representing a specific phonetic realization of a syllable) directly activates the appropriate motor plan and sensory state for the realization of a syllable. Thus, the information concerning a motor plan and its appropriate sensory (auditory and somatosensory) states is stored completely within the synaptic link weights of the mappings between the phonetic map and the sensory state maps.

Modeling speech learning: babbling and imitation

The organization of the phonetic map is based on the link weight values of the neural mappings between the phonetic map and the sensorimotor state maps. These link weight values result from learning or training and directly reflect the acquired knowledge and the acquired speech skills. In our modeling approach it is assumed that the phonetic map is a self-organizing map (SOM; also called Kohonen-Map). Further, our approach assumes that the adjustment of synaptic link weights between the SOM and the state maps is established mainly within the early stages of speech acquisition, meaning in the first years of life, but that it can be modified and further developed over the lifetime.

Overview: babbling and imitation

It is well known that babies start to imitate facial expressions and communicative gestures within the first year of life (e.g. [39,40]). In parallel, they start to *imitate speech items*, which they hear from a caretaker, relatively early [14]. Later on, if a toddler is already capable of communicating with the caretaker, the toddler actively asks for specific words, e.g. by pointing at an object, and then looking to the caretaker (e.g. [41,42]). At this stage of learning to speak, the toddler receives (i) an acoustic realization of a speech item uttered by the caretaker and (ii) information concerning the meaning of the speech item. The first point is modeled in our approach by activating the auditory realization, while the second point is modeled by synchronously activating the phonemic representation for the relevant speech item (see red dashed arrows in Figure 2 in the area of "communication with caretaker"). Thus, the caretaker produces an acoustic realization which activates an auditory state within the auditory state map of the toddler (i.e. of the model). This auditory state can now be imitated by the toddler in order to generate an appropriate motor plan state.

But how does the toddler imitate a speech item produced by the caretaker? There are two major problems. Firstly, at the beginning of the learning procedure, the toddler does not know how to generate a motor plan state which could result in an auditory state similar to that produced by the caretaker. Secondly, the toddler does not possess the adult vocal tract and therefore cannot imitate the fundamental frequency or formant structures of a speech item produced by an adult: a toddler normally produces higher formants as well as higher fundamental frequency. This second problem is referred to as the "speaker normalization problem" (e.g. [43]).

These problems are not addressed in this modeling study for several reasons. Firstly, they may be solved in part by the caretaker, since caretakers are able to use a specific child-directed way of speaking ("motherese", see e.g. [14]). This way of speaking normally involves the adult trying to adapt the formant frequency and fundamental frequency targets of their speech sounds towards those targets which can be produced by the child. Additionally, the problem may be solved in part by the toddler herself: Even if the auditory result produced by the toddler differs from the original that was produced by the caretaker, the caretaker can reward the toddler for each correct or at least understandable realization of a word ("reinforcement learning"). Thus, the toddler is able to establish an initial association between his own realizations and adult realizations of the same word (e.g. the model stores two or more realizations for each speech item; one or more child versions and one or more adult versions).

A further, theoretically more important solution for these problems is that before the imitation state, the toddler (or the neural model) should already have some experience or knowledge of how to generate a motor plan state for a given auditory state. This knowledge is normally acquired by the toddler during the "babbling" phase [7,10,44]: The toddler (or the model) starts with randomly generated motor plan states (see red dashed arrow in Figure 2, "motor initiation"), then executes these motor plans (i.e. generation of an articulatory and acoustic speech signal by the vocal tract model), and finally performs the sensory processing of these signals, so that the appropriate auditory and somatosensory states are generated for all motor plan states (see Figure 2). Thus the model develops a set of "sensorimotor associations" by "exploring the acoustic and articulatory states of its own vocal tract". This set of associations is stored at the level of the phonetic map and its mappings towards the motor plan and sensory maps. This learning or training is called babbling training. Babbling always precedes imitation, but the two can also be somewhat interwoven due to the reasons outlined in this section. Clear language-specific speech production starts at around 10 months of age). Indeed, the importance of babbling training in speech acquisition, particularly the sensory processing of the signals produced in babbling, is made clear by the fact that hearing children progress from babbling sounds to babbling syllables, while deaf children do not progress to the syllable stage (e.g. [45]).

Adjustment of synaptic link weights

Next we will describe how the adjustment of synaptic link weights between the state maps and the self-organizing phonetic map takes place in our model during training (learning). We assume that a *training data set* comprises an amount of D *training items*, i.e. D state representations d (d = 1, ..., D) with a model neuron activation pattern a_{id} (i = 1, ..., N_i), comprising a motor plan state, sensory state, and a phonemic state activation pattern, and that each training item is applied to the model C times (C = *number of training cycles*) where the succession of training items varies randomly per training cycle d(t). The *synaptic link weights* w_{ij} between *state map neurons* n_i and *phonetic map neurons* n_j were updated (i.e. were changed incrementally) from training step t to training step t + 1 (T = C*D, t = 1, ..., T training steps in total) by applying the following equation:

$$w_{ij}(t+1) - w_{ij}(t) = H_j(t)^* L(t)^* \left(a_{id}(t) - w_{ij}(t)\right), \tag{1}$$

where i = 1, ..., N_i is the index for all state map model neurons and where j = 1, ..., N_j is the index for all phonetic map model neurons. Here, $H_j(t)$ denotes the Gaussian neighborhood kernel around the best matching model neuron j (i.e. winner neuron), if the phonetic map is activated by the training item (see Equation 4), and where L denotes the learning rate. The learning rate decreases over time exponentially:

$$L(t) = L(0)^* \exp(-0.00001^* t), \tag{2}$$

where L(0) = 0.9 is the initial learning rate. The radius σ(t) of the Gaussian neighborhood kernel (Equation 3) also decreases exponentially during training (Equation 4):

$$H_j(t) = \exp\left\{-0.5^* \left(n_k - n_j\right)^2 / (s(t))^2\right\}. \tag{3}$$

Here, n_k denotes any neuron within the self-organizing map and $|n_k - n_j|$ denotes the distance of a neuron n_k to the winner neuron n_i defining the center of the neighborhood kernel. σ(t) is defined here arbitrarily as the radius of the neighborhood kernel. The neighborhood factor H_j, which is 1 for n_j, declines below a value of 0.6 if the neuron distance exceeds σ(t) and below 0.14 if the neuron distance exceeds 2σ(t). The temporal decrease of the neighborhood radius is:

$$s(t) = s(0)^* \exp(-0.00001^* t), \tag{4}$$

where σ(0) = 5 neurons. Thus, 2σ(0) is 40% of the overall length of the phonetic map (the overall length is 25 neurons). The winner neuron n_j, specifying the center of the neighborhood kernel, is calculated by looking for the maximally activated neuron j within the phonetic map. The activation pattern $a_j(t)$ of the phonetic map during the application of the training item $a_i(t)$ is calculated as

$$a_j(t) = \text{sum}_{(i=1,\ldots,Ni)}\{w_{ij}(t)^* a_i(t)\}. \quad (5)$$

The incremental change of synaptic link weights, quantified by the equations above, describes *Hebbian learning*, because synaptic link weights w_{ij} mainly change for those synaptic connections for which phonetic map neurons as well as state map neurons are strongly activated. At least synaptic link weights w_{ij} increase towards a_i, i.e. link weights adapt to the currently applied activation pattern a_i ($i = 1, \ldots, N_i$). Thus, this learning is *non-supervised*: No ideal activation pattern is known in advance for the phonetic map. Learning always tries to approximate the activation pattern of the currently applied training item. Links with link weights w_{ij} grow in both direction, i.e. from state maps towards phonetic map as well as vice versa. This bidirectionality is compatible with the idea of Hebbian learning.

We started babbling training with a random link weights initialization for w_{ij}, taking values between 0 and 0.5 in order to model moderate neural interconnectivity at the beginning of training. Our criterion for stopping imitation training is defined as follows: A predefined amount of neurons of the *phonemic* map (e.g. 95%) should indicate a strong synaptic connection with one neuron of the phonetic map (e.g. $w_{ij} > 0.8$). That implicates that a predefined percentage of syllables is learned by the self-organizing network in a way that at least one phonetic realization exists for that syllable. From our experience, this indicates that the associated phonetic states already have a correct association between motor plans and sensory states for that syllable.

The situation is more complex in the case of babbling training, since in that case no simple criterion can be given for reaching a specific level in sensorimotor knowledge or sensorimotor skill learning. Since babbling training is always guided by imitation, in order to train appropriate regions within the multidimensional space of motor parameters, the criterion used in our modeling approach is that babbling training will be continued as long as a speech item under imitation is not reproduced "distinguishably". It should be noted that all babbling trials performed by our model for producing an imitation training item (this can be referred to as "guided babbling") are controlled aurally by a supervisor. It is up to this person to decide subjectively whether an item is "distinguishable" or not. In the case of this study, babbling knowledge was sufficiently acquired after 500 training cycles.

In our experience, babbling training requires a great deal more training cycles than imitation training, since the organization of the phonetic map emerges during babbling training and is only refined during imitation training (up to 500 training cycles are required during babbling training, while 50 to 150 training cycles are sufficient in many imitation training situations; see [44]). This decrease in training cycles towards imitation training may result from the fact that – based on our experience with our simulation experiments – the emergence of phoneme regions never indicates a reorganization of the sensorimotor structure of the phonetic map with respect to the phonetic features that have already been acquired.

The learning equations (1)-(5) given above are based on Kohonen's learning equations [21]. However, Kohonen used just one state map and thus did not subdivide state maps with respect to different modalities. Our model adapts and expands Kohonen's approach to the areas of speech acquisition and speech processing by considering the sensory, motor, and phonemic modalities.

Furthermore, Kohonen's approach calculates synaptic link weights only in one direction, e.g. from the sensorimotor state maps towards self-organizing map. In contrast, we assume that synaptic connections with strength w_{ij} arise not only from model neurons n_i of the state maps towards model neurons of the self-organizing phonetic map n_j, but also in the opposite direction, from the phonetic map towards the state maps with the same link weight values w_{ij}. This *bidirectionality* within the mappings of our model (Figure 2) is important especially in modeling not just speech acquisition and speech perception, but speech production as well. This modification is a further contribution of the present paper in adapting Kohonen networks to the areas of speech acquisition and speech processing.

Results

Description of simulation experiments

Six groups of simulation experiments were performed in order to feed speech knowledge into our model (Table 1). Because imitation always requires some previous sensorimotor knowledge (see above), in our model, babbling training is performed before imitation training. But it is important to note that the language-specific speech items spoken by the caretaker normally guide babbling: the set of all possible motor plan states which could be generated randomly would be infinitely huge and thus could not be trained by the model. For this reason, babbling training is always directed towards the specific language being acquired (e.g. [46]).

In order to feed first sensorimotor knowledge into our model, we started with protovocalic babbling (Table 1). In a first group of simulation experiments, the tongue position of the vocal tract model was randomly positioned and the corresponding auditory state was calculated for each training item. After babbling training, the synaptic link weights between the motor plan map and the auditory map are tuned in a way that (i) the motor plan states and the appropriate sensory states (as produced by the sensorimotor processing loop) are *associated* and that (ii) these sensorimotor states are *ordered* with respect to two important *vocalic phonetic features* "front-back" and "low-high" [44].

In a second group of babbling experiments, motor plan states were initiated which start from different labial, apical, and dorsal closures and which end in different protovocalic states (protoconsonantal babbling, Table 1). After training, the *consonantal phonetic feature* "place of articulation" (labial, apical, dorsal) are learned. These are in addition to the vocalic phonetic features, which had previously been learned. Three compact regions appear within the resulting self-organizing map. These regions represent the three different places of articulation [44,47].

Because initial sensorimotor knowledge is gained after performing these babbling training simulations (i.e. the associations between the motor plan and the sensory states, along with the ordering of these sensorimotor states at the level of the phonetic map with respect to phonetic features), the motor plan states can now be roughly estimated based on the auditory states. Thus, the model now is ready for vowel and consonant imitation. We continued with imitation training of a model language comprising five vowels V =/i, e, a, o, u/(see third group of simulation experiments, Table 1) and 15 syllables comprising all combinations of these five vowels with the consonants/b, d, g/as CV-syllables (consonant-vowel-syllables, see fourth group of simulation experiments, Table 1). This imitation training does not increase the number of phonetic features, which have already been learned by the model during babbling, but rather achieves a labeling of specific sensorimotor states with respect to phonemic

Table 1 Groups of simulation experiments for speech acquisition

Group name	Specification of training items	Number of training items	Size of SOM	Phonetic features learned	Published
Protovocalic babbling (prelinguistic)	Variation of tongue position	1076	15 × 15	Front-back; low-high;	[44]
Protoconsonantal babbling (prelinguistic)	Variation of initial closure: labial, apical, or dorsal	279	15 × 15	Place of articulation;	[44,47]
Vowel imitation (part of model language)	5 vowels (V): /i, e, a, o, u/	500	15 × 15	Front-back; low-high;	[44]
Consonant imitation (part of model language)	15 CV syllables: 5 vowels (V) with 3 consonants (C): /b, d, g/	465	15 × 15	Place of articulation;	[44,47]
Syllable imitation (model language: V, CV, CCV)	70 syllables: 5 V + 5 V with 9 single C: /b, d, g, p, t, k, m, n, l/ as CV + 5 V with 6 clustered CC: /bl, gl, pl, kl/ as CCV	600	25 × 25	Place and manner; voiced-voiceless; type of syllable;	[36,48]
Imitation of frequent syllables (real language)	Standard German; 200 most frequent syllables;				
	Male speaker; one realization per syllable	200	15 × 15	Stressed-unstressed; vowel type;	[22]
			20 × 20	Same;	
			25 × 25	Same;	
	Female speaker; up to 27 realizations per syllable	703	25 × 25	Same;	This study

categories. Due to the inclusion of phonemic states during imitation training (in contrast to babbling training), here there is also an association established between the sensorimotor states and the phonemic states. Thus, many neurons of the phonetic map can be labeled phonemically. Because phonemic states usually represent similar sensorimotor states, *phoneme regions* occur for each phonemic vowel state and for each phonemic CV-syllable state at the level of the phonetic map. A model neuron within the phonetic map is a member of a phoneme region if there are strong synaptic connections between this model neuron and a specific model neuron within the phonemic map which represents that phoneme or phonemic state [44,47].

After imitation training of V- and CV-syllables with V =/i, e, a, o, u/and C =/b, d, g/, a fifth group of simulation experiments was performed. Within this group of simulation experiments, the set of consonants was extended (C =/b, d, g, p, t, k, m, n, l/) and first CCV-syllables (i.e. syllables with initial double-consonant clusters/bl, gl, pl, gl/) were allowed and consequently trained. Within this training step – which included guided babbling with the new groups of consonants undergoing imitation training – new consonantal phonetic features result from the further ordering of sensorimotor items at the level of the phonetic map, i.e. "voiced-voiceless", and "manner of articulation" (plosive, nasal, lateral; see Table 1). Also, an appropriate ordering of phoneme regions is achieved at the level of the phonetic map for all 70 phonemic syllable states (see [36,48]).

Finally, a sixth group of simulation experiments was performed in order to train a natural language, in our case Standard German (see Table 1). First, we assembled a corpus of Standard German on the basis of 40 children's books. These books were targeted for children up to six years of age. We transcribed 6513 sentences in total, leading to 70512 words in total, comprising 4763 different syllables [22]. The 200 most frequent syllables were extracted from this database and used for training the model. In contrast to model language training (simulation experiments 3, 4, and 5 in Table 1), where phoneme realizations within the training set were generated in a synthetic way through a rule-based version of our articulatory synthesizer [25], here we used natural acoustic realizations of syllables which were uttered by native speakers, one male (33 years old) and one female (27 years old), both without any known anomalies in speech or hearing.

In a first simulation experiment of this sixth group of experiments (Table 1), only one sentence was uttered by the male speaker for each frequent syllable. This syllable was extracted from the acoustic signal, and the appropriate auditory state was generated and imitated by the model. Thus, just one realization was trained per syllable, but the training already took into account the differences in frequency of occurrence for the 200 most frequent syllables (for details, see [22]). In the present paper, we report on a further simulation of imitating of these 200 most frequent syllables by using a refined training data set comprising more than one realization per syllable (see Results section).

We now provide a summary of the babbling training results. Babbling training always resulted in (i) an *association of motor plan and appropriate sensory states*, and (ii) an *ordering of these sensorimotor states with respect to sensorimotor or phonetic features*. Subsequent imitation training always resulted in an additional labeling of specific sensorimotor states – represented by specific model neurons within the phonetic map – with respect to phonemic categories. Thus, *phoneme regions* appear within the phonetic map, since similar sensorimotor states are usually represented by adjacent model neurons within the phonetic map (see e.g. Figure 8, Appendix). The emergence of phoneme regions implies that a phonemic state can be represented by more than one neuron at the level of the phonetic map. These neurons at the level of the phonetic map, representing the same phonemic state, are referred to as *exemplars* in the context of our modeling approach; see also [49] for a comparable use of this term. Thus, a phonemic state can be represented by *different learned or trained phonetic realizations*, referred to as exemplars of that phonemic state. For example, the three model neurons in the upper left corner of the phonetic map shown in Figure 8 (see Appendix) represent three (stored or learned) exemplars of the Standard German syllable/mIt/("with"). A closer inspection of these three exemplars indicates that there are slightly different sensorimotor or phonetic features for each state (see motor plan states and auditory states of these exemplars, presented in Figures 9 and 10 in the Appendix).

Imitation training of frequent syllables

A female speaker (27 years old; with no known anomalies in speech or hearing) uttered up to 27 realizations of the 200 most frequent syllables of our children's book corpus [22]. Syllable realizations were produced in proportion to the frequency of occurrence of a syllable in the corpus (Table 2). Thus, 703 sentences were recorded in total (D = 703).

The syllable under consideration was marked within each sentence and a motor plan state was generated for each of these syllables using our resynthesis procedure [50]. Neural representations of all motor plan states and auditory states were calculated for each syllable. The resulting training items were applied to the neural model.

In total, 120 training cycles were sufficient in order to represent 95% of the 200 most frequent syllables of our corpus (T = C * D = 120 * 703 = 84360 training steps in total). The criterion for terminating the training was defined as follows: A strong synaptic connection ($w_{ij} > 0.8$) was established for 95% of the model neurons within the phonemic map (phonemic states) and at least one model neuron within the phonetic map for each of these phonemic states. Thus, at least one exemplar must exist for 95% of all phonemic states (190 of 200 syllables). The learning curve resulting from training our model is presented in Figure 5. It can be seen that a strong increase in syllable learning occurs above training cycle 60. A saturation effect is reached after approximately 120 training cycles. We checked to make sure that the 5% of syllables which were not learned were below rank 150 for the 200 most frequent syllables.

The neurons of the phonetic map which exhibit a strong synaptic link towards a phonemic state (i.e. all exemplars) are marked in Figure 8 (Appendix) by their phonemic transcription. The link weight distributions for all model neurons of the phonetic map towards the motor plan and auditory state map are given in Figures 9 and 10 (Appendix). From these displays of the organization of the phonetic map, which is

Table 2 Frequency of occurrence of a specific syllable in the children's book corpus and in the training set

Rank of syllable (with respect to frequency)	Frequency of occurrence in corpus	Number of training items
1	2367	27
20	692	8
50	390	4
100	193	2
200	88	1

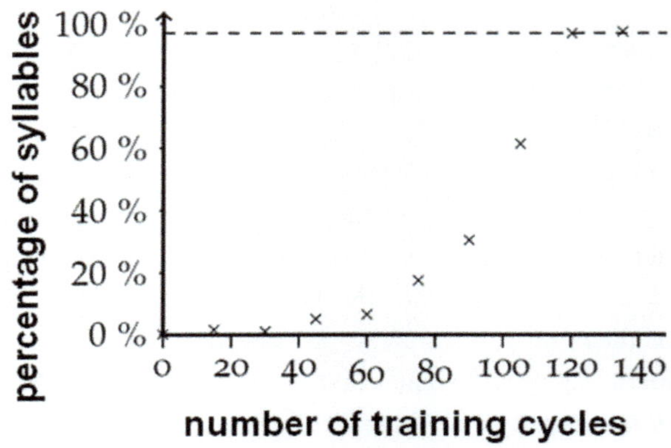

Figure 5 Learning curve resulting from simulation of imitation training for our children's book corpus. X-axis: number of training cycles; y-axis: percentage of syllables already acquired.

Figure 6 25 × 25 phonetic map after training, representing phonemic link weights (see also Figure 8 in the Appendix). Separate regions emerge within this map, representing the syllable features stressed vs. unstressed and the vocalic features reduced vs. nonreduced. Nonreduced vowels can be short vowels, long vowels and diphthongs in Standard German.

present after imitation training, it can be clearly seen that neighboring states at the level of the phonetic map represent similar sensorimotor features. Thus, the distribution of states at the level of the self-organizing phonetic map can be labeled as quasi-continuous. This in part reflects the ordering of sensorimotor states with respect to phonetic features. The organization of the phonetic map for the experiment described here is discussed in more detail in Discussion section.

A careful inspection of the distribution of the phonemic transcriptions over the phonetic map (Figure 8, Appendix) indicates an ordering of syllables with respect to the syllable feature *stressed-unstressed* and with respect to the vocalic features *nonreduced-reduced*, where nonreduced vowels in Standard German can be short vowels, long vowels, or diphthongs (Figure 6). Also, a loose ordering with respect to other vocalic and consonantal phonetic features can be seen.

Because our training data set is designed in such a way that frequent syllables occur with more realizations than less frequent syllables (see Table 2), syllables are processed with different frequency during training. After training, this is reflected in the resulting organization of the phonetic map. Here, the number of neurons representing a syllable with strong synaptic connection to a phonemic state (i.e. the number of exemplars per syllable, or the size of the phoneme region) is proportional to the frequency of occurrence of that syllable in the training set (Figure 7). This means that the more often a syllable is processed by the model, the larger the area is within the phonetic map which represents this syllable. (Or, the more often a syllable is processed by the model, the higher the number of exemplars which are stored for that phonemic state is and the bigger the appropriate phoneme region is).

Moreover, we checked whether the phonetic variability of the exemplars for a specific frequent syllable, represented in the phonetic map, is comparable to the phonetic variability within the different realizations of that syllable within the training set. This was

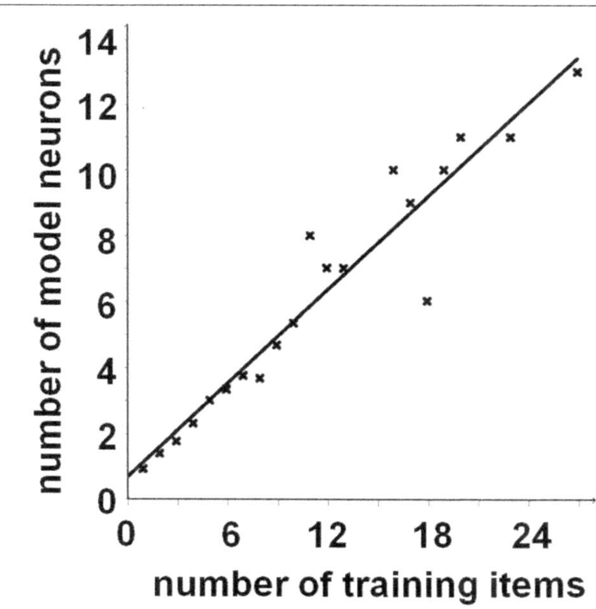

Figure 7 Number of model neurons representing a syllable at the level of the phonetic map as a function of the number of training items occurring for that syllable within the training set.

done by comparing the mean variance of the activation pattern of the neural auditory states resulting from the realizations of the training set for a specific syllable to the mean variance of the activation patterns of neural auditory states which are co-activated from neurons of the phonetic map, assuming that that neuron of the phonetic map represents an exemplar of the same syllable (exemplars which had already been learned, represented by link weights. It can be seen that the phonetic token variability within the training set is comparable to that of exemplars for some syllables, while it is smaller for most syllables (Table 3).

Speech processing

Because speech production and speech perception are already a part of speech acquisition (babbling and imitation are perception of a caretakers utterance, followed by several production trials of the toddler or model), the performance of speech production and perception increases during the continuously ongoing process of speech acquisition. In our modeling approach, we defined a specific landmark within the process of acquisition at which the performance in production and perception was tested. Thus, speech acquisition was defined as (nearly) complete if 95% of the syllables occurring in the training set have been acquired. It should be noted that this criterion is a direct reflection of the fact that our training sets and self-organizing maps are of fixed sizes. The remaining 5% of syllables are easily acquired if the training set is widened, e.g. if in an augmented training set and an augmented phonetic map are used for further training at this landmark for speech acquisition. Thus, testing the quality of performance of the model's production and perception capabilities at a landmark of speech acquisition is also a check on the quality of speech acquisition accomplished so far.

Bidirectional cortical mappings

It is important to keep in mind that self-organizing maps as defined by [19] only use mappings from that map, which represents the training items (i.e. the sensorimotor state maps in our case) towards the self-organizing map (the phonetic map in the case

Table 3 Training results for 10 most frequent syllables

Syllable	Number of model neurons (exemplars)	Mean variance of exemplars	Mean variance of training items
'?Unt	13	0,004114	0,004037
'di:	11	0,001899	0,004117
'zi:	10	0,002830	0,007564
t@	10	0,002761	0,003878
'dE6	10	0,001903	0,008438
'?E6	8	0,002896	0,013582
n@	7	0,002053	0,005684
g@	6	0,001418	0,003845
n@n	5	0,002184	0,005175
b@	3	0,001906	0,001881

First column: phonetic transcription (SAMPA) of the 10 most frequent syllables of our children's book corpus; second column: number of neurons representing all realizations of these syllables within the phonetic map (number of exemplars); third column: mean variance of the auditory states of all realizations by activating specific neurons within the phonetic map, representing exemplars for a syllable; fourth column: mean variance of the auditory states of all realizations activated by the training items directly.

of our modeling approach). But Hebbian learning in principle allows the emergence of neural connections between synchronously activated neurons in both directions. Thus, in our modeling approach we assume that the neural mapping between state maps and phonetic map is *bidirectional*. Thus, neural connections occur and increase in connectivity in both directions, i.e. from state maps towards the phonetic map and vice versa. The increase in connectivity directly reflects the degree of activation of the model neurons within each training step.

This existence of bidirectional mapping with the same link weight values for the synaptic neural connections in both directions is important for the other working modes of the model. Aside from speech acquisition, the model comprises the working modes of speech production and speech perception.

In the case of speech production, it can be assumed that the production of a word or utterance leads to a successive activation of phonemic syllable states at the level of the phonemic map. Thus, if the production of a specific syllable starts, we expect the full activation of that neuron, which represents that specific syllable at the level of the phonemic map. This directly leads to a co-activation of neurons at the level of the phonetic map, initiated by the neural mapping from the phonemic state map towards the phonetic map (Figure 2). Subsequently, the strongest co-activated model neuron at the level of the phonetic map leads to a co-activation of the appropriate motor plan and sensory states via the neural mapping from the phonetic map towards motor plan and sensory state maps. On the basis of the motor plan activation pattern, the syllable can be articulated by activating the feedforward execution path. Finally the resulting auditory state, activated by a subsequent sensory processing of the articulatory-acoustic vocal tract model output via the sensory processing pathway, can be compared with the auditory state of that syllable, already activated from the neural mapping between the phonetic map and the auditory state map. In this case, two states must be compared at the level of the auditory state map, which is not included in the version of our model introduced here. However, a concept for the comparison of sensory states and for the calculation of sensory error signals has already been introduced in [7].

In the case of speech perception (e.g. perception of the caretaker), an auditory state is activated at the level of the auditory state map (Figure 2). This leads to a co-activation of the phonetic map via the mapping from the auditory state to the phonetic state map and then to the selection of a winner neuron at the level of the phonetic map. Subsequently, a phonemic state is activated at the level of the phonemic map if the winner neuron at the level of the phonetic map is part in the relevant phoneme region.

It is possible that due to the context within a specific communication situation, competing candidates (competing syllable states) are already activated at the level of the phonemic map. This could lead to top-down effects in perception and influence the neural activation pattern at the level of the auditory map. These top-down effects in speech perception are not included in the version of our model presented in the present paper.

Speech production

The quality of the 50 most frequent syllables produced after speech acquisition (i.e. after the adjustment of synaptic link weights of the mappings between the phonetic map and all state maps as described in above) was tested by performing a combined acoustic perception and vocal tract action identification test. A phonetic expert (BK, male, 53 years old, no known speech or hearing anomalies) performed this test. In this test, the 50 most frequent syllables, which were already acquired during simulation of speech acquisition, were generated by the model. A winner neuron was calculated at the level of the phonetic map by activating a specific phonemic state (a specific syllable). Subsequently, the appropriate motor plan state was co-activated for that winner neuron, and then the articulatory and acoustic signal was generated using the feed-forward execution path including our vocal tract model (see Figure 2).

In a first step of this test, the expert was advised to transcribe all 50 acoustic speech stimuli. Stimuli were presented twice each in random order. In the case of differences between both transcriptions of the same acoustic stimulus (as was the case for 8 stimuli in this experiment), the stimulus was presented a third time and the expert was advised to choose one of both transcriptions as the nearest transcription.

An initial evaluation of the results of this transcription process was done by comparing the transcriptions with the phonemic transcriptions defining the phonemic states during the speech acquisition process. The overall rate of correct transcriptions was 78%. A close inspection of the resulting transcriptions showed that transcription errors mainly resulted from confusions in the place of articulation for nasals and for voiced plosives. Our impression is that this results primarily from the lack in acoustic quality of our vocal tract model rather than from errors in controlling articulation (errors in generating motor plan states).

Thus, in the case of those speech items which showed transcription errors, we checked in a second step whether the correct vocal tract action was produced at the level of the articulatory signal. This was done by inspecting the articulatory movement pattern generated by the vocal tract model. This resulted in a rate of correct syllable production of 94%, which we assume means that the model is capable of articulating well after speech acquisition.

The remaining errors (3 of 50 syllables) resulted from an incorrect perceptual categorization of the phonetic feature "stressed vs. unstressed", which may result from the fact that the settings of the pulmonary and laryngeal actions was fixed within the current version of our model. Thus, the main acoustic cue for identifying the stressed vs. unstressed condition, which is available in our model, is the length of the whole syllable and the length of single sound segments within each syllable.

Speech perception

Speech perception (i.e. the speech recognition rate) was tested by calculating the recognition rate for the 200 most frequent syllables, spoken three times by the same person (female, 27 years old). This person was the same person whose voice was recorded for the training items. These test items were fed into the model and resulted in the activation of a winner neuron at the level of the phonetic map and subsequently to an activation of one neuron at the level of the phonemic map (the identification process). The rates for correct identification are listed in Table 4. It can be seen in Table 4 that the recognition rate decreases with decreasing frequency of occurrence of a syllable within the children's book corpus.

Discussion and conclusions

A quantitative model is introduced in this paper which is capable of simulating the basic processes of speech acquisition, speech production, and speech perception. This

Table 4 Recognition rate of test items of 1 to N with N = 5 to N = 200 most frequent syllables

Number of most frequent syllables	Recognition rate in % without frequency correction	Recognition rate in % with frequency correction
5	100,0	100,0
10	97,2	97,4
20	91,1	93,2
50	90,2	92,7
100	88,9	92,3
200	85,8	91,8

The recognition rate without frequency correction (second column) shows the mean recognition rates for all N syllables, each syllable weighted equally. The frequency correction (third column) shows recognition rates for all N syllables, now weighted with respect to the frequency of occurrence of each syllable within the corpus (i.e. highly frequent syllables occur more often and also indicate higher recognition rates).

approach does not include higher-level linguistic processes such as the acquisition of semantic or higher-level grammatical knowledge. Thus, modeling of the mental lexicon (cf. [28]) is beyond the scope of our approach. Our model focuses on the sensorimotor aspects of speech acquisition, production, and perception. The speech knowledge, including speech production skills, acquired by our model may be comparable to the knowledge which is theoretically assumed to constitute the mental syllabary [51]. But in contrast to the model presented in [51], our model is strictly quantitative; it is capable of processing acoustic and articulatory data; and it is based on computational and theoretical neuroscience.

Thus our approach for modeling sensorimotor aspects of speech should be seen as a counterpart to the quantitative computational neurolinguistic approach developed by [5,38] for modeling the mental lexicon. Initial ideas as to how to combine these two approaches are discussed in [22,42]; see the discussion concerning P-Map and S-Map in these papers. Consequently, our modeling approach focuses on the early phases of speech acquisition (mainly first 18 months of life), while other approaches are focused on the rapid growth of the mental lexicon (the vocabulary spurt, starting at around 18 months; e.g. [52]).

As mentioned in the Introduction of the present paper, our approach should be seen as complementary and not as contradictory to the sensorimotor model introduced by Guenther (DIVA model [7,10,11,35]). On the one hand, while the DIVA model and our model both comprise phonemic, motor and auditory and sensorimotor neural state representations, our approach additionally introduces the phonetic map as an intermediate map between the sensorimotor and linguistic parts of the model. This intermediate level could be added to the DIVA model as well. On the other hand, a quantitative modeling of the processing of sensorimotor mismatch – as is done by Guenther and colleagues [ibid.] using specific "error maps" – is not currently implemented in our approach but could be added easily. A difference between the approaches is that our approach explicitly assumes a motor planning level (cf. [2]), while the DIVA approach directly associates its speech sound map with its primary motor map.

The modeling approach for speech acquisition and speech processing (production and perception) as introduced in the present paper bases the ability to simulate behavioral phenomena, such as the production and perception of speech items and learning to speak, upon brain-related principles such as Hebbian learning and self-organization.

This goal is achieved by using an activation rate model. In contrast, if we were to use fine-grained "neuromicroscopic" approaches such as spiking neuron models (e.g. [15]), we would not be able to simulate "macroscopic" large-scale behaviors such as speech acquisition, production and perception. Thus, our approach is a feasible compromise between simplicity in modeling neural "microfunctions" on the one hand and the ability to simulate complex "macroscopic" behavioral phenomena on the other. Moreover, our approach provides explanatory power for understanding the behavioral phenomena of speech processing from a neurofunctional point of view. At this point, it should be noted that our model not only gives quite good results in the performance of production and perception after training (see Discussion section), but also is capable of simulating complex behavior such as categorical speech perception [44].

A basic functional neural structure is hypothesized in this paper and is introduced as a preliminary neurofunctional model for speech acquisition, speech production, and speech perception. Three hypotheses form the basis of the model's structure: (i) We assume that there are three neural maps representing higher-level motor plans and higher-level auditory and somatosensory states. The activation patterns of syllabic speech items are manifested at this level during the time period of processing (i.e. producing or perceiving) of that speech item. (ii) In parallel, a phonemic (i.e. abstract linguistic) state representation is assumed to be activated for each speech item at the level of the phonemic map. The activation of a syllable state at the level of the phonemic map occurs at the beginning of the production process as well as at the end of the perception process. (iii) *Only one* intermediate neural map is assumed in order to associate motor plan, sensory, and phonemic states. This neural map is postulated to be a self-organizing map. This map is referred to as the phonetic map because it is supramodal, i.e. it is above the motor and sensory modalities. At the level of this map, phonetic features result from the ordering of sensorimotor speech states during neural self-organization.

Two hypotheses are assumed for speech acquisition, i.e. for training the model: (i) Hebbian learning (cf. [53]) occurs during speech acquisition: "Wire" two neurons if they are activated within the same time period, i.e. if these neurons "fire" within the same time interval. More specifically, one of these neurons must be part of the phonetic map, while the other neuron is part of the sensorimotor state maps. Because many neurons, especially at the level of the state maps, may be activated in parallel, the "wiring" process (i.e. the increase in synaptic link weights, also referred to as the increase in synaptic association) occurs in parallel for many neurons between the sensorimotor state maps and the phonetic map. (ii) Training can be subdivided into babbling training and imitation training, where babbling training comprises only motor and sensory information because the goal of babbling training is to learn sensorimotor relations. Learning sensorimotor relations means associating sensory and motor plan states. This is easily accomplished during babbling because sensory states are produced directly from motor plan states by using the peripheral vocal tract model, which is included in our model (Figure 2). Additionally, imitation training requires phonemic information in order to categorize sensorimotor states acquired previously at the level of the phonetic map. Imitation training leads to the emergence of phoneme regions.

The results of the imitation training experiment described in this study, which used the 200 most frequent syllables of Standard German on the basis of a children's book sentence corpus, indicate that the phonetic map can be subdivided into two major areas: stressed and unstressed syllables. In addition, in the case of stressed syllables, some sub-areas can be found which reflect different types of vowels (long, short, diphthong; see Figure 6; frequent unstressed syllables in our corpus are mainly produced with a reduced vowel). Other imitation experiments indicate that sensorimotor states are ordered at the level of the phonetic map with respect to different phonetic features such as front-back, low-high, place and manner of articulation, etc. (see Table 1). This result is not affected if different random initializations for link weights are used or if motor and sensory representations are weighted in a different way. These modifications mainly cause rotations or a mirroring of the phonetic map, but they do not affect the resulting phonetic ordering.

Moreover, the simulation experiment discussed in this paper indicates that the more often a training item is processed by the model, the larger the resulting phoneme region is, i.e. the more model neurons represent exemplars (or realizations) of that phonemic state at the level of the phonetic map (see Figure 7). In addition, it should be noted that in our approach, imitation training requires fewer training cycles than babbling training. The number of training cycles is approximately 600 for babbling training and around 150 training cycles for imitation [44]. This reflects the fact that the phonetic map and especially the ordering of phonetic states is in large part established during pre-linguistic babbling training. This ordering is refined during imitation training with respect to language-specific demands. Even the emergence of phoneme regions does not lead to a reorganization of the phonetic map, which is already established during babbling training.

There are still some shortcomings in our approach which we will address by further refining the model in future studies. Firstly, simulation experiments for babbling and imitation were done step-by-step and not – as it would be more realistic – as one continuously ongoing simulation experiment. The main reason for that is that we did not use a growing self-organizing map (GSOM, see [5]) but rather SOMs of fixed sizes (see Table 1). A further reason is that we are currently not able to guide babbling by imitation. The current version of the model is not able to imitate a given speech item directly and thus is not able to generate motor plans which are related to a potentially successful motor plan for imitating the speech item. This work is done manually in the current version of the model. It is for this reason that we currently establish babbling and imitation training sets manually.

Thus, a further step in improving our model would be the implementation of a module for babbling training guided by imitation. In addition, the introduction of GSOMs is an important future step and will allow us to avoid predefining the size of the phonetic map and to allow an ongoing continuous acquisition of speech knowledge including sensorimotor speech skills. This augmented model should also include the acquisition of prosodic features, such as the basic intonation contours of the target language, as well as the acquisition of basic stress patterns of words and utterances.

A second shortcoming of the current version of our model is the problem of speaker normalization [43]. Even in the case of imitation training, the imitation trials produced by the model are evaluated aurally by an expert before they are used as training items. In the current version of the model, it would be possible to store both auditory realizations (the one pre-produced by the external speaker or caretaker) and is the one successfully imitated by the model or toddler. Both groups of auditory realizations could be associated with the phonemic states as well as with motor plan states. In the case of imitating more than one external speaker, we potentially could integrate two or more "speakers" into the phonetic map into future versions of our approach, which would allow us to augment the phonetic map to include the phonetic dimension of "different speakers" or "different types of speakers" and thus to address the problem of adaptation to a new speaker in our framework.

A third shortcoming of our approach is that we start with sensorimotor (i.e. phonetic) information during babbling training, and directly thereafter we extend this to phonemic information for imitation training. From the viewpoint of natural speech acquisition, it would be more realistic to start imitation on the basis of face-to-face communication scenarios, where the model can associate words with semantic categories rather than directly with a (perfect) phonemic representation. There is some evidence showing that phonemic information is not available at the earliest developmental stages, but rather emerges during speech acquisition [42]. Thus, we will use sensorimotor and semantic information rather than sensorimotor and phonemic information for future training experiments. This information could be used for speech acquisition if our model comprised a semantic level

mIt	'mIt	'hat	'hat		'?Ist	'?Ist	'?Es	'?Es	'?Es	'vas	'vas	'das	'das	'das	'das	'gants		'dEn	'de:n	'de:	'de:	'de:m	'kan	'kOmt
'mIt		'hat	'hat		'?Ist	'?Ist	'?Es	'?Es		'bIs		'das	'das	'das			'dan	'dan		'de:n	'de:m	'de:n		'ta:k
	'gIpt	'rYk	'?Et		'?IC	'?I	'?ap	'?ap	d@t	'bEt	'dOx	'dOx		'za:k	'za:k	'vEn	'vEn	'man		'de:n		'mo:nt	'mo:nt	
'nICt	'nICt		lIC		'?IC	'?IC	vas					've:k	'zi:t		'fra:kt		n@n	'man	'hant		'na:x	'ma:l		'SnEl
'nICt		'?Ist	lIC			'?IC	'di:	'di:	'du:	'di:		'ge:t		'So:n	'fOn	'fOn		'vIl		'ja	'ma:l	'da:		'tsvaI
	'mUs		n@s	'zIC	'zIC		'bE:	'di:	'di:	'di:		'gro:	'fe:		'tsUm	'tsUm		'vIl	'vOl	'da:	'da:	'kaI		
'jEtst	'maxt	'nOx		'zIC	'zIC		'di:	'di:	'di:	'di:	'vo:	'gro:		'fy6	'fraU		'vI6t		'va6	'va6	'dE6	'dE6	'ga6	baI
				'zIC		'vi:	'di:	'du:	'du:	'vi:	'vi:		'fo6	'fraU			'vI6t	'vi6	'dE6	'bE6	'dE6	'dE6	'bE6	'dE6
'tsu:	'tsu:	'Spi:	'fi:		'le:	'li:				'zi:		'zi:	'fo6		'klaI			'dE6	'dE6	'dE6	'vi6	'm6		'dE6
	'tsu:	tsIm	'fo:	f6		'blu:	'zi:	'zi:	'zi:	'zi:	'zi:	'zi:		'klaI	'klaI	'fa:	'va6	'vE6	'vE6		'?i6		'?i6	'?E6
ts@	tsu:	ts@n		f@l	'fIn		'zi:	'zi:		'zo:			'laN	'ja		'zaI	'zaI	'da:			'?E6	'?E6		'?E6
f@			f@n		C@n	C@n		'ze:	'zo:	'zo:	'zOn	'jUN	'mE:t	'me6	'hOI			'ra:	d6	'?E6	'?E6		'?aIn	
s@	S@		f@n	f@n	C@n	x@n	x@n				'kOn		ma		'ha:		'vaI			'?E6	'?E6	'?aIn	'?aIn	'?aIn
	fE6	z@		z@n	z@n			m@n	l@n	n@m	N@n		ma:l		'ha:	'?a:	'?a6	'?aI	'?aI	'?aI	'?aI	'?aIn	'?a	'?aIn
z@	fE6	z@				m@n	m@n			n@	N@n	g@n	dEl	'tE		'?aI	'?aI	'?a	'?aI	'?aI	'?aU	'?a	'?aUf	'?a
		l@		'ma	'nOk	'nOk		r@n	r@n	n@n	n@n	k@n	t@l	t6	pa:		'?a:	'?a:	'?a:	'?a	'?aU	'?aUf	'?aUf	'?aUf
n@		l@	ma:	'ma		m6		r@n	r@n	n@n	n@n		g@l	g6						'?aUx	'?aUx	'?aUs	'?aUs	'?aUf
n@	n@	n@		ma	ma:		n6		l@n	n@n		d@n	g@l		'?al		'ri:	'?i:	'ri:		'?a	'?alts	'?a	'haUs
	n@	n@		hE		m6	ma:	ma:		b@n	b@n	d@n		g@l	'?al		'ri:	'?i:		'?Ent	'?an	'?a	'?alts	
l@		m@	n@				l6		g@n	g@n	g@n	d@n	t@t	t@		'?o:	'?o:		'?Im	'?En	'?an	'?an	'?a	k@l
jo:	jo:	w@		ni:	di:	'da	'da	b6	b6	g@n	k@n	d@n	t@	t@		'?y:		'?Im	'?In	'hIn	'?an	'?an	'?am	'hIn
n@	l@	li	r@		r@	b6		d6		b@n	k@n		t@	t@			'?Un		'?In	'?I	'?I	'?i:n	'?In	'?i:m
g@			d@	d@		b6		d6	g6		k@	t@	t@	t@	t@t	'kat		'?Unt	'?Unt		'?Unt	'?Um	'?Um	'?i:
g@	b@					d6		t6	t6		p@	k@		t@	ts@	'kOn	t@l		'?Unt	'?Unt	'?Unt	'?Unt	'?Unt	'?Unt
g@	b@	b@	g@	g@	g@	d6		t6	t6	'pa	pi:	pi:	t@	t@		'kO	'kIn	'kIn		'?Unt	'?Unt	'?Unt	'?Unt	'?Ins

Figure 8 25 × 25 phonetic map after imitation training as described in Results section. A model neuron is marked with a phonemic transcription if the synaptic link weight between that model neuron of the phonetic map and a model neuron of the phonemic map is strong (link weight > 0.8). An apostrophe at the beginning of the syllable indicates that the syllable is stressed. Transcriptions are given in SAMPA notation (Speech Assessment Methods Phonetic Alphabet).

in addition to the sensorimotor level described in this paper (ibid.). Additionally, this would expand our model towards word and utterance processing and thus more realistic performance tests for production and perception could be performed, especially considering that speech perception is difficult to check at syllable level.

A fourth shortcoming of our approach is that we have no explicit modeling of time as exists in spiking neuron models or more detailed rate models such as the neuroanatomically grounded Hebbian learning model developed by [8]. Time modeling would allow us to avoid the zero-paddings which currently occur in our syllable-based representations of motor plans and neural spectrograms. Furthermore, these approaches would allow the processing of whole utterances rather than isolated syllables. It is a major goal to introduce time as an explicit parameter into future versions of our model.

Appendix

Figures 8, 9, and 10 display the synaptic link weights of the *same* 25 × 25 phonetic map resulting from training the 200 most frequent syllables of our children's book corpus as

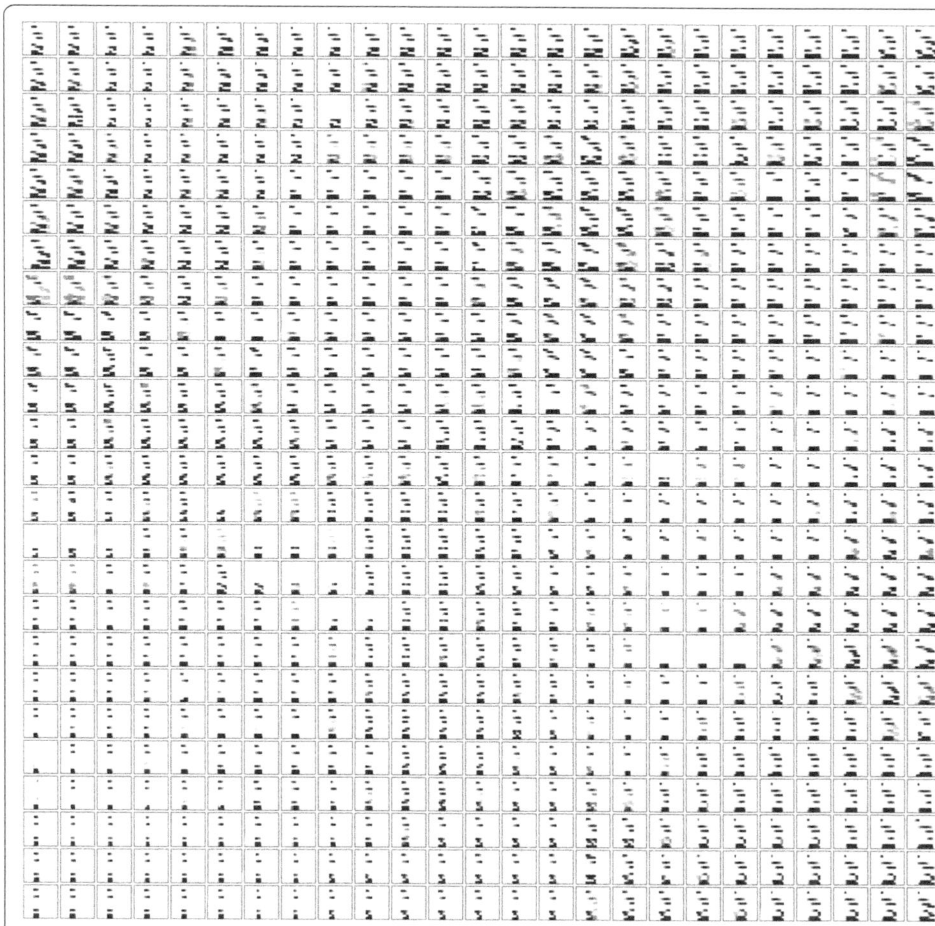

Figure 9 25 × 25 phonetic map after imitation training as described in Results section. Each model neuron within this phonetic map displays all synaptic link weights towards the motor state map (white: no connectivity, $w_{ij} = 0$; black; high connectivity, $w_{ij} = 1$). Thus, if one model neuron within the phonetic map is maximal activated, the resulting motor plan state activation pattern is displayed here in the box representing that model neuron.

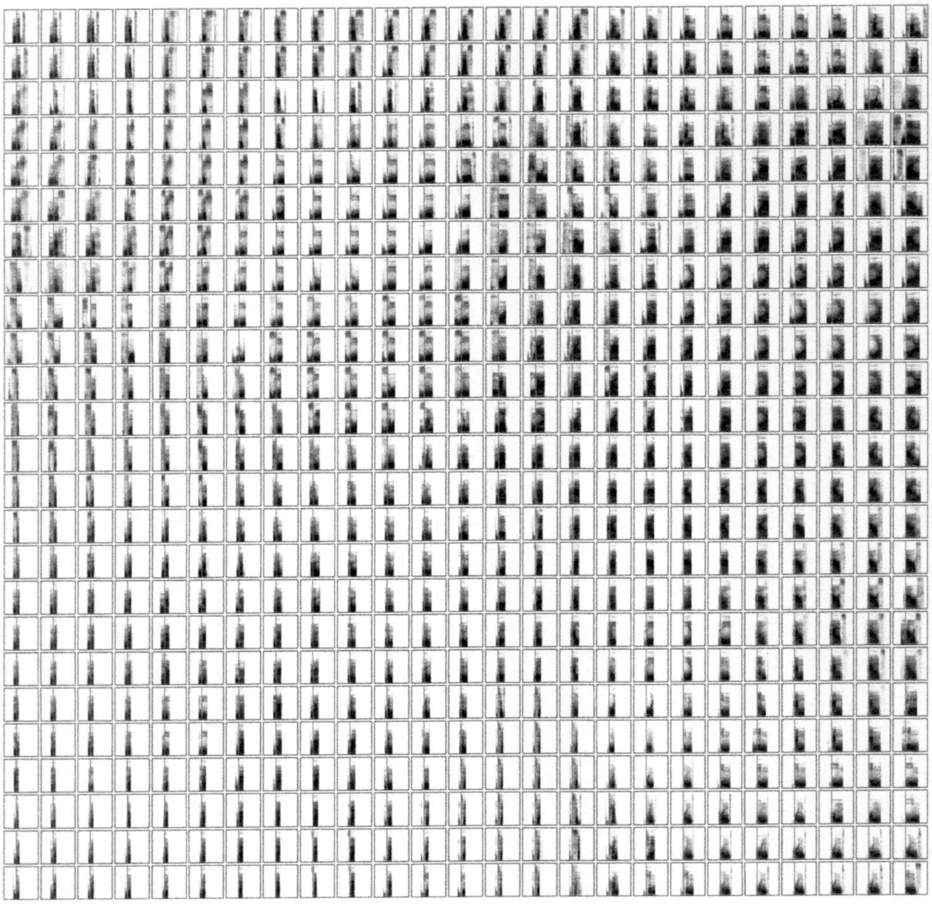

Figure 10 25 × 25 phonetic map after imitation training as described in Results section. Each model neuron within this phonetic map presents all synaptic link weights towards the auditory state map (white: no connectivity, $w_{ij} = 0$; black; high connectivity, $w_{ij} = 1$). Thus, if one neuron within the phonetic map is maximal activated, the resulting auditory state activation pattern is displayed here in the box representing that model neuron.

described in above. All three figures represent the speech knowledge stored as synaptic link weights between the phonetic map and the phonemic map (Figure 8), between the phonetic map and the motor plan state map (Figure 9), and between the phonetic map and the auditory state map (Figure 10). Thus, Figure 8 presents all model neurons within the phonetic map which represent a realization of a specific syllable (phoneme regions), Figure 1B presents the knowledge concerning the production of a phonetic realization (motor state for each realization) and Figure 1C indicates the knowledge of how that realization sounds (auditory state for each realization).

Competing interests

The authors declare that they have no competing interests.

Authors' contributions

BJK and EK developed the overall structure of the model and the design for the learning experiments. JK carried out the computer implementation and run the simulations. All authors read and approved the final manuscript.

Acknowledgements
This work was supported in part by German Research Council DFG, project number KR 1439/15-1, and in part by EU-COST action 2102. We thank Peter Birkholz for providing us with an articulatory-acoustic vocal tract model, and we thank Cornelia Eckers for lending her voice for the corpus.

Author details
[1]Neurophonetics Group, Department of Phoniatrics, Pedaudiology, and Communication Disorders, Medical School, RWTH Aachen University, Aachen, Germany. [2]Cognitive Computation and Applications Laboratory, School of Computer Science and Technology, Tianjin University, Tianjin, China. [3]Education and Rehabilitation of the Deaf and Hard of Hearing, Department of Special Education, Faculty of Human Sciences, University of Cologne, Cologne, Germany.

References

1. Wise RJS, Greene J, Buechel C, Scott SK: **Brain regions involved in articulation.** *Lancet* 1999, **353**:1057–1061.
2. Riecker A, Mathiak K, Wildgruper D, Erb M, Hertrich I, Grodd W, Ackermann H: **fMRI reveals two distinct cerebral networks subserving speech motor control.** *Neurology* 2005, **64**:700–706.
3. Hickok G, Poeppel D: **The cortical organization of speech processing.** *Nat Rev Neurosci* 2007, **8**:393–402.
4. McClelland JL, Elman JL: **The TRACE model of speech perception.** *Cogn Psychol* 1986, **18**:1–86.
5. Li P, Farkas I, MacWhinney B: **Early lexical development in a self-organizing neural network.** *Neural Netw* 2004, **17**:1345–1362.
6. Westermann G, Miranda ER: **A new model of sensorimotor coupling in the development of speech.** *Brain Lang* 2004, **89**:393–400.
7. Guenther FH: **Cortical interaction underlying the production of speech sounds.** *J Commun Disord* 2006, **39**:350–365.
8. Garagnani M, Wennekers T, Pulvermüller F: **A neuroanatomically grounded Hebbian-learning model of attention-language interactions in the human brain.** *Eur J Neurosci* 2008, **27**:492–513.
9. Wennekers T, Garagnani M, Pulvermüller F: **Language models based on Hebbian cell assemblies.** *J Physiol Paris* 2006, **100**:16–30.
10. Guenther FH, Ghosh SS, Tourville JA: **Neural modeling and imaging of the cortical interactions underlying syllable production.** *Brain Lang* 2006, **96**:280–301.
11. Guenther FH, Vladusich T: **A neural theory of speech acquisition and production.** *J Neurolinguistics* 2012, **25**:408–422.
12. Perkell JS: **Movement goals and feedback and feedforward control mechanisms in speech production.** *J Neurolinguistics* 2012, **25**:382–407.
13. Arbib MA, Erdi P, Szentagothai J: *Neural Organization.* Cambridge, MA: The MIT Press; 1998.
14. Kuhl PK: **Early language acquisition: cracking the speech code.** *Nat Rev Neurosci* 2004, **5**:831–843.
15. Gerstner W, Kistler W: *Spiking Neuron Models.* Cambridge, UK: Cambridge University Press; 2002.
16. Kasabov N: **To spike or not to spike: A probabilistic spiking neuron model.** *Neural Netw* 2010, **23**:16–19.
17. Oberauer K, Lewandowsky S: **Modeling working memory: a computational implementation of the Time-Based Resource-Sharing theory.** *Psychon Bull Rev* 2011, **18**:10–45.
18. Bednar JA, Kelkar A, Miikkulainen R: **Scaling self-organizing maps to model large cortical networks.** *Neuroinformatics* 2004, **2**:275–301.
19. Kohonen T: **The self-organizing map.** *Proc IEEE* 1990, **78**:1464–1480.
20. Kohonen T: **Things you haven't heard about the self-organizing map.** In *Proceedings of IEEE International Conference on Neural Networks.* USA: ICNN; 1993:1147–1156.
21. Kohonen T: *Self-Organizing Maps.* 3rd edition. Berlin: Springer; 2001.
22. Kröger BJ, Birkholz P, Kannampuzha J, Kaufmann E, Neuschaefer-Rube C: **Towards the acquisition of a sensori-motor vocal tract action repository within a neural model of speech processing.** In *Analysis of Verbal and Nonverbal Communication and Enactment: The Processing Issues (LNCS 6800).* Edited by Esposito A, Vinciarelli A, Vicsi K, Pelachaud C, Nijholt A. Berlin, Germany: Springer; 2011:287–293.
23. Kröger BJ, Kopp S, Lowit A: **A model for production, perception, and acquisition of actions in face-to-face communication.** *Cogn Process* 2010, **11**:187–205.
24. Kröger BJ, Birkholz P, Kannampuzha J, Eckers C, Kaufmann E, Neuschaefer-Rube C: **Neurobiological interpretation of a quantitative target approximation model for speech actions.** In *Studientexte zur Sprachkommunikation: Elektronische Sprachsignalverarbeitung 2011.* Edited by Kröger BJ, Birkholz P. Dresden, Germany: TUDpress; 2011:184–194.
25. Kröger BJ, Birkholz P: **A gesture-based concept for speech movement control in articulatory speech synthesis.** In *Verbal and Nonverbal Communication Behaviours (LNAI 4775).* Edited by Esposito A, Faundez-Zanuy M, Keller E, Marinaro M. Berlin: Springer; 2007:174–189.
26. Birkholz P, Jackel D, Kröger BJ: **Construction and control of a three-dimensional vocal tract model.** In *Proceedings of the International Conference on Acoustics, Speech, and Signal Processing.* USA: ICASSP; 2006:873–876.
27. Birkholz P, Jackel D, Kröger BJ: **Simulation of losses due to turbulence in the time-varying vocal system.** *IEEE Transactions on Audio, Speech, and Language Processing* 2007, **15**:1218–1225.
28. Levelt WJM, Roelofs A, Meyer A: **A theory of lexical access in speech production.** *Behav Brain Sci* 1999, **22**:1–75.
29. Elman JL: **An alternative view of the mental lexicon.** *Trends Cogn Sci* 2004, **8**:301–306.
30. Cisek P: **Neural representations of motor plans, desired trajectories and controlled objects.** *Cogn Process* 2005, **6**:15–24.
31. Saltzman E, Munhall KG: **A dynamical approach to gestural patterning in speech production.** *Ecol Psychol* 1989, **1**:333–382.
32. Kröger BJ: **A gestural production model and its application to reduction in German.** *Phonetica* 1993, **50**:213–233.

33. Kröger BJ, Schröder G, Opgen-Rhein C: **A gesture-based dynamic model describing articulatory movement data.** *J Acoust Soc Am* 1995, **98:**1878–1889.
34. Pasley BN, David SV, Mesgarani N, Flinker A, Shamma SA, Crone NE, Knight RT, Chang EF: **Reconstructing speech from human auditory cortex.** *PLoS Biol* 2012, **10:**e1001251. doi:10.1371/journal.pbio.1001251.
35. Golfinopoulos E, Tourville JA, Guenther FH: **The integration of large-scale neural network modeling and functional brain imaging in speech motor control.** *Neuroimage* 2010, **52:**862–874.
36. Kröger BJ, Birkholz P, Lowit A: **Phonemic, sensory, and motor representations in an action-based neurocomputational model of speech production (ACT).** In *Speech Motor Control: New developments in basic and applied research.* Edited by Maassen B, Van Lieshout P. New York: Oxford: University Press; 2010:23–36.
37. Kröger BJ, Birkholz P, Kannampuzha J, Neuschaefer-Rube C: **Modeling sensory-to-motor mappings using neural nets and a 3D articulatory speech synthesizer.** In *Proceedings of the 9th International Conference on Spoken Language Processing.* USA: ICSLP & Interspeech 2006; 2006:565–568.
38. Li P, Zhao X, Mac Whinney B: **Dynamic self-organization and early lexical development in children.** *Cognit Sci* 2007, **31:**581–612.
39. Meltzoff AN, Moore MK: **Explaining facial imitation: a theoretical model.** *Early Development and Parenting* 1997, **6:**179–192.
40. Knapp ML, Hall JA: *Nonverbal Communication in Human Interaction.* 7th edition. Wadsworth, USA: Cengage Learning; 2010.
41. Tomasello M: *Origins of Human Communication.* Cambridge, MA: The MIT Press; 2008.
42. Kröger BJ, Birkholz P, Neuschaefer-Rube C: **Towards an articulation-based developmental robotics approach for word processing in face-to-face communication.** *PALADYN Journal of Behavioral Robotics* 2011, **2:**82–93.
43. Johnson K: **Speaker normalization in speech perception.** In *The Handbook of Speech Perception.* Edited by Pisoni DB, Remez RE. Oxford, UK: Blackwell; 2008:ch15.
44. Kröger BJ, Kannampuzha J, Neuschaefer-Rube C: **Towards a neurocomputational model of speech production and perception.** *Speech Comm* 2009, **51:**793–809.
45. Oller DK, Eilers RE: **The role of audition in infant babbling.** *Child Dev* 1988, **59:**441–449.
46. De Boysson-Bardies B, Sagart L, Durand C: **Discernible differences in the babbling of infants according to target language.** *J Child Lang* 1984, **11:**1–15.
47. Kröger BJ, Kannampuzha J, Lowit A, Neuschaefer-Rube C: **Phonetotopy within a neurocomputational model of speech production and speech acquisition.** In *Some Aspects of Speech and the Brain.* Edited by Fuchs S, Loevenbruck H, Pape D, Perrier P. Berlin: Peter Lang; 2009:59–90.
48. Kröger BJ, Miller N, Lowit A, Neuschaefer-Rube C: **Defective neural motor speech mappings as a source for apraxia of speech: Evidence from a quantitative neural model of speech processing.** In *Assessment of Motor Speech Disorders.* Edited by Lowit A, Kent R. San Diego, CA: Plural Publishing; 2011:325–346.
49. Pierrehumbert JB: **Exemplar dynamics, word frequency, lenition and contrast.** In *Frequency Effects and Emergent Grammar.* Edited by Bybee J, Hopper P. Amsterdam: John Benjamins; 2001:137–158.
50. Bauer D, Kannampuzha J, Kröger BJ: **Articulatory Speech Re-Synthesis: Profiting from natural acoustic speech data.** In *Cross-Modal Analysis of Speech, Gestures, Gaze and Facial Expressions (LNAI 5641).* Edited by Esposito A, Vich R. Berlin: Springer; 2009:344–355.
51. Levelt WJM, Wheeldon L: **Do speakers have access to a mental syllabary?** *Cognition* 1994, **50:**239–269.
52. Plunkett K: **Lexical segmentation and vocabulary growth in early language acquisition.** *J Child Lang* 1993, **20:**43–60.
53. Hebb DO: *The Organization of Behavior.* New York: Wiley and Sons; 1949.

5

Intrinsic synergistic-topological mechanism versus synergistic-topological matrix in microtubule self-organization

Vlado A Buljan[1,2,3*], R M Damian Holsinger[1,4], Brett D Hambly[5], Vangelis Kanellis[1], Elie Matar[1], Xanthe Larkin[1], Guo Jun Liu[3], John J Bohorquez-Florez[1] and Richard B Banati[1,3,6]

* Correspondence:
vlado.buljan@sydney.edu.au
[1]Brain and Mind Research Institute, Sydney Medical School, The University of Sydney, Sydney, NSW 2050, Australia
[2]Department of Physiology, Sydney Medical School, The University of Sydney, Sydney, NSW 2006, Australia
Full list of author information is available at the end of the article

Abstract

Background: In this body of work we investigate the *synergistic-topological* relationship during self-organization of the microtubule fiber *in vitro*, which is composed of straight, axially shifted and non-shifted, acentrosomal microtubules under crowded conditions.

Methods: We used electron microscopy to observe morphological details of ordered straight microtubules. This included the observation of the differences in length distribution between microtubules in ordered and non-ordered phases followed by the observation of the formation of interface gaps between axially shifted and ordered microtubules. We performed calculations to confirm that the principle of summation of pairwise electrostatic forces act between neighboring microtubules all their entire length.

Results: We have shown that the self-organization of a microtubule fiber imposes a variety of topological restrictions onto its constituting components: (a) tips of axially shifted neighboring microtubules are not in direct contact but rather create an *interface gap*; (b) fibers are always composed of a restricted number of microtubules at given solution conditions; (c) the average length of microtubules that constitute a fiber is always shorter than that of microtubules outside a fiber; (d) the length distribution of microtubules that constitute a fiber is narrower than that of microtubules outside a fiber and this effect is more pronounced at higher GTP-tubulin concentrations; (e) a *cooperative* motion of fiber microtubules due to actualization of the summation principle of pairwise electrostatic forces; (f) appearance of local GTP-tubulin *depletion* immediately in front of the tips of fiber microtubules.

Conclusion: Overall our data indicate that under crowded conditions *in vitro*, the self-organization of a microtubule fiber is governed by an *intrinsic synergistic-topological* mechanism, which *in conjunction* with the *topological* changes, GTP-tubulin *depletion*, and *cooperative* motion of fiber constituting microtubules, may generate and maintain a *synergistic-topological matrix*. Failure of the mechanism to form biologically feasible microtubule synergistic-topological matrix may, *per se*, precondition tumorigenesis.

Background

Microtubules are functionally and critically involved in the major events of intracellular organization including cell morphology [1], cellular motility [2], cell division [3], intracellular transport [4], signal transduction [5] and memory encoding [6].

In order to perform their roles accurately and efficiently, it is essential for microtubules to undergo highly coordinated behaviour, which implicates diverse forms of crosstalk phenomena including crosstalk between microtubules and other cytoskeletal components, between two or more individual microtubules or between different domains of single microtubules [3,7-15]. It has been recognized that crass-talk between cytoskeletal components (including microtubules) is *conditio sine qua non* for synergy to occur in the system [16]. Therefore, it is not surprising that in the presence of highly correlated motion of microtubules, combined with crosstalk, the presence of different modes of synergistic mechanism may be observed [17-19].

It has been shown that the mitotic spindle is a self-made machine, i.e., it can self-organize under biologically poor conditions (in the absence of centrosome and chromatin) [20-22]. To explain these findings it has been proposed that the physical state of the cytoplasm surrounding chromosomes is changed in a way that favors microtubule growth and self-organization into the spindle [21]. Indeed, it has recently been shown that self-organization of anastral spindles are guided by the *synergy* of dynamic instability, autocatalytic microtubule production and spatial signaling gradients [19]. Taken together these works may support the idea that some synergy operating mode(s) may be inherent to microtubule systems and local solution conditions.

The idea that the physical conditions of the cytoplasm may favor microtubule mitotic spindle self-formation is similar to the concept of a spindle matrix. The concept of the spindle matrix was introduced in order to explain the mechanics of complex movement of microtubules in mitotic spindles and its highly correlated behavior in relation to chromosome movement during cellular division [23]. Recent elaboration of the matrix concept in the study of spatio-temporal control of complex reorganization of microtubules in mitotic spindles during mitosis have shown that the concept requires further development to include molecular details, pliable structural dynamic features and cross-linking features [17,24,25]. As noted above, cross-linking may endow the system with cross-talk ability, which is indeed a synergistic feature. Therefore, it follows that the concept of the spindle matrix should also be bestowed with appropriate *synergistic* aspects. It has been suggested that the matrix exhibits elastic gel properties [26]. On the other hand, the high resistivity of the self-organized pattern in microtubules (*in vitro*) to disruptive forces may suggest that a similar matrix is formed, which, in association with the spatio-temporal pattern of microtubules *in vitro*, exhibits properties of an elastic gel [27]. Therefore, it would be beneficial to extend the study of the microtubule spindle matrix to the analogous matrix of any self-organized spatio-temporal microtubule pattern, especially those that include a far simpler pattern that is formed under solution conditions *in vitro*.

To investigate the possibility that intrinsic synergistic mechanisms in conjunction with the topology of spatio-temporally self-organized microtubules may form, we define this as microtubule synergistic-topological matrix . Under biologically reduced solution conditions *in vitro*, we have used a simple system comprising GTP-tubulin and a small pool of classical microtubule associated proteins (see Methods).

Under the crowded conditions, in terms of GTP-tubulin concentration, we have observed self-organized spatio-temporal patterns of microtubules (fibers). The crowded conditions, if they are strong enough, may reduce the degree of freedom of motion of protein molecule resulting in anisotropic motion [28].

In order to solve the problem of blind mason, during its addition to the tip, GTP-tubulin has to have significant degree of freedom of motion (rotational and translational diffusion) [29]. Therefore, by limiting the accommodating space for free rotation and translation of GTP-tubulin molecule, the highly crowded conditions are in conflict with solving the problem of blind mason during microtubule assembly. However, we have shown that crowded conditions in our experiments were not too strong so that the problem of blind mason is properly solved.

The fibers have always imposed some topological restrictions onto the constituent microtubules and have self-generated local *depletion* of GTP-tubulin in the immediate vicinity of the tips of microtubules. Obviously, affecting the shape of the fiber by acting at the microtubule tips, the local *depletion* of GTP-tubulin determines the dynamic of microtubule fiber in axial direction. In addition, there is strong mutual interdependence between the topology of microtubules and local solution conditions. Therefore, the spatio-temporal pattern of microtubules could be treated as an integrated composite of microtubule patterns itself and locally associated specific solution conditions. In an analogy to the spindle matrix, the nonlinear physics of integrated composites could be mathematically modeled by means of some sort of matrix.

An important implication of our data is that the self-organized microtubules are *co-localized, interconnected* and *interdependent*. These three features are critical conditions for crosstalk and consequently for *intrinsic synergistic* mechanisms to operate in the system as well [16]. The physical basis of these features is the principle of summation of electrostatic pairwise forces, which acts between neighboring microtubules all the way along their lengths [30]. Our calculations show that under the crowded conditions of our experiments the pairwise electrostatic forces follow the principle of summation. Therefore, the shape and the dynamic of fiber along its perpendicular direction are determined by the action of the principle of the forces summation across the constitutional microtubules. In addition, the principle endows constitutional microtubules with *cooperative* behavior.

Taken together the *depletion* of GTP-tubulin and the action principle of pairwise forces summation are critical and independent ingredients in the *synergistic* mechanism and the formation of *synergistic-topological matrix*.

As for the rationale of our data we propose that, under crowded conditions *in vitro*, the self-organization of microtubule fibers is governed by an *intrinsic synergistic* mechanism that *in conjunction* with the *topological* changes in microtubules may generate a *synergistic-topological matrix*.

Synergistic-topological mechanisms are pivotal in the formation and maintenance of a matrix. Failure of intrinsic synergistic-topological mechanisms to form and maintain biologically feasible microtubule synergistic-topological matrices may precondition the system to decline into biologically non-feasible events such as aneuploidy during mitosis that may in turn precondition the system for tumorigenesis [15,31,32].

Methods

The systems tested

Unless stated otherwise all experiments were performed with microtubules in MES buffer containing (in mM): 100 MES, 1 EGTA, 0.5 $MgCl_2$, 1.0 GTP, pH 6.6. Microtubules were reconstituted at different concentrations of microtubule protein (MTP). MTP refers to 95% tubulin +5% MAPs [33].

Microtubule assembly was initiated by the addition of 1 mM GTP at 36?C. A stationary state of assembly was reached in 30 minutes. Microtubule protein was prepared using bovine brain as described elsewhere [33].

In order to observe the effect of calcium on microtubule morphology and their ordering, in some experiments we added an excess of 1 mM free calcium (Ca^{2+}).

Electron microscopy: the morphology of ordered rudimentary microtubule phases on a micrometer scale

Our goal was to observe morphological details of ordered straight microtubules. This included the observation of the differences in length distribution between microtubules in ordered and non-ordered phases followed by the observation of the formation of interface gaps between axially shifted and ordered microtubules. For this purpose the specimens were viewed using an analytical Philips CM12 transmission electron microscope (TEM) at the Australian Key Centre for Microscopy and Microanalysis, at the University of Sydney. EM images were obtained using a Gatan multiscan CCD camera and an exposure time of 1.51 s.

In order to be viewed by EM, microtubules were prepared according to Langford s method [34]. Briefly, samples of microtubules containing different MTP concentrations were left to assemble for 60 min at 36?C. Microtubules were stabilised by adding 50% sucrose, fixed and stained with 2% uranyl acetate on 200 mesh grids. Samples were left overnight at room temperature prior to microscopic examination.

Microtubule lengths were measured using the Prism Image Analysis System (Analytical Vision Inc, CA). The length of microtubules (both, curved and straight) has been measured as the average contour microtubule length. In further text the average contour microtubule length will be referred to as microtubule length. We were primarily interested in microtubules of straight morphology. For this purpose our buffer did not contain any excess Ca^{2+}. However, for the purpose of comparison we prepared curved microtubules by adding an excess of 1 mM Ca^{2+} to solutions prior to microtubule assembly [35].

Microtubule samples were applied to coverslips (No 1 Superior Micro Cover Glasses W. Germany) using the procedures described in [36].

Results

Using transmission electron microscopy, we have studied the basic parameters of self-organized microtubule fibers under crowded conditions *in vitro*. The basic fiber parameters include: fiber size (the total number of constituent microtubules), the average length and length distribution of microtubules that constitute the fiber. We have also observed numerous, nanometer size, microtubule free space (which we termed *interface gap*) between the tips of approaching microtubules within fibers. In addition, our calculations have revealed that total depletion of GTP-tubulin can occur in the immediate vicinity of the tips of microtubule fibers, that is more pronounced versus more crowded conditions.

Symmetry breaking *per se* indicates the presence of a synergistic mechanism(s)

Formation of self-organized microtubule fibers and their spatial networks are accompanied by symmetry breaking, which *per se* indicates the presence of a synergistic mechanism(s) [37,38].

Straight microtubule fibers

We have observed straight microtubules spontaneously organize into straight fibers at 36?C in the presence of 7 mg/ml MTP and calcium chelated by 1 mM EGTA (Figure 1A). Strictly speaking the fiber in Figure 1A is not constituted purely by the straight microtubules. A very few slightly curved microtubules can be seen as apparently part of the fiber structure. However, bearing in mind that the fiber s geometrical, physical and self-organizational features are critically determined by the majority of straight microtubules, we will continue to use term straight microtubule fibers .

Obviously, straight fibers reflect distinct topological relations among the constituent microtubules. Single microtubules were axially shifted relative to their neighbors in the fiber. In this manner, fibers might extend by several average microtubule lengths. Furthermore, multitudes of straight microtubule-based fibers may spontaneously create a network, i.e. they can form a discrete spatial pattern of a higher topological complexity (Figure 1B) at higher MTP concentrations (10 mg/ml). If microtubules are predominantly straight, then they can maintain a partial parallel ordering pattern at some locations, while they cross each other multiple times elsewhere, forming star-like patterns. Additionally, some locations maybe free of microtubules while they densely populate other locations. Consequently, the initially homogenous space of microtubule multitudes become heterogeneous through symmetry breaking, i.e. differentiated into three regions: (a) microtubule free regions, (b) regions consisting of star-like microtubule patterns and (c) the railroad regions of parallel microtubules. Admittedly, massive spatio-temporal correlations must be involved in this symmetry breaking phenomena that *per se* provides evidence of the action of intrinsic synergy.

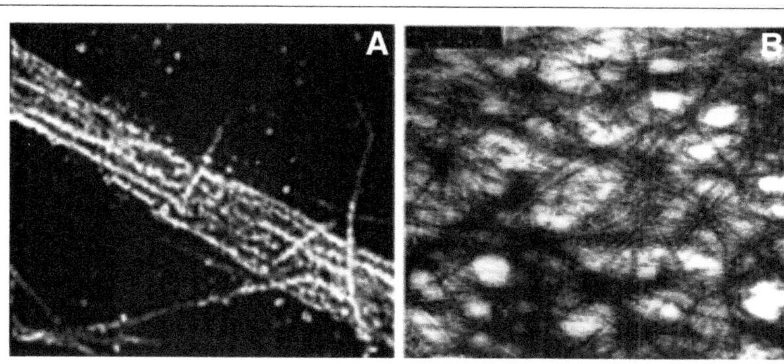

Figure 1 Self-organisation of straight microtubules, at crowded conditions in vitro is guided by breaking the symmetry of initially homogenous space, that reflects the presence of massive spatio-time correlations, i.e. immanent principle of synergy. A. Fibers constituted by straight microtubules. Electron-microscopy (EM), filtered, image at 12500?in vitro at 7 mg/ml of microtubule protein concentration, and 360ℂ. **B**. Straight fibers spontaneously self-organize into a spatial patterns of higher complexity (network) in vitro, at higher microtubule protein (MTP) concentration (10 mg/ml). Predominantly straight microtubules retain partial parallel ordering pattern in some locations, they are multiple crossing each other at other locations, while microtubule free locations are formed as well.

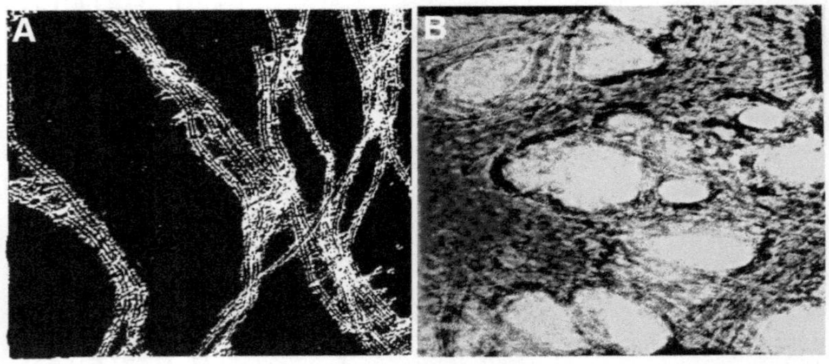

Figure 2 Self-organisation of curved microtubules, at crowded conditions in vitro, is guided by breaking the symmetry of initially homogenous space, that reflects the presence of massive spatio-time correlations, i.e. immanent principle of synergy. Electron-microscopy (EM), filtered image of microtubules fibers at 12500? **A**. Several curved and entangled fibers constituted by curved microtubules, at 7 mg/ml MTP and 1 mM Ca2+ in excess. **B**. Curved fibers spontaneously self-organize into a spatial patterns of higher complexity (network) in vitro, at higher microtubule protein (MTP) concentration (10 mg/ml), and 1 mM Ca2+ in excess. Curved fibers are constituted of partially parallel curved microtubules, they are multiple crossing each other at other locations, while microtubule free locations are formed as well.

Curved microtubule fibers

Although curved as a whole, microtubules may retain partial parallel ordering, thus forming partially curved fibers. An example of curved fibers is seen in Figure 2A, at 7 mg/ml MTP in the presence of excess calcium. At a higher MTP concentration (10 mg/ml) curved fibers spontaneously create a spatial network (Figure 2B). Here, curved microtubules retain a parallel ordering in some locations, while crossing each other elsewhere. Topological relations are different here in comparison to a fiber formed by straight microtubules. Furthermore, some spatial domains are filled with microtubules while others are microtubule free. Curved microtubules, like straight microtubules, are also increasingly breaking the spatial symmetry of the initially homogenous space, making it increasingly more heterogeneous and therefore organizationally (i.e. synergistically) more complex.

More details about self-organization of straight and curved microtubule forms

When the curved microtubules are dominant constituent of the curved fiber such as in the case shown in Figure 2A, then their average curvature corresponds to the curvature of the fiber itself within the standard deviation shown below. Using the circle approximation, we found that the average curvature of the curved fiber microtubules in Figure 2A is 0.0015 ? 0.00028 nm^{-1}.

This situation is different when the curved microtubules are the minority among the dominant population of the straight microtubules constituting the fiber in Figure 1A. We have estimated that due to presence of impurity, constituted of a few curved microtubules, the width of the whole fiber in Figure 1A is enlarged on average by the factor 1.21. In terms of microtubule diameter and the distance between neighboring microtubules, this enlargement is equivalent to the sum of outer diameter of individual microtubules and the distance between neighboring microtubules. However, this enlargement does not appear to affect the fiber straightness. Also, it does not significantly affect the ratio between its length and width. Thus, the fiber still has straight and distinctly extended form.

The other population of rarely found curved and straight microtubules, located randomly across the fiber and eventually near the fiber, we did not consider to be a part of fiber structure. However, we assume them to be a part of non-ordered microtubule population. This population cannot be self-organized because after the fibers formation, the concentration of the rest of microtubules population is decreased below the critical concentration for self-organization.

To illustrate the complexity and the strength of segregation and alignment of straight and curved microtubule forms we use a very simple example of hybrid microtubules (Figure 3). Hybrid microtubule is composed of intrinsically curved and straight parts. Depending on the relative size of its parts the following may happen, straight parts for example are driven into simple alignment, while some portion of the curved parts are separated from straight parts and undergo their own alignment, while the remaining curved parts, probably due to sterical reasons, may be entangled (Figure 3). This case of hybrid microtubules illustrates how strong are spontaneous tendencies for segregation of morphologically different microtubule parts (straight from curved parts) while

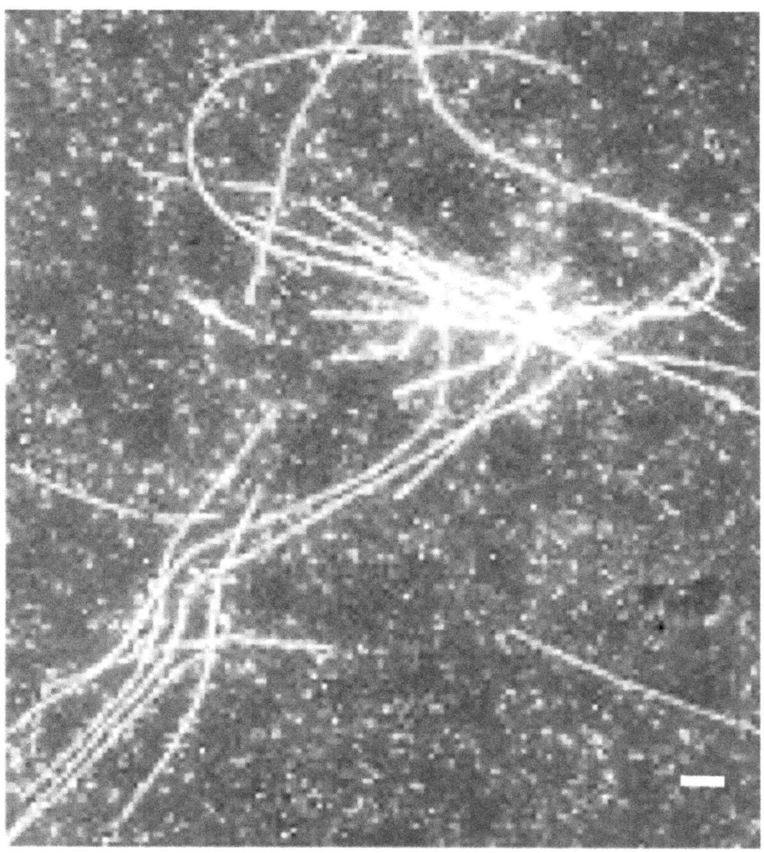

Figure 3 The segregation (demix) and alignment of straight and curved parts of hybrid microtubules, on the basis of morphological criteria. Hybrid microtubule is composed of morphologically different regions: straight and curved. Morphologically different regions tend to be spontaneously segregated along the same microtubule, while morphologically corresponding regions tend to be aligned between the neighboring hybrid microtubules. Bar is 0.1 μm.

there is also strong isomorphic alignment (curved with curved and straight with straight parts).

Since the curved microtubules were minority in our experiments their effects on the direction of self-organization process could be negligible. However, we estimated the average curvature by applying circle approximation. Therefore we have calculated the curvature as inverse of relevant circle radius. The obtained average curvature of the curved microtubules, which are a minority in the straight microtubules dominated fiber (Figure 1A), is 0.001 ? 0.00039 nm^{-1}, while the obtained value of curvature for the curved microtubules outside the fiber is 0.006 ? 0.0008 nm^{-1}. Therefore, in the straight microtubules dominated fiber, the average curvature of the curved microtubules (minority) is smaller by factor 6 than that of the non-ordered microtubules, and by factor 1.5 than the curvature of the curved microtubules of curved microtubules dominated fiber, as shown in Figure 2A.

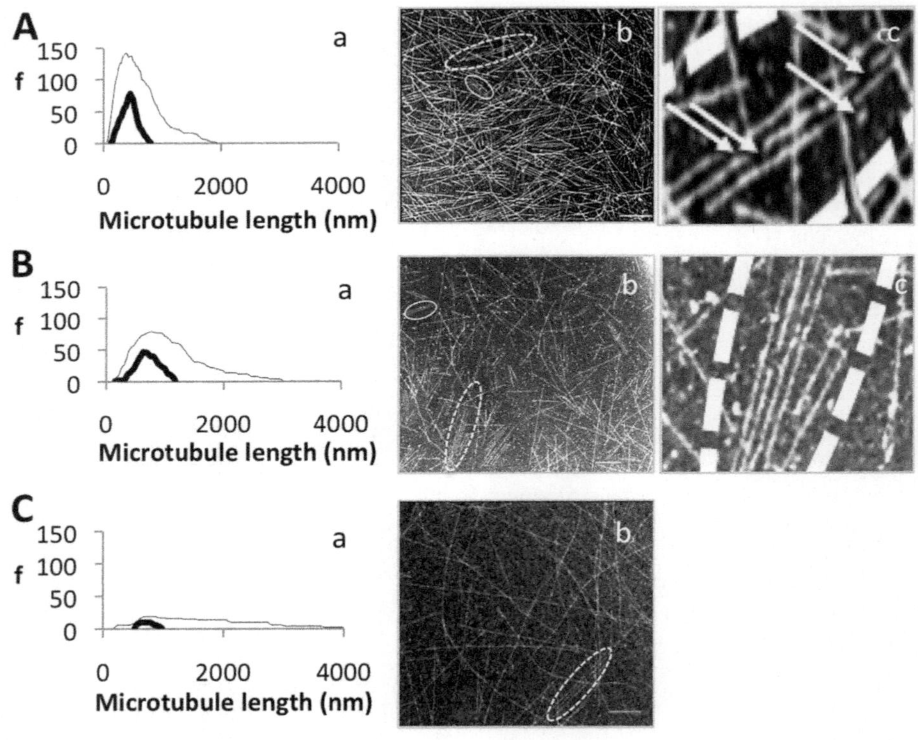

Figure 4 The self-ordered microtubule phase differs from non-ordered phase in terms of microtubule length distributions parameters and interface gaps. The length distribution in ordered (▬), and disordered phase (▬), at 36C, and MTP concentration: 5 mg/ml **A**(a); 2.0 mg/ml **B**(a). Then at 26C, and MTP concentration 5 mg/ml **C**(a). A typical examples of electron microscopy images of ordered and disordered microtubules phase, taken at stationary state of microtubule assembly, at 36C, and MTP concentration: 5 mg/ml **A**(b); 2.0 mg/ml **B**(b). Then at 26C, and MTP concentration 5 mg/ml (**C**(b). White dotted ellipse marks the fiber constituted by staggered microtubules. Figures **A**(c), and B(c) are details taken from Figures **A**(b) and Figure **B**(b) respectively; arrows indicate interface gap. White ellipse marks the fiber constituted by non-staggered microtubules. Bars are 0.5 μm.

The basic topological features of straight microtubule fibers

Our results show that microtubule fiber patterns, *in vitro,* manifest in two configurations: *axially shifted* (staggered) and *non-shifted* microtubules. Fibers consisting of axially shifted microtubules are shown in Figure 1A, Figure 4A(b), B(b) and C(b) (white dotted ellipse). The pattern consisting of non-shifted neighboring microtubules is visible in Figure 4A(b) and B(b) (white ellipse). The fiber consisting of non-shifted neighboring microtubules cannot be extended.

The axial microtubule shift may extend fibers for several average microtubule lengths. Microtubules that constitute the extended fiber are laterally separated by microtubule-associated proteins (MAPs) by approximately 15 nm, while they are axially separated by irregular nanometer size microtubule free spaces i.e. the *interface gaps*. The interface gaps have irregular lengths that on average are commensurate to the average lengths of MAPs.

There are numerous interface gaps randomly located throughout the fiber. As mentioned, the size of the interface gap varies from location to location across the microtubule fiber but is always at a nanometer scale (Figure 4A(b), B(b), C(b)). The interface gap is laterally semi-confined by the walls of neighboring (parallel) microtubules and thus, is partially and unevenly separated from the bulk solution. The variability of the size of the interface gap is determined by the growth/shrinkage stage of microtubules that constitute the axial semi-confinement.

Fiber imposes topological restrictions onto its constitutive microtubules

The length distribution of microtubule self-organization into a fiber is narrower than that of microtubules outside the fiber

In this section, we compare the distribution of microtubule length in the ordered versus non-ordered phase (Figure 3). The exemplars of typical electron microscopy images of ordered and non-ordered microtubules have been obtained during a stationary state of microtubule assembly, at 36?C and MTP concentration of 5.0 mg/ml (Figure 3A(b)) and 2.0 mg/ml (Figure 3B(b)) as well as at 26?C and a MTP concentration of 5.0 mg/ml (Figure 3C(b)).

Different degrees of microtubule self-organization (doublets, triplets etc.) are seen at both temperatures (36?C and 26?C) and at concentrations of MTP above 1.0 mg/ml. Statistical parameters of these data are shown in Figure 5A. The average microtubule length increased as the concentration of MTP decreased. The average length dramatically increased at a lower temperature (26?C) (at the MTP concentration of 5.0 mg/ml).

When we compared length distributions of ordered versus non-ordered microtubules (Figure 4A(a), B(a), C(a) and Figure 5A), we obtained interesting results. The length distribution of ordered microtubules was always significantly narrower than that of non-ordered microtubules. In case of an MTP concentration of 5.0 mg/ml, the length distribution of ordered microtubules was approximately 3-fold narrower than that of non-ordered microtubules. In case of an MTP concentration of 2.0 mg/ml, the length distribution of ordered microtubules is approximately 2.5 times narrower than that of non-ordered microtubules. When the temperature was decreased to 26?C and at a concentration of 5.0 mg/ml MTP, the length distribution of ordered microtubules was approximately 9-fold narrower than that of non-ordered microtubules.

A

Experimental conditions		Statistics						
		NOT ORDERED MICROTUBULES						
Temperatur-e [C°]	MTP Concentrati-on [mg/ml]	Average length [nm]	Median [nm]	Standard Deviation	Variance [nm]	Minimal length [nm]	Maximal length [nm]	Distribution width [nm]
	5	474	405	333	18	50	1925	1875
36	1	967	788	691	26	150	3000	2850
	0.5	1070	1013	667	26	100	3200	3100
25	5	1576	1496	1052	32	125	4100	4075
		ORDERED MICROTUBULES						
	5	423	429	150	12	125	800	675
36	1	725	724	235	15	150	1175	1025
	0.5							
25	5	677	604	151	12	525	975	450

MTP - (microtubule protein (tubulin + MAPs))

B

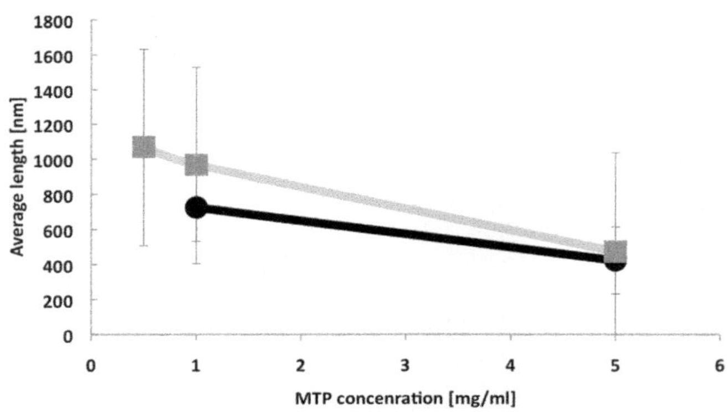

Figure 5 The analysis of the length distribution and average length of ordered and non-ordered microtubules. A. Shows statistical parameters of length distributions. **B**. The average lengths of ordered (▬) and non-ordered (▬) microtubules decrease linearly versus MTP concentration increase.

The average length of self-organized microtubules in a fiber is always shorter than those outside a fiber

The average length of self-organized microtubules in a fiber is 1.12?, 1.33? and 2.33? shorter than the average length of non-organized microtubules at 5.0 mg/ml MTP and 2.0 mg/ml MTP at 36?C and 5.0 mg/ml MTP at 26?C respectively (Figure 5A).

The average length of both microtubule populations decreases linearly versus MTP concentration see graph in Figure 5B. It is interesting to note that the range of standard deviation see graph in Figure 5B, and the related variance, see Figure 5A of ordered microtubules inside the fiber is much more narrower than that of the non-ordered microtubules outside the fiber.

The total number of microtubules that constitute an individual fiber is always restricted

Under our experimental conditions, we found that fibers comprising non-shifted microtubules consisted of approximately ~2 to ~5 microtubules (Figure 4A(b), B(b)). However, fibers comprising axially shifted microtubules consisted of a minimum of 2 and a maximum of 12 microtubules (Figure 1A, Figure 4A(b), B(b), C(b)).

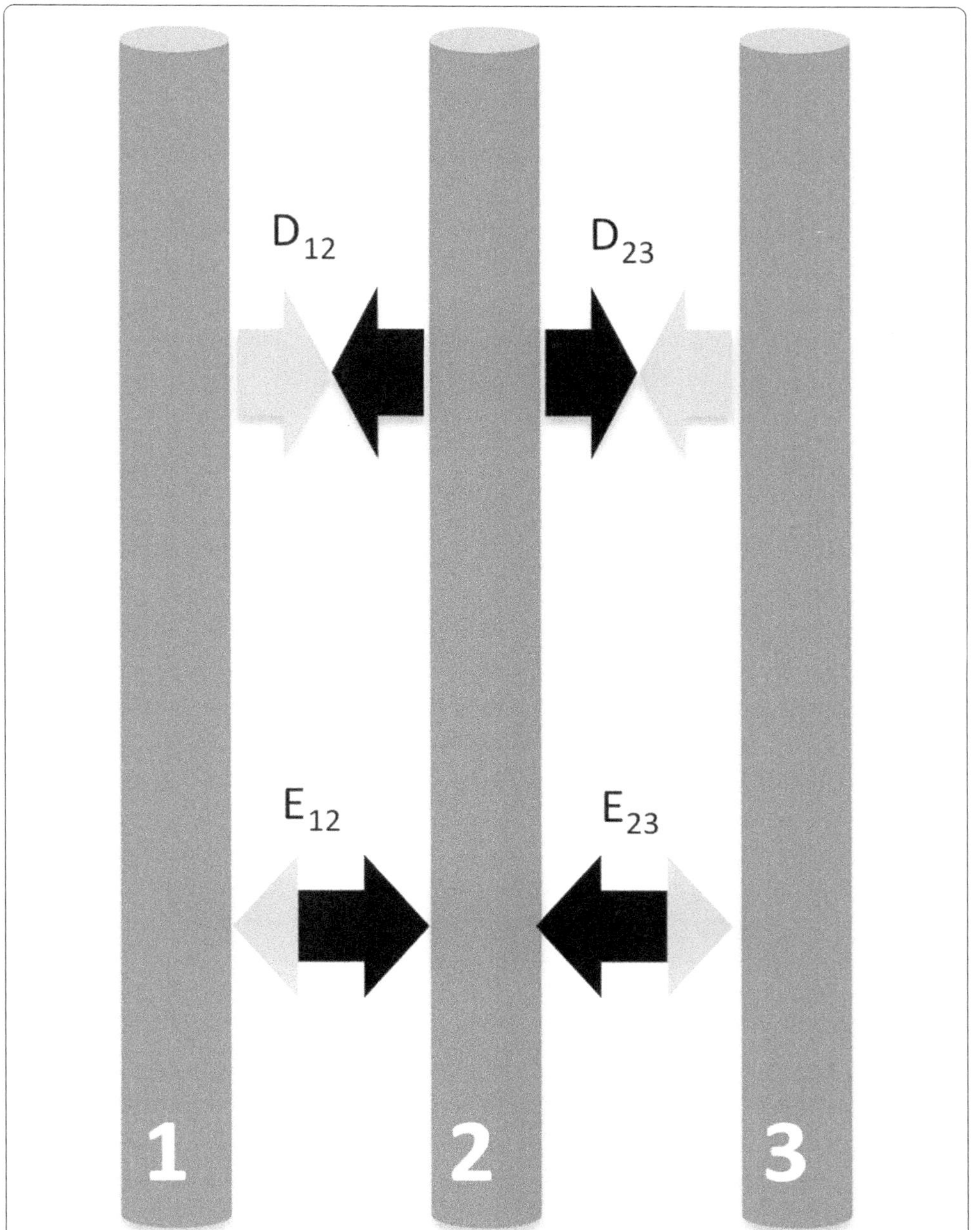

Figure 6 Bundle of three parallel microtubules (1, 2, 3): an illustration of the principle of summation of pairwise forces. Regarding the microtubule labeled as (2) this principle can be written as follows: $\vec{D}_{12}\quad \vec{D}_{23}\quad \vec{E}_{12}\quad \vec{E}_{23}\quad 0$ (3) ➡⬅D ATRACTIVE dipole-dipole pairwise interactions between two parallel microtubules ⬅➡E REPULSIVE electrostatic pairwise interactions between two parallel microtubules.

The pairwise electrostatic forces follow the principle of summation under the crowded conditions in our experiments

Once entropically brought close enough (such as in Figure 1A, Figure 2A, Figure 3, Figure 4A(b), B(b), C(b)) straight microtubules or straight part of curved microtubules may fall within the range of action of a variety of pairwise electrostatic forces. Due to their principle of summation these forces further tune ordering of pattern, increase its stability and endow constituting microtubules with cooperativeness as illustrated through the following text and in Figure 6; eq. 3. The microtubule related electrostatic forces have two sources from which they originate: (a) highly charged αβ-tubulin which in addition possesses permanent dipole moment [39,40], and (b) salty (aqueous based) medium which is necessary for microtubule growth (in vitro as well as in vivo). Salt in aqueous medium is dissolved into its constituting ions and this concentration is represented in terms of ionic strength. As far as ionic strength is concerned microtubules undergo spatial self-ordering in the range of physiological ionic strength (0.1 - 0.2 M) and somewhat lower [35,41]. Finally, in order to have as much as possible satisfactorily picture about interactions which guide the microtubules spatial pattern formation in our microtubule system, it is important to bear in mind that: (a) ionic strength was ~0.1 M, and (b) αβ-tubulin concentration was above the threshold considered as critical for microtubule crowded conditions. Under these solution conditions, variety of interactions may occur in the system, but microtubules self-organization in particular may be affected mainly due to: microtubule surface charge (repulsive electrostatic interactions), dipole-dipole attractive interactions between two neighboring microtubules, screening microtubules negative charges by salt counterions, charge-fluctuation forces between microtubules, and the principle of summation of pairwise electrostatic interactions. However, under conditions of our experiments (salt concentration 0.1 M), charge-fluctuations interactions are at least about an order of magnitude smaller than electrostatic interactions [30]. Therefore, we continue our analysis having in mind that electrostatic repulsive interactions and dipole-dipole attractive interactions are critical in ruling the formation and maintenance of microtubule pattern under our experimental conditions.

Let us now consider interactions which straight microtubules experience by being organized in parallel fashion of spatio-temporal pattern, and immersed in salty medium. Due to negatively charged surface, the neighbouring microtubules will experience strong electrostatic repulsive force. This force is weakened significantly due to the screening with the salt counterions. It is calculated that if two straight parallel microtubules are at the axis-to-axis distance of ~40 nm, then the repulsive electrostatic force between them is about ~9.0 pN [42]. This calculation was made under the assumptions that: (a) the average microtubule length is ~5 μm, (b) ionic strength of salty solution is ~0.1 M (therefore the effect of screening was included). Our experimental data show that at the same ionic strength (~0.1 M) bundled neighboring microtubules of average length of 5 μm were kept approximately at the distance of ~40 nm. This obvious agreement between the theoretical predictions and our experimental data indicates per se that repulsive electrostatic force between charged microtubules is critical for bundle formation and maintenance. If so, then the immediate question is what is the force that may counterbalance this repulsive electrostatic force (force between the two charged microtubule)? It is quite natural to speculate that attractive dipole-dipole force is potential candidate to counterbalance this repulsive force. Under the similar solution conditions, mentioned above, dipole-dipole attractive interactions between the two neighboring microtubules were found to be about ~330 pN, which per se is far stronger than repulsive electrostatic force [42]. Why then two neighboring microtubules, in the bundle, do not collapse onto each other? If this happens it would lead to a quite static structure. However, from the biological point of view, bundle has to be and it is indeed highly dynamic spatio-temporal pattern [27]. Fortunately, for microtubule in the bundle the situation is more complex than for two parallel microtubules outside of bundle. This is due to the principle of summation of pairwise forces, which is taking place under crowded conditions of our experiments. Indeed, because of a particular parallel geometry of bundled straight microtubules, the same kind of force may be self-attenuated. Let us illustrate how this principle may act in the case of a simple bundle constituted by three straight and parallel microtubules (Figure 6). In this case the microtubule, which is placed between its two exterior neighbors is continually under the influence of pairwise forces generated between it and its neighbors. In other words, the interior microtubule is in synchronous fashion pulled outward to the left and the right sides by dipole-dipole attractive forces. The net effect of summation of these pairwise forces taken together is their self-attenuation. On the other hand, the same microtubule is pushed inward to the left and the right directions by repulsive electrostatic forces generated by its exterior neighbours. Again, the principle of summation of pairwise repulsive forces will lead to their net self-attenuation. Obviously, the principle of summation of pairwise forces opens the ample possibilities of different combinations of forces that may support dynamic equilibrium of the microtubule self-organized pattern.

Let us see now if the principle of pairwise forces summation is working under conditions of our experiments. For that purpose we employed the criterion already developed in the work of [30]. The criterion is mathematically expressed in the following inequality

$$n_m >> \frac{1}{b\pi\left(R^2 \quad a^2\right)}. \qquad 2$$

In the above inequality $\boldsymbol{n_m}$ is the density of mobile ions in solution, $\boldsymbol{R}$ is the neighboring microtubules central axis-to-central axis distance, $\boldsymbol{a}$ is microtubule half diameter, and b is the length per unit charge along microtubule surface. For the salt concentration of 0.1 M we obtain the following density of mobile ions

$$\boldsymbol{n_m} \quad 6.02 \times 10^{25} \mathrm{m}^{-3}.$$

To calculate the right side of inequality (2) we use the microtubule distance R = 40 nm, microtubule half-diameter $\boldsymbol{a}$ = 12.5 nm, and b = 8?. So for the right side of inequality (2) we obtain

$$\frac{1}{b\pi\left(R^2 \quad a^2\right)} \quad 0.00246 \times 10^{25} m^{-3}$$

The left side of inequality (2) is far greater than its right side. Therefore criterion (2) is satisfied, and consequently the principle of pairwise forces summation is operating under our experimental conditions. Consequently, the principle of pairwise electrostatic forces summation leads to fine tuning (cooperation) of constituting microtubule motion within the fiber and its additional stabilization.

The local GTP-tubulin depletion may occur in the immediate vicinity of the tips of microtubules

The set of topological restrictions imposed on microtubules self-organized into a fiber is an important observation as it may indicate that an inherent mechanism regulates self-formation and maintenance of the fiber. It was shown that the *depletion* of GTP-tubulin in a region located in the immediate vicinity of the tips of microtubules might control the number of microtubules organized into a microtubule aster [43]. We assume that, similar to the case of a microtubule aster, local depletion of GTP-tubulin may be a part of local solution conditions that lead to a restricted number of microtubules that constitute a fiber.

Odde s calculations indicated that there are three factors that are critical in order to obtain total local depletion of GTP-tubulin near the tip of a single free (i.e. not bound in any spatial pattern) growing microtubule [44]. These factors are: the bulk GTP-tubulin concentration, the diffusivity and the rate of microtubule assembly. The calculations are based on a steady-state mean-field model for the concentration of GTP-tubulin and the assumption that there is only one GTP-tubulin binding site per microtubule that effectively contributes to microtubule assembly.

According to these calculations, total depletion of GTP-tubulin near the tip of a single microtubule may occur when the bulk GTP-tubulin concentration is 1 mg/ml (the threshold at which crowded conditions start to appear), diffusivity (intermediate) is 6.4 ? 10^{-12} m^2/s and the assembly rate of an individual microtubule is 65 μm/min [44]. Experimentally measured, the real maximal microtubule assembly rate is 20 μm/min [45]. Odde therefore concluded that, under the given conditions, the GTP-tubulin depletion could not be created in front of a tip of a single growing individual microtubule.

The factors critical for total GTP-tubulin depletion in our experiments are different to those employed by Odde and favor the occurrence of depletion. The bulk GTP-tubulin concentration used in our experiments is approximately 10 times higher than those used by Odde. Lower GTP-tubulin diffusivity may favor the creation of GTP-tubulin depleted regions. Odde used an intermediate value for diffusivity. In our system, diffusivity is expected to be much lower due to a strong influence of boundary conditions imposed by the presence of a multitude of crowded, single microtubules as well as microtubule spatio-temporal structures (such as fibers). Therefore, the significantly higher degree of crowded conditions in our system may lead to the significantly lower diffusivity than an intermediate one employed by Odde [44].

Conversely, under highly crowded conditions (GTP-tubulin concentration approximately 7 mg/ml), the *local* reaction diffusion conditions in front of a set of microtubule tips organized into a fiber may be quite different as compared to those of the single free microtubule, which in turn may favor the creation of small GTP-depleted regions in the immediate vicinity of the tips of microtubule fibers [46]. Let us assume that each individual microtubule (within a fiber) grows *independently*, so that we may assume that they *independently* take up GTP-tubulin from the common pool located in the immediate vicinity of their tips. According to Odde s calculations, a single free microtubule, in order to create a total local depletion of the GTP-tubulin, should grow at a rate of 65 μm/min. Therefore, to reach the same degree of local depletion in front of the tips of a fiber comprising two microtubules, each microtubule should grow at an assembly rate of 32.5 μm/min. Consequently, in a fiber consisting three microtubules, each microtubule should grow at a rate of 21.67 μm/min in order to produce a total local depletion (Figure 7).

If we accept Odde s assumption that there is only one GTP-tubulin binding site per microtubule, then (see the graph in Figure 7), total local depletion of the GTP-tubulin will certainly occur in close proximity of a fiber terminal, which is constituted by at least four microtubules (where the assembly rate of each microtubule is 16.25 μm/min). However, in our experimental conditions, we found an abundance of fibers comprising only two and three microtubules (Figure 4), which may indicate that there is more than one GTP-tubulin binding site per individual microtubule that constitutes a fiber. Thus, under highly crowded

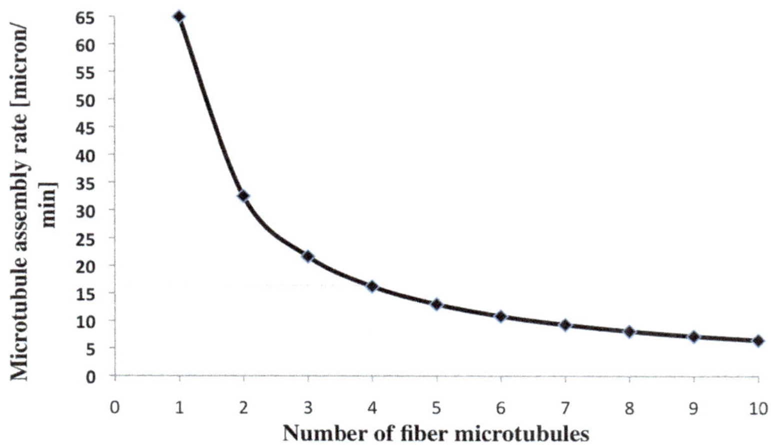

Figure 7 Theoretical estimation of effective assembly rate of each individual microtubule within a fiber at which the total depletion of GTP-tubulin in the immediate vicinity of fiber microtubule tips may be induced. At crowded conditions (see the text), the total local depletion of GTP-tubulin at the tip of individual, out of fiber, microtubule will occur if microtubule grow at 65 microns/min (Odde [44]). However, maximal microtubule growing rate is 20 microns/min (Shelden and Wadsworth [45]). Therefore total depletion of GTP-tubulin cannot be induced in front of individual, out of fiber, microtubule. However, each individual fiber microtubule *independently* contributes to the depletion. Since their effect on depletion is cumulative, they may induce depletion each growing at rate less than 65 microns/min. In the case of fiber constituted by the four microtubules, depletion can be induced. Indeed, if each of four microtubule grow at ~16.5 microns/min, then in terms of depletion efficiency; $16.5 \cdot 4 = 66$ microns/min, total local depletion may be achieved. At number of fiber microtubules higher than four, depletion will be certainly induced.

conditions, a total local depletion may occur even if the fiber is made up by less than four microtubules (Figure 7).

The extension of the GTP-tubulin depletion zone versus the solution of blind mason problem

As we have mentioned above, apart from the tubulin concentration, its degree of freedom of motion manifested as translational and rotational diffusivity plays important role in the extension of GTP-tubulin depletion zone. In order to be docked in the appropriate location of GTP-cap tip the incoming tubulin emerges from the bulk solution. If emerging tubulin is endowed with full degree of freedom of motion, this would be the most favorable (energetically and entropically) situation. In other words, its subsequent manoeuvres and placement in the location with any orientation would be equally probable. Therefore, the blind mason problem would be directly solvable. However, crowded conditions, depending of their degree of crowdedness, may impair full freedom of tubulin motion by imposing anisotropic conditions.

Consequently this may impair the solution of blind mason problem. Nominally, we are working at and above tubulin concentration of 1 mg/ml. This concentration is considered as threshold above which crowded conditions arise [43,47-49]. A propos of blind mason, it is reasonable to ask the following question: does tubulin has enough space to perform necessary manoeuvres by means of free rotation and translation at the critical concentration of microtubule self-organization and higher? As long as incoming tubulin has enough space to perform free rotation and translation, the problem of blind mason is solved. Concerning tubulin, as far as we know, there are no appropriate data that can indicate the changes (reduction) of degree of freedom of motion versus crowded conditions. However, recently it was studied the emergency of anisotropy of motion of lysosome versus its concentration within the range which includes crowded conditions (Table three in [28]). The volume of lysozyme molecule, approximated by cylinder is ~141.3 nm^3 [50]. The volume of αβ-tubulin molecule, approximated by cylinder is ~464.5 nm^3 [51]. To extrapolate the observed data of emergency of anisotropy of lysozyme motion and apply it to the case of tubulin, we calculate the accommodating space factor for both lysozyme and αβ-tubulin. The term accommodating space factor refers to the number of times the accommodating space is greater than the volume of the protein. Accommodating space of protein is the space, which accommodates protein free rotation and translation. The results (Table 1) show that anisotropy of lysozyme motion emerged at highly crowded condition, i.e. at lysozyme concentration of 150 mg/ml. At that concentration the accommodation space factor for lysozyme is 1.12. Since at the same concentration, the accommodating space factor for tubulin (2.61) is very close to the lysozymes accommodating space factor, we speculate that anisotropy of tubulin motion should emerge at approximately close concentration of lysozymes one. Therefore, in our working concentration range (1-10 mg/ml) crowded conditions are too weak to impair overcoming blind mason problem. From the dynamical point of view, in order to perform subsequent positional-orientational adjustments by means of translational and rotational diffusion tubulin is endowed with additional local space - this will be done at the cost of depletion zone extension.

Mutual interdependence of microtubule pattern and associated local solution conditions

Through all our data we see that the spatio-temporal pattern of microtubules are a sensitive function of solution conditions. For example in the presence of excess calcium, curved microtubules and therefore curved fibers are formed (Figure 2A,B; Figure 3). Furthermore, the concentration of GTP-tubulin affects parameters of microtubule self-organization. The occurrence of fibers is more probable at more crowded conditions. The length distribution of patterned microtubules changes differently in higher GTP-tubulin concentrations compared to those of non-patterned microtubules. In the presence of high GTP-tubulin concentrations, the width of the length distribution of patterned microtubules is shorter compared to those of non-patterned microtubules (Figure 4A,B,C).

Our calculations have shown that total local depletion of GTP-tubulin may occur in the immediate vicinity of microtubule tips due to pattern growth dynamics (Figure 7).

Table 1 Emergency of anisotropy of αβ-tubulin molecule motion in the bulk solution

	LYSOZYME		TUBULIN	
Concentration [mg/ml]	**Fas**	**Anisotropy**	**Fas**	***Anisotropy**
1	169.85	NO	394	NO
10	16.99	NO	39.4	NO
150	1.12	YES	2.61	YES

The emergency of anisotropy of lysozyme molecule motion may qualitatively indicate the emergency of anisotropy of αβ-tubulin motion under corresponding solution conditions. F_{as} is the accommodating space factor which says how many times the accommodating space is greater than the volume of one protein molecule. Accommodating space is the space which accommodates free rotation and translation of single protein molecule. The volume of lysozyme molecule, approximated by cylinder is ~141.3 nm^3 (Matthews and Remington [50]). The volume of αβ-tubulin molecule, approximated by cylinder is ~464.5 nm^3 (Nogales, et al. [51]). *Anisotropy is the anisotropy of αβ-tubulin motion obtained by extrapolation from lysosyme data under corresponding solution conditions (Table three in Dlugosz and Antosiewicz [24]).

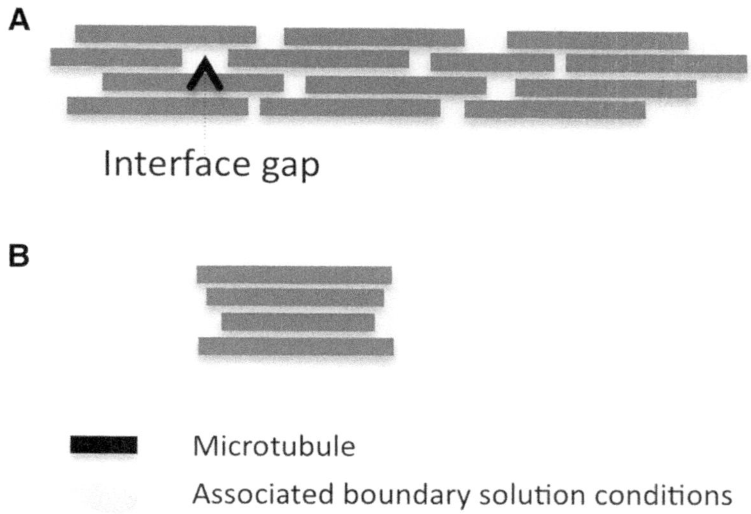

Figure 8 Synergistic-topological matrix model*. This model summarizes our findings that indicate that there is strong mutual interdependence between the pattern formation and the local boundary solution conditions, which is driven by intrinsic synergistic-topological mechanism (see Figure 9). For example calcium, GTP-tubulin concentration, the vicinity of other microtubules etc. may affect the morphology of pattern formation and vice versa the pattern affects the local solution condition (e.g. by fiber microtubule dynamic instability and generating depletion). **A**. Synergistic-topological matrix* constituted by the pattern (fiber) of axially shifted (staggered) microtubules, and associated specific boundary solution conditions. **B**. Synergistic-topological matrix* constituted by the pattern (fiber) of non-shifted (non-staggered) microtubules, and associated specific boundary solution conditions. *Mutual interdependence of microtubule pattern and associated local specific solution conditions may be physico-mathematically modeled as synergistic-topological matrix that is formed through intrinsic synergistic mechanism (see Figure 9). Associated local solution conditions include: local cross diffusion, local GTP-tubulin fluctuations (including GTP-tubulin depletion), and different molecular species (such as proteins, and ions). The matrix concept specifically related to microtubule mitotic spindle system can be found elsewhere (Lince-Faria, et al. [24] and Johansen, et al. [26]).

Therefore, collectively, these data indicate that there is strong mutual interdependence between the pattern morphology dynamics and local solution conditions that can be physico-mathematically modeled by a matrix that is driven by intrinsic synergistic-topological mechanisms (Figure 8A,B; Figure 9A,B,C).

Discussion

Existing data indicate that synergism is an essential element involved in biological complexity of cytoskeletal (including microtubules) function. However, the mechanism(s) by which synergism guides the self-organization of the microtubule system is yet to be understood. For this reason it would be useful to study synergistic mechanism(s) related to microtubule multitudes under minimalistic but crowded (in terms of GTP-tubulin concentration) biological conditions, *in vitro*.

Entropically driven spontaneous demix and self-organization of curved and straight microtubules

We have observed microtubule self-organization under crowded conditions where straight microtubules were dominant. However, to get rationale for this particular case it is helpful to bear in mind what happen if the concentration numbers of straight and

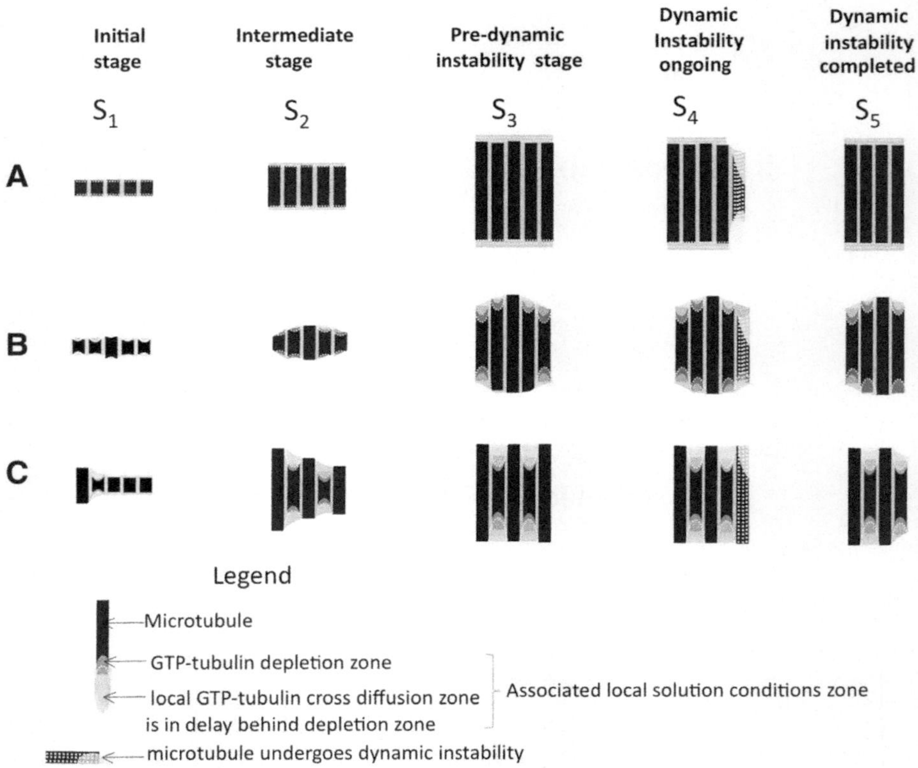

Figure 9 The model of three classes of synergistic-topological matrix shown as shaped by the synergistic-topological mechanism. Here we use the example of quintet fiber and illustrate, three main classes of fiber end-configurations obtained in cases: **(A)** The effects of local GTP-tubulin depletion and local cross diffusion are equilibrated. This case is relevant for biological system such as polarized epithelial cell (Schroer et al. [57]) (see Figure 12A). **(B)** Depletion effects are overwhelmed by stabilization effects of cross diffusion. This case is relevant for biological systems such as rod shape cell of the fission yeast *Schizosaccharomyces pombe* (Siegrist and Doe [72]) (see Figure 12B). **(C)** Local cross diffusion effects are overwhelmed by depletion effects. This case is relevant for biological system of *fibroblast*, i.e. in formation of leading edge cortical polarity during fibroblast migration (Siegrist and Doe [72]) (see Figure 12C). The zone of associated local solution conditions as part of the matrix are determined by the extension of depletion and cross diffusion. As the number of fiber microtubules increases, its topology becomes more complex by exerting stronger effects on local events. Furthermore, in addition to the initial stage (S1) (which is ruled prevalently by intrinsic microtubule stability), synergy between GTP-tubulin local depletion, local cross diffusion, and dynamic instability, begin to play prominent role in establishing the matrix of the fiber end-configuration. Depletion destabilizes microtubule end and inhibits microtubule growth. Cross diffusion stabilizes microtubule end and stimulates microtubule growth. Dynamic instability eliminates microtubule, partially or totally due to depolymerisation. Dynamic instability may occur at any stage, but for clarity we illustrate it as commencing at stage S4. It is assumed that external microtubules, as the most unstable are the most likely prone to dynamic instability. It may start at any external side of fiber or at both.

curved microtubules are approximately equal. In addition, the system contains minority of microtubules whose curvature is at the critical threshold, which enables them to undergo alignment (we call these microtubules slightly curved).

If so, then at sufficiently crowded conditions, the system is unstable and undergoes spontaneous phase segregation (demix) of coexisting phases, while slightly curved microtubules join the population of straight microtubules [52]. Also when the straight microtubules become much longer than the distance between neighboring microtubules, then the spontaneous orientational ordering is initiated [53]. At the end of the process (composed of demix and orientational ordering) the total gain of entropy for straight,

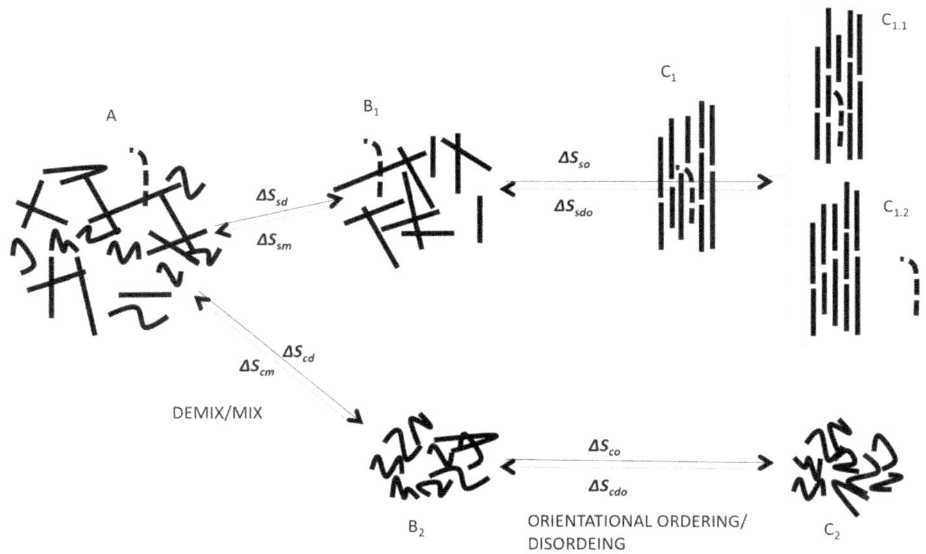

Figure 10 Spontaneous (entropically driven) segregation (demix) of straight and curved microtubules and final orientational ordering under sufficiently crowded conditions. Here we illustrate the self-organization of microtubule system which is initially homogeneous, i.e., it is constituted by approximately equal amount of straight and curved microtubules. In addition, the system contains minority of curved microtubules (- - - -) whose curvature is at the critical threshold for alignment (we call them slightly curved microtubules). Under crowded conditions, the distance between neighboring microtubules becomes rapidly shorter than the average microtubule length; the system is unstable and undergoes spontaneous phase segregation (demix) of coexisting phases (curved and straight), subsequently the separate final patterns are formed by spontaneous orientational ordering. A. Homogenous mix of straight and curved microtubules. B1. Demix of straight microtubule phase with minority of slightly curved microtubules. (***ΔSsd*, *ΔSsm*** - entropy of demix and mix, respectively). B2. Demix of curved microtubules phase. (***ΔScd*, *ΔScm*** entropy of demix, and mix, respectively). C1 Orientational ordering of straight microtubules with slightly curved microtubules. (***ΔSso*, *ΔSsdo*** - entropy of orientation and disorientation, respectively). C1.1 To be finally accommodated the curvature of curved microtubule should be made smaller than the critical threshold. This could be done by forced microtubule depolymerisation/repolymerisation that may be induced by mechanism of summation of pairwise electrostatic forces of neighboring microtubules. C1.2 If curvature of the curved microtubule cannot be reduced, curved microtubule is rejected. C2. Curved microtubules orientational partial ordering. (***ΔSco*, *Δscdo*** - entropy of partial orientation and disorientation, respectively). Spontaneity of demix and orientational ordering of either population is entropically driven, i.e.:

$$\begin{array}{l} \Delta Ssd \quad \Delta Sso > \Delta Ssm \quad \Delta Ssdo \\ \Delta Scd\, \Delta Sco > \Delta Scm \quad \Delta Scdo \end{array} \quad (1)$$

curved, and slightly curved microtubule populations will be higher than the total loss of entropy during mix and orientational disordering (Figure 10A, B1, B2, C1, C2; inequality (1)) [52,53]. Therefore, the total process constituted of phase demix and final pattern formation due to orientational ordering, is entropically driven.

In practice slightly curved and straight microtubules can be often mixed in the self-organized pattern, such as the fiber in Figure 1A. Here, the main question is about the mechanism how the slightly curved microtubule got into the fiber structured of dominantly straight microtubules? We have no data regarding the direct history of how the slightly curved microtubule got into the fiber (Figure 1A). However, the data regarding the average curvature indicate that the average curvature of non-ordered microtubules is greater than the curvature of slightly curved microtubules built in ordered structure of the fiber. We can speculate that there are two possible scenarios. The first scenario is that the entropically driven processes of demix and alignment may temporarily bring slightly curved microtubule into the fiber structure (Figure 10C1). Once microtubules have been brought close enough to each other by entropical drive making (loose) fiber (Figure 10C1), the pairwise electrostatic forces in particular the action principle of pairwise forces summation further tune and stabilize the fiber [30,39,40,42]. Consequently, the curvature of a slightly curved microtubules may become even smaller due to the complex neighboring microtubule electrostatic interactions within the fiber, in particular due to the actualization of the principle of summation of pairwise electrostatic forces which may cause the cycle of microtubule disassembly/reassembly and/or their direct bending. Therefore, the slightly curved microtubule may be straightened and stabilized i.e. accommodated as part of fiber structure (Figure 10C1.1).

The second possible scenario is that the fiber structure may simply reject microtubules whose curvature is above the certain threshold as energetically and entropically costly in order to be accommodated within the fiber structure (Figure 10C1.2).

Concerning segregation and alignment phenomena in the system of hybrid microtubules (Figure 3) one can explain it on the basis of Onsager s principle: curved microtubules, with the certain degree of straightness and rigidity, undergo partial parallel ordering and in part just entanglement [53] (Figure 3).

Spontaneous symmetry breaking during microtubule fiber formation and its network self-formation provides evidence for the presence of intrinsic synergy

Although simple, our chosen system enables spontaneous formation of different forms of microtubule fibers comprised of straight and curved microtubules, which further undergo spontaneous segregation and symmetry breaking, thereby forming more complex spatial networks (Figure 1A and Figure 2A; Figure 1B, and Figure 2B). Symmetry breaking is obviously accompanied by an abundance of spatially and temporally correlated microtubule multitudes. Therefore, spontaneous symmetry breaking during microtubule self-organization reflects the presence of intrinsic synergy mechanism(s) [37,38].

The fiber imposes topological restrictions on constituting microtubules

An exemplar of different classes of fibers formed under different solution conditions are illustrated in Figure 4A(b), B(b), and C(b) and marked by a dotted white ellipse. We have found that under a given solution condition the self-organization of a microtubule fiber imposes a set of topological restrictions onto microtubules that constitute the fiber. We have found variable, nanometer-size microtubule free spaces located between fiber microtubules that are approaching/receding (i.e. tip-to-tip), along the same fiber track (Figure 4A(c)), B(c), see also an illustration of the interface gap in Figure 8A). As already mentioned, we termed these nano-spaces *interface gaps*. The nanometer-size microtubule free spaces, between tip-to-tip fiber microtubules are also observed in a microtubule MAPs free system, where fibers are ordered by strong external magnetic fields [54]. In our system, MAPs are present and fibers are spontaneously self-organized. It was also shown that microtubules undergo spontaneous self-organization in the form of liquid nematic crystals under solution conditions commensurable with ours [55]. Importantly, the pattern of nematic liquid crystals implicates the existence of microtubule free spaces between tip-to-tip of axially approaching microtubules. The microtubule free spaces are also present in the fibers of axonal microtubules that have an axially shifted configuration [56,57]. Therefore, the presence of the microtubule *interface gap*, under a variety of experimental conditions, may perhaps suggest that it has a regulatory role in events that may be transmitted along individual fiber microtubules such as mechanical force, energy and information processing [26,39,58-61]. This may be critical in cellular wound healing [62,63].

In relation to topological restrictions in fibers, the following are pertinent: (a) the total number of microtubules that constitute a fiber is always restricted at given solution conditions; (b) the average length of microtubules within a fiber is consistently shorter than the average length of free microtubules; (c) the length distribution of fiber microtubules is significantly narrower in comparison to free microtubules (Figure 4A(a), B(a), C(a)). The associated numerical data are summarized in Figure 4A,B. In addition, we believe that the differences in standard deviations between ordered and non-ordered microtubules as shown in Figure 5B graph, and the differences in variance intervals (Figure 5A) may come, not from the quality of measurement, but from a different nature (in terms of freedom of motion) of these two microtubules populations. Non-ordered microtubules grow freely, stochastically and independently of each other, while ordered microtubules are subjected to the local structural packing constriction conditions, which may prevent significant differences in their lengths. In our view, these differences in standard deviations and variances, as such may per se indicate that particular mechanism, that is, synergy mechanism may be behind the microtubules self-ordering event.

By being inherent to the fiber microtubule topology, these restrictions drive the fiber self-organization. To explain how these restrictions function, we propose the existence of an intrinsic synergistic-topological mechanism.

The synergistic-topological mechanism

It is very well known that microtubules capacity to undergo self-ordering is critical for their biological function. Due to its biological function, a stable self-ordered microtubule pattern has also to retain a certain degree of freedom of motion (flexibility). This extraordinary dynamic feature of microtubule self-organized patterns is realized by means of the principle of summation of pairwise electrostatic forces [30]. This principle plays fiber integrative role, and in the same time it endows constituting microtubules with the certain degree of cooperativeness. Once entropically brought close enough (such as in Figure 1A, Figure 2A, Figure 3; Figure 4A(b), B(b), C(b)) straight microtubules or straight part of the curved microtubules may fall into the range of action of a complex variety of pairwise electrostatic forces. While these forces stabilize the pattern, they also, due to the principle of their summation, endow constituting microtubules with certain degree of cooperative motion (Figure 6). This is of essential biological importance. We have shown that under our experimental crowded conditions the principle of summation of pairwise forces is working (inequality (2) is satisfied). Although it is not limited to the triplet of microtubules, the principle of forces summation can be fully understood in the case of microtubules triplet (Figure 6 eq. (3)). Indeed, left neighbor (labeled with number 1) of intermediate microtubule (labeled with 2) knows what exactly does its right neighbor (labeled as 3), and responds with appropriate reaction according to eq. (3). This is the essence of elementary cooperativeness among fiber constituting microtubules. Rooted in the principle of summation of pairwise electrostatic forces, the cooperativeness of fiber constituting microtubules is an important factor in the synergistic-topological mechanism about which we will talk later.

Substantial differences between the topology of self-organized microtubules (where the principle of summation of pairwise forces is working) and free single microtubules, under the *same* bulk solution conditions, suggest that some *local* solution factor(s) and mechanism(s) may be responsible for the above-mentioned topological restrictions in fiber microtubules. Based on the analysis of our data, we propose that the key solution factor responsible for the set of the above-mentioned topological restrictions in fiber microtubules is the *local depletion* of GTP-tubulin. While depletion occurs locally in the immediate vicinity of the tips of fiber microtubules, and therefore acts on their dynamic directly, the principle of summation of pairwise electrostatic forces operates between neighboring microtubules all along their lengths - across the fiber (Figure 6).

It was shown that the *local depletion* of GTP-tubulin cannot occur in the immediate vicinity of the tip of only one (single) free growing microtubule [44]. Our calculations revealed that at given solution conditions the *local depletion* of the GTP-tubulin may occur in the immediate vicinity of the tips of multitude of self-organized fiber microtubules at a certain critical number of fiber microtubules (Figure 6).

The question that arises from this is what would be the mechanism by which local GTP-tubulin depletion may operate to control the size of microtubule fiber? During fiber growth each fiber microtubule grows *independently*. Therefore, the tips of all fibers constituting microtubules are *independently* taking up GTP-tubulin from the *same* small pool that is located in the immediate vicinity of fiber terminals, that is, within a nanometer distance from the tips of growing/shrinking microtubules. Consequently, all of the microtubule tips from the same fiber may strongly compete for incoming GTP-tubulin. The competition intensifies with increased number of growing microtubules within the fiber. Each individual fiber microtubule contributes *independently* to the eventual local GTP-tubulin depletion, as each microtubule, of the same fiber, grows *independently*. Therefore, we may assume that the total contribution to depletion may be calculated as a linear sum of all individual fiber microtubule contributions, sometimes referred to as collective reactivity [64]. If the number of growing microtubules, which constitute a fiber, is high enough, then they may take up more GTP-tubulin from the pool than can be supplied by local diffusion. Therefore, local *depletion* of the GTP-tubulin may effectively and instantly occur in the immediate vicinity of a growing fiber terminal (Figure 6).

The local GTP-tubulin *depletion* may randomly and non-uniformly lead to the exhaustion of the addition of GTP-tubulin to the tips of different individual microtubules and thus, may randomly slow down or even halt the growth of some individual fiber microtubules. The instant halting of the growth of an individual fiber microtubule, in conjunction with the eventual burst of GTP-hydrolysis at the terminal GTP-tubulin subunits of that particular microtubule, creates conditions that instantaneously switch on *dynamic instability* of the same fiber microtubule [65]. Once the *dynamic instability* is switched on, the microtubule may shrink, or it may be catastrophically and completely destroyed. In this way the number of growing fiber microtubules may be reduced due to *dynamic instability* in conjunction with the dynamic of *depletion* impact.

Once the number of growing microtubules is reduced, the *depletion* may start to decline due to a constant supply of fresh GTP-tubulin by local *cross diffusion,* which is efficiently empowered by crowded conditions. In this way, the growth of some of fiber microtubules, whose growth was stopped due to depletion, may resume again while the others may continue to grow at an optimal rate according to the given solution conditions. The spatial variation in local GTP-tubulin depletion can be understood as the nonthermal and nonequilibrium *GTP-tubulin local fluctuations* [66,67]. Under the given conditions, these fluctuations may be maintained as long as they are coupled to *dynamic instability* and *cross diffusion* of GTP-tubulin. This tight cooperation between *dynamic instability, GTP-tubulin local fluctuations and cross diffusion*, is the essence of *synergy* mechanisms, which operate at microtubule tips of the fiber terminals.

Finally, the action of the principle of pairwise forces summation is constantly implicated in all of these events, and can be understood as essential part of a synergy mechanism, that is its modulator.

Apart from the apparent randomness of local GTP-tubulin *depletion*, there are two stereo-specific aspects of a microtubule GTP-cap tip that may *topologically* condition the rate of GTP-tubulin *depletion*, and the extension of GTP-tubulin *depletion* zone. The first of these is the burst of GTP-hydrolysis at the microtubule GTP-cap tip, and the second is related to the necessity of blind mason problem solving.

The burst of GTP-tubulin hydrolysis

The burst of GTP-tubulin hydrolysis is *topologically* conditioned by intrinsic structural parameters of the GTP-cap of each individual fiber microtubule [65]. In turn, the GTP-hydrolysis burst may switch on dynamic instability of a particular individual microtubule and cause its subsequent partial or total depolymerization. Therefore, under the given solution conditions, the elimination pattern of an individual fiber microtubule by dynamic instability is topologically conditioned by intrinsic GTP-cap stereo-selective parameters. This stereo-selectivity may lead to different end-configuration topologies of fiber terminals (Figure 9A,B,C).

GTP-tubulin depletion zone is conditioned by the necessity of blind mason problem solving

It is well documented that tubulins structural arrangement in any microtubule lattice (known so far) strictly follows unique symmetry pattern manifested as the left-handed (chirality) helicity [68-70]. It was proposed that since tubulin, as the main building block of microtubule wall, possesses well defined particular asymmetric shape, it can be docked with precision by appropriate packing constraints of symmetry pattern of microtubule lattice, and this should be the most direct and the easiest way to solve the problem of blind mason during microtubule growth [29]. Therefore in dynamic picture tubulin docking needs *additional* effective local space to perform necessary spatial adjustments due to rotational and translational movement. In turn, this will condition the size of the GTP-tubulin *depletion* zone. We believe that the proper solution to the blind mason problem lies within the frame of nonlinear dynamic. Before its docking begins in the proper location of GTP-cap tip lattice, the incoming tubulin has to emerge from the bulk solution. In order to initiate and complete the docking, it is obviously preferable (energetically and entropically) for tubulin to have on disposition the full degree of freedom of its rotational and translational motion. In this way, all the subsequent positional and orientational adjustments will be almost equally possible. Anisotropy in tubulin motion occurs due to crowdedness, and in turn it (anisotropy) may impair, energetically and entropically, the tubulin docking.

Consequently, in this situation the capacity to overcome the problem of blind mason is reduced. However, by extrapolating data for lysosyme, we have shown in Figure 7 that in vitro experiments, where only tubulin is present in the solution, anisotropy in tubulin motion in the bulk may appear only under the strong crowded conditions (the bulk concentration of tubulin ≥150 mg/ml) [28]. Therefore, in our experiments, the degree of crowdedness is enough low to allow the problem of blind mason to be solved at the cost of extended zone of GTP-tubulin depletion.

The final acts of tubulin docking are governed by the complexity of physical interactions. Firstly, due to the left-handed helical lattice symmetry of microtubule wall the docking tubulin interacts with tubulins, which already constitute the GTP-cap tip by means of spatially configured electrostatic interactions. Secondly, it also interacts with the other (competing) docking tubulins, and here are included electrostatic interactions, and excluded volume. Obviously tubulin, before its final addition to the GTP-cap tip, is undergoing strong nonlinear regime of cooperative interactions. Both, the strong interaction cooperativeness and nonlinearity phenomena are necessary ingredients in symmetry breaking during new structure formation [71]. In fact, single microtubule, as well as complex microtubules spatial patterns, manifests distinct symmetry features. On the other hand, from the thermodynamic point of view, since microtubule system is inherently open system, then so-called, spontaneous processes of microtubule growth and self-organization within the system are driven by the production of entropy, more specifically entropy decrease. Taking into account entropy decrease, breaking the symmetry, nonlinearity and cooperativeness of relevant physical interactions, and specific geometry of tubulin molecule, they all together endow the system as the whole with the power to resolve the enigma of blind mason by navigating the tubulin through adjusting its energy, position and proper orientation in order to be added to the GTP-cap tip.

In Figure 9, we have shown three possible classes of synergistic-topological patterns shaped by synergistic-topological mechanism. The first class (Figure 9A) is produced when the effects of local GTP-tubulin depletion and local cross diffusion are equilibrated. Dynamic instability occurs as a synergistic-topological function of those two. The corresponding biological example for this case may be seen in a polarized epithelial cell [57] (Figure 11A). The second class of fiber end-configuration topology is developed when the cross diffusion stabilization effects overwhelm the depletion effects (Figure 9B). This case is relevant for biological systems such as rod shaped cells of the fission yeast *Schizosaccharomyces pombe* [72] (see Figure 11B). The third class is generated when depletion effects overwhelm the local cross diffusion effects. This case is relevant for biological system of *fibroblast,* i.e. in formation of leading edge cortical polarity during fibroblast migration [72] (see Figure 11C).

Overall, intrinsic synergy mechanism in conjunction with topological features of the interface gap may control the gap size. Briefly, it was shown that when microtubules grow in a semi-confined volume, the parameters of microtubule dynamic instability and therefore the overall dynamic of microtubule growth, might be strongly influenced by subtle changes of the local GTP-tubulin concentration [30,73]. The multitude of neighboring microtubules imposes strong and uneven physical constraints on the GTP-tubulin influx into interface gap space. The impaired (tunable; single file) diffusion, i.e. cross diffusion as a particular kind of active diffusion is most likely the only mechanism by which GTP-tubulin may be brought, with a relative retardation,

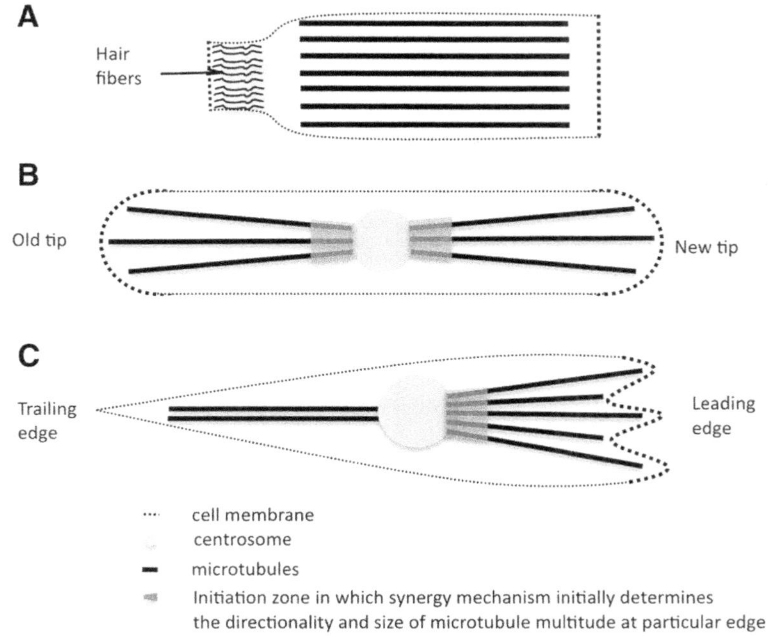

Figure 11 A biological examples of intrinsic control of size, shape and directionality of microtubule spatio-temporal pattern by synergistic-topological mechanism. **A**. Polarized epithelial cell (Schroer et al. [57]); local GTP-tubulin depletion and local cross diffusion are equilibrated. **B**. Microtubules intrinsic induction of cortical polarity in fission yeast (*Schizosaccharomyces pombe*) (Siegrist and Doe [72]); the cross diffusion stabilization effects overwhelm the depletion effects. **C**. Microtubules intrinsic induction of cortical polarity during *fibroblast* migration (Siegrist and Doe [72]); depletion effects overwhelm the local cross diffusion effects.

into a space of the interface gap [66,74-78]. Unevenly retarded GTP-tubulin *cross-diffusion* may influence internal *local fluctuations* of GTP-tubulin concentration within the gap. This local GTP-tubulin fluctuation may in conjunction with a topological stage of growing microtubule (within the gap) instigate a burst of GTP-hydrolysis, which in turn leads to *dynamic instability*. Consequently, the microtubule may start to shrink back and thus enlarge the interface gap. If the gap is enlarged to a certain size, the influx of fresh GTP-tubulin, by means of cross diffusion, may recommence and the microtubule growth may resume.

In conclusion, in the case of the interface gap, it is possible that tight correlation, i.e. synergy between *dynamic instability, GTP-tubulin local fluctuations and cross*

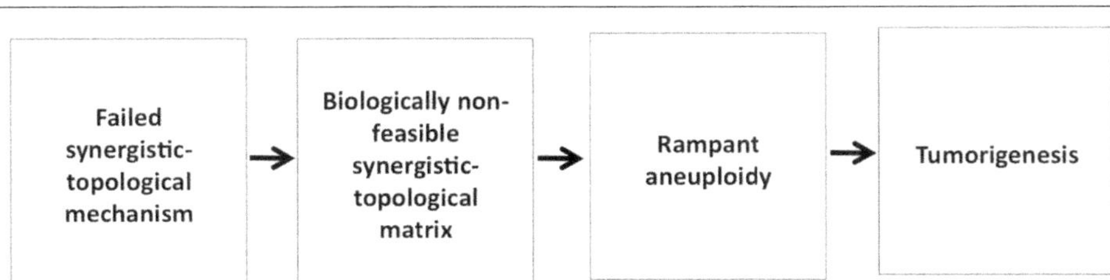

Figure 12 The possible role of microtubule synergistic-topological mechanism in tumorigenesis initiation. Failed synergistic-topological mechanism may initiate formation of biologically non-feasible microtubule synergistic-topological matrix. Non-feasible matrix mayenable rampant aneuploidy, that *per se* may preconditioned tumorigenesis to start (Duesberg, et al. [31]; Li, et al. [32]).

(active) diffusion in conjunction with a *topological* stage of the interface gap microtubule controls the size of the gap. Thus, synergy mechanisms in conjunction with interface gap topology may be an important aspect of the controlling mechanism of force and energy transmission as well as signal processing along fiber microtubules [26,39,58-61].

Possible biological relevance of a synergistic-topological mechanism

At biologically reduced conditions *in vitro*, we observed strong mutual interdependence of spatio-temporal microtubule pattern formation and locally associated solution conditions that may be explained by the existence of particular matrix. This matrix would be an integrated composite of microtubule spatio-temporal pattern and its specific locally associated solution conditions. The concept of a matrix related to the microtubule system of a mitotic spindle has already been proposed [23]. In our case, we consider that a synergistic-topological matrix is generated and maintained by the synergistic-topological mechanism including its modulator (the principle of summation of pairwise electrostatic forces between constituting microtubules of spatio-temporal microtubule pattern). If synergistic-topological mechanisms fail then it enables formation of biologically non-feasible synergistic-topological matrices. This may further enable rampant aneuploidy that may *per se* precondition tumorigenesis [15,31,32] (Figure 12).

Competing interests

The authors declare that they have no competing interests.

Authors contributions

VAB conceptualized and designed the study, performed the experiments and wrote the manuscript; RMDH assisted with some of the experiments and assisted with the preparation of the manuscript; BDH critically evaluated the manuscript and assisted with the preparation of later versions; VK assisted with some of the experiments; EM assisted with some of the experiments; XL assisted with some of the experiments; GJL assisted with obtaining measurements described within the manuscript; JJB-F assisted with obtaining measurements described within the manuscript and RBB was involved in the discussions of the project concepts, evaluated the methodologies and assisted with the preparation of the manuscript. All authors read and approved the final manuscript.

Acknowledgments

We would like to express our sincere gratitude to Professor Maxwell Bennett, AO (Professor of Neuroscience, University Chair and Scientific Director, Brain and Mind Research Institute, The University of Sydney) for his genuine interest in this work and continued support.

We thank to microscopist Dr. Ian Kaplin (Australian Key Centre for Microscopy and Microanalysis, The University of Sydney) for technical assistance.

My personal (VB) sincere thanks to Mrs. Maria and Mr. Con for their kind help in the accomplishment of this work.

Author details

[1]Brain and Mind Research Institute, Sydney Medical School, The University of Sydney, Sydney, NSW 2050, Australia. [2]Department of Physiology, Sydney Medical School, The University of Sydney, Sydney, NSW 2006, Australia. [3]Australian Nuclear Sciences and Technology Organisation (ANSTO), The Bragg Institute and the Australian Synchrotron, Sydney, Australia. [4]Discipline of Biomedical Science, School of Medical Sciences, Sydney Medical School, University of Sydney, Lidcombe, NSW 1825, Australia. [5]Discipline of Pathology, School of Medical Sciences, Sydney Medical School, Bosch Institute, University of Sydney, Sydney, NSW 2006, Australia. [6]Discipline of Medical Radiation Sciences, Faculty of Health Sciences, University of Sydney, Lidcombe, NSW 1825, Australia.

References

1. Neumann B, Hilliard MA: **Loss of MEC-17 leads to microtubule instability and axonal degeneration.** *Cell Rep* 2014, **6:**93 103. doi:10.1016/J.Celrep.2013.12.004.
2. Etienne-Manneville S: **Actin and microtubules in cell motility: which one is in control?** *Traffic* 2004, **5:**470 477. doi:10.1111/J.1600-0854.2004.00196.X.
3. Tamura N, Draviam VM: **Microtubule plus-ends within a mitotic cell are moving platforms with anchoring, signalling and force-coupling roles.** *Open Biol* 2012, **2:**Artn 120132. doi:10.1098/Rsob.120132.
4. Gerlitz G, Reiner O, Bustin M: **Microtubule dynamics alter the interphase nucleus.** *Cell Mol Life Sci* 2013, **70:**1255 1268. doi:10.1007/S00018-012-1200-5.

5. Mitsuyama F, Kato K, Hirosawa K, Mikoshiba K, Okuya M, Karagiozov K, Kato Y, Kanno T, Sanoe H, Koide T: **Redistribution of microtubules in dendrites of hippocampal CA1 neurons after tetanic stimulation during long-term potentiation.** *Ital J Anat Embryol* 2008, **17**:27.
6. Craddock TJ, Tuszynski JA, Hameroff S: **Cytoskeletal signaling: is memory encoded in microtubule lattices by CaMKII phosphorylation?** *PLoS Comput Biol* 2012, **8**:e1002421. doi:10.1371/journal.pcbi.1002421.
7. Akhshi TK, Wernike D, Piekny A: **Microtubules and actin crosstalk in cell migration and division.** *Cytoskeleton* 2014, **71**:1 23. doi:10.1002/Cm.21150.
8. Bouissou A, Verollet C, de Forges H, Haren L, Bellaiche Y, Perez F, Merdes AB: **Raynaud-Messina: gamma-Tubulin ring complexes and EB1 play antagonistic roles in microtubule dynamics and spindle positioning.** *Embo J* 2014, **33**:114 128. doi:10.1002/Embj.201385967.
9. Zanic M, Widlund PO, Hyman AA, Howard J: **Synergy between XMAP215 and EB1 increases microtubule growth rates to physiological levels.** *Nat Cell Biol* 2013, **15**:688. doi:10.1038/Ncb2744.
10. Janke C, Kneussel M: **Tubulin post-translational modifications: encoding functions on the neuronal microtubule cytoskeleton.** *Trends Neurosci* 2010, **33**:362 372. doi:10.1016/J.Tins.2010.05.001.
11. Stoppin-Mellet V, Fache V, Portran D, Martiel JL, Vantard M: **MAP65 coordinate microtubule growth during bundle formation.** *Plos One* 2013, **8**:ARTN e56808. doi:10.1371/journal.pone.0056808.
12. Mitchison T, Wühr M, Mitchison P, Nguyen K, Ishihara A, Groen CM: **Field: growth, interaction and positioning of microtubule asters in extremely large vertebrate embryo cells.** *Cytoskeleton (Hoboken)* 2012, **69**:738 750.
13. Wuhr M, Dumont S, Groen AC, Needleman DJ, Mitchison TJ: **How does a millimeter-sized cell find its center?** *Cell Cycle* 2009, **8**:1115 1121.
14. Smurova AABKM, Verin AD, Alieva IB: **Microtubule system in endothelial barrier dysfunction: disassembly of peripheral microtubules and microtubule reorganization in internal cytoplasm.** *Cell Tissue Biol* 2008, **2**:45 52.
15. Sayas CJ, Avila J: **Crosstalk between axonal classical microtubule-associated proteins and end binding proteins during axon extension: possible implications in neurodegeneration.** *J Alzheimers Dis* 2013, **x**:xx.
16. Gavin RH: **Synergy of cytoskeleton components - cytoskeletal polymers exhibit both structural and functional synergy.** *Bioscience* 1999, **49**:641 655. doi:10.2307/1313440.
17. Johansen KM, Johansen J: **Recent glimpses of the elusive spindle matrix.** *Cell Cycle* 2002, **1**:312 314.
18. Teng J, Takei Y, Harada A, Nakata T, Hirokawa N: **Synergistic effects of MAP2 and MAP1B knockout in neuronal migration, dendritic outgrowth, and microtubule organization.** *Mol Biol Cell* 2001, **12**:433A.
19. Clausen T, Ribbeck K: **Self-organization of anastral spindles by synergy of dynamic instability, autocatalytic microtubule production, and a spatial signaling gradient.** *Plos One* 2007, **2**:ARTN e244. doi:10.1371/journal.pone.0000244.
20. Schuh M, Ellenberg J: **Self-organization of MTOCs replaces centrosome function during acentrosomal spindle assembly in live mouse oocytes.** *Cell* 2007, **130**:484 498. doi:10.1016/J.Cell.2007.06.025.
21. Karsenti E, Vernos I: **Cell cycle - the mitotic spindle: a self-made machine.** *Science* 2001, **294**:543 547. doi:10.1126/Science.1063488.
22. Brunet S, Polanski Z, Verlhac MH, Kubiak JZ, Maro B: **Bipolar meiotic spindle formation without chromatin.** *Curr Biol* 1998, **8**:1231 1234. doi:10.1016/S0960-9822(07)00516-7.
23. Pickettheaps JD, Tippit DH, Porter KR: **Rethinking Mitosis.** *Cell* 1982, **29**:729 744. doi:10.1016/0092-8674(82)90435-4.
24. Lince-Faria M, Maffini S, Orr B, Ding Y, Florindo C, Sunkel CE, Tavares A, Johansen J, Johansen KM, Maiato H: **Spatiotemporal control of mitosis by the conserved spindle matrix protein Megator.** *J Cell Biol* 2009, **184**:647 657. doi:10.1083/Jcb.200811012.
25. Johansen KM, Johansen J: **Cell and molecular biology of the spindle matrix.** *Int Rev Cytol* 2007, **263**:155. doi10.1016/S0074-7696(07)63004-6.
26. Johansen KM, Forer A, Yao CF, Girton J, Johansen J: **Do nuclear envelope and intranuclear proteins reorganize during mitosis to form an elastic, hydrogel-like spindle matrix? (vol 19, pg 345, 2011).** *Chromosome Res* 2011, **19**:683. doi:10.1007/S10577-011-9217-4.
27. Buljan VA, Holsinger RMD, Brown D, Bohorquez-Florez JJ, Hambly BD, Delikatny EJ, Ivanova EP, Banati RB: **Spinodal decomposition and the emergence of dissipative transient periodic spatio-temporal patterns in acentrosomal microtubule multitudes of different morphology.** *Chaos* 2013, **23**:Artn 023120. doi10.1063/1.4807909.
28. Dlugosz M, Antosiewicz JM: **Evaluation of proteins rotational diffusion coefficients from simulations of their free brownian motion in volume-occupied environments.** *J Chem Theory Comput* 2014, **10**:481 491. doi:10.1021/Ct4008519.
29. Hunyadi V, Chretien D, Flyvbjerg H, Janosi IM: **Why is the microtubule lattice helical?** *Biol Cell* 2007, **99**:117 128. doi:10.1042/Bc20060059.
30. Podgornik R, Parsegian VA: **Charge-fluctuation forces between rodlike polyelectrolytes: pairwise summability reexamined.** *Phys Rev Lett* 1998, **80**:1560 1563. Doi: 10.1103/Physrevlett.80.1560.
31. Duesberg P, Li RH, Fabarius A, Hehlmann R: **The chromosomal basis of cancer.** *Cell Oncol* 2005, **27**:293 318.
32. Li RH, Yerganian G, Duesberg P, Kraemer A, Willer A, Rausch C, Hehlmann R: **Aneuploidy correlated 100% with chemical transformation of Chinese hamster cells.** *P Natl Acad Sci USA* 1997, **94**:14506 14511. doi:10.1073/Pnas.94.26.14506.
33. Shelansk M, Gaskin F, Cantor CR: **Microtubule assembly in absence of added nucleotides.** *P Natl Acad Sci USA* 1973, **70**:765 768. doi10.1073/Pnas.70.3.765.
34. Langford GM: **Length and appearance of projections on neuronal microtubules invitro after negative staining - evidence against a crosslinking function for maps.** *J Ultra Mol Struct R* 1983, **85**:1 10. doi:10.1016/S0022-5320(83)90111-9.
35. Buljan V, Ivanova EP, Cullen KM: **How calcium controls microtubule anisotropic phase formation in the presence of microtubule-associated proteins in vitro.** *Biochem Bioph Res Co* 2009, **381**:224 228.
36. Turner DC, Chang CY, Fang K, Brandow SL, Murphy DB: **Selective adhesion of functional microtubules to patterned silane surfaces.** *Biophys J* 1995, **69**:2782 2789.

37. Anderson PW: **More is different - broken symmetry and nature of hierarchical structure of science.** *Science* 1972, **177:**393. doi:10.1126/Science.177.4047.393.
38. Corning PA: **Synergy and self-organization in the evolution of complex-systems.** *Syst Res* 1995, **12:**89 121. doi:10.1002/Sres.3850120204.
39. Tuszynski JA, Hameroff S, Sataric MV, Trpisova B, Nip MLA: **Ferroelectric behavior in microtubule dipole lattices - implications for information-processing: signaling and assembly disassembly.** *J Theor Biol* 1995, **174:**371 380.
40. Tuszynski JA, Malinski W, Carpenter EJ, Luchko T, Huzil JT, Ludena RF: **Tubulin electrostatics and isotype specific drug binding.** *Can J Phys* 2008, **86:**635 640. doi:10.1139/P07-199.
41. Barton JS, Vandivort DL, Heacock DH, Coffman JA, Trygg KA: **Microtubule assembly kinetics - changes with solution conditions.** *Biochem J* 1987, **247:**505 511.
42. Tuszynski JA, Brown JA, Crawford E, Carpenter EJ, Nip MLA, Dixon JM, Sataric MV: **Molecular dynamics simulations of tubulin structure and calculations of electrostatic properties of microtubules.** *Math Comput Model* 2005, **41:**1055 1070. doi:10.1016/J.Mcm.2005.05.002.
43. Zimmerman SB, Minton AP: **Macromolecular crowding - biochemical, biophysical, and physiological consequences.** *Annu Rev Bioph Biom* 1993, **22:**27 65. doi:10.1146/Annurev.Bb.22.060193.000331.
44. Odde DJ: **Estimation of the diffusion-limited rate of microtubule assembly.** *Biophys J* 1997, **73:**88 96.
45. Shelden E, Wadsworth P: **Observation and quantification of individual microtubule behavior invivo - microtubule dynamics are cell-type specific.** *J Cell Biol* 1993, **120:**935 945. doi:10.1083/Jcb.120.4.935.
46. Tabony J: **Morphological bifurcations involving reaction diffusion processes during microtubule formation.** *Science* 1994, **264:**245 248. doi:10.1126/Science.8146654.
47. Hall D, Minton AP: **Macromolecular crowding: qualitative and semiquantitative successes, quantitative challenges.** *Bba-Proteins Proteom* 2003, **1649:**127 139. doi:10.1016/S1570-9639(03)00167-5.
48. Minton AP: **The influence of macromolecular crowding and macromolecular confinement on biochemical reactions in physiological media.** *J Biol Chem* 2001, **276:**10577 10580.
49. Minton AP: **Effects of excluded surface area and adsorbate clustering on surface adsorption of proteins: II. Kinetic models.** *Biophys J* 2001, **80:**1641 1648.
50. Matthews BW, Remingto SJ: **3 dimensional structure of lysozyme from bacteriophage-T4.** *P Natl Acad Sci USA* 1974, **71:**4178 4182. doi:10.1073/Pnas.71.10.4178.
51. Nogales E, Whittaker M, Milligan RA, Downing KH: **High-resolution model of the microtubule.** *Cell* 1999, **96:**79 88. doi:10.1016/S0092-8674(00)80961-7.
52. Herzfeld J: **Crowding-induced organization in cells: spontaneous alignment and sorting of filaments with physiological control points.** *J Mol Recognit* 2004, **17:**376 381.
53. Onsager L: **The effects of shape on the interaction of colloidal particles.** *Ann Ny Acad Sci* 1949, **51:**627 659. doi:10.1111/J.1749-6632.1949.Tb27296.X.
54. Liu YF, Guo YX, Valles JM, Tang JX: **Microtubule bundling and nested buckling drive stripe formation in polymerizing tubulin solutions.** *P Natl Acad Sci USA* 2006, **103:**10654 10659. doi:10.1073/Pnas.0510381103.
55. Hitt AL, Cross AR, Williams RC: **Microtubule solutions display nematic liquid-crystalline structure.** *J Biol Chem* 1990, **265:**1639 1647.
56. Baas PW, Ahmad FJ: **Beyond taxol: microtubule-based treatment of disease and injury of the nervous system.** *Brain* 2013, **136:**2937 2951. doi:10.1093/Brain/Awt153.
57. Schroer TA, Sheetz MP: **Functions of Microtubule-Based Motors.** *Annu Rev Physiol* 1991, **53:**629 652. doi:10.1146/Annurev.Physiol.53.1.629.
58. Sekulic DL, Sataric BM, Tuszynski JA, Sataric MV: **Nonlinear ionic pulses along microtubules.** *Eur Phys J E* 2011, **34:**Artn 49. doi:10.1140/Epje/I2011-11049-0.
59. Das M, Levine AJ, MacKintosh FC: **Buckling and force propagation along intracellular microtubules.** *Epl-Europhys Lett* 2008, **84:**Artn 18003. doi:10.1209/0295-5075/84/18003.
60. Sataric MV, Tuszynski JA, Zakula RB: **Kink-like excitations as an energy-transfer mechanism in microtubules.** *Phys Rev E* 1993, **48:**589 597. doi:10.1103/Physreve.48.589.
61. Hameroff SR, Watt RC: **Information-processing in microtubules.** *J Theor Biol* 1982, **98:**549 561.
62. Rodriguez OC, Schaefer AW, Mandato CA, Forscher P, Bement WM, Waterman-Storer CM: **Conserved microtubule-actin interactions in cell movement and morphogenesis.** *Nat Cell Biol* 2003, **5:**599 609. doi:10.1038/Ncb0703-599.
63. Kirschner M, Mitchison T: **Beyond self-assembly - from microtubules to morphogenesis.** *Cell* 1986, **45:**329 342.
64. Tabony J, Vuillard L, Papaseit C: **Biological self-organisation and pattern formation by way of microtubule reaction diffusion processes.** *Adv Complex Syst* 1999, **02:**221 276. doi:10.1142/S0219525999000138.
65. Buljan VA, Holsinger RMD, Hambly BD, Banati RB, Ivanova EP: **Intrinsic microtubule GTP-cap dynamics in semi-confined systems: kinetochore-microtubule interface.** *J Biol Phys* 2013, **39:**81 98. doi:10.1007/S10867-012-9287-3.
66. Brangwynne CP, Koenderink GH, MacKintosh FC, Weitz DA: **Intracellular transport by active diffusion.** *Trends Cell Biol* 2009, **19:**423 427. doi:10.1016/J.Tcb.2009.04.004.
67. Brangwynne CP, Koenderink GH, MacKintosh FC, Weitz DA: **Nonequilibrium microtubule fluctuations in a model cytoskeleton.** *Phys Rev Lett* 2008, **100:**Artn 118104. doi10.1103/Physrevlett.100.118104.
68. Chasey D: **Left-handed subunit helix in flagellar microtubules.** *Nature* 1974, **248:**611 612. doi:10.1038/248611a0.
69. Amos LA, Klug A: **Arrangement of subunits in flagellar microtubules.** *J Cell Sci* 1974, **14:**523 549.
70. Mandelkow EM, Schultheiss R, Rapp R, Muller M, Mandelkow E: **On the surface lattice of microtubules - helix starts, protofilament number, seam, and handedness.** *J Cell Biol* 1986, **102:**1067 1073. doi:10.1083/Jcb.102.3.1067.
71. Ebeling W: *Strukturbildung bei irreversiblen prozessen.* Leipzig: B.G. Teubner Verlagsgesellschaft; 1976.
72. Siegrist SE, Doe CQ: **Microtubule-induced cortical cell polarity.** *Gene Dev* 2007, **21:**483 496. doi:10.1101/Gad.1511207.
73. Janulevicius A, van Pelt J, van Ooyen A: **Compartment volume influences microtubule dynamic instability: a model study.** *Biophys J* 2006, **90:**788 798. Doi: 10.1529/Biophysj.105.059410.

A multi-compartment pharmacokinetic model of the interaction between paclitaxel and doxorubicin

Kenneth JE Vos[1†], Angela G Martin[1], Maxine G Trimboli[1], Lindsay Forestell[1], Khaled Barakat[2,3] and Jack A Tuszynski[2,4*†]

*Correspondence: jack.tuszynski@gmail.com
†Equal contributors
[2]Department of Physics, University of Alberta, T6G 2J1 Edmonton, Canada
[4]Department of Oncology, Faculty of Medicine and Dentistry, University of Alberta, T6G 2J1 Edmonton, Canada
Full list of author information is available at the end of the article

Abstract

Background: In this paper the interactions between paclitaxel, doxorubicin and the metabolic enzyme CYP3A4 are studied using computational models. The obtained results are compared with those of available clinical data sets. Analysis of the drug-enzyme interactions leads to a recommendation of an optimized paclitaxel-doxorubicin drug regime for chemotherapy treatment.

Methods: A saturable multi-compartment pharmacokinetic model for the multidrug treatment of cancer using paclitaxel and doxorubicin in a combination is developed. The model's kinetic equations are then solved using standard numerical methods for solving systems of nonlinear differential equations. The parameters were adjusted by fitting to available clinical data. In addition, we studied the interaction of each drug with the metabolic enzyme CYP3A4 through blind docking simulations to demonstrate that these drugs compete for the same metabolic enzyme and to show their molecular mode of binding. This provides a molecular-level justification for the introduction of interaction terms in the kinetic model.

Results: Using docking simulations we compared the relative binding affinities for the metabolic enzyme of the two chemotherapy drugs. Since paclitaxel binds more strongly to CYP3A4 than doxorubicin, an explanation is given why doxorubicin has no apparent influence upon paclitaxel, while paclitaxel has a profound effect upon doxorubicin. Finally, we studied different time sequences of paclitaxel and doxorubicin concentrations and calculated their AUCs.

Conclusions: We have found excellent agreement between our model and available empirical clinical data for the drug combination studied here. To support the kinetic model at a molecular level, we built an atomistic three-dimensional model of the ligands interacting with the metabolic enzyme and elucidated the binding modes of paclitaxel and doxorubicin within CYP3A4. Blind docking simulations provided estimates of the corresponding binding energies. The paper is concluded with clinical implications for the administration of the two drugs in combination.

Keywords: Paclitaxel; Doxorubicin; Metabolism; Combination chemotherapy; Molecular dynamics; Docking binding energy; Multi-compartment model; Pharmacokinetics; Pharmacodynamics

Background

World Health Organization (WHO) statistics show that 12.5% of the deaths worldwide result from cancer related malignant neoplasms [1]. Thus, a critical need exists for effective treatments against these illnesses. Chemotherapy is the treatment of cancer using an anti-neoplastic drug or a combination of such drugs. There are more than one hundred distinct chemotherapy drugs approved for use in North America and Europe that can be divided into approximately ten categories. The first group consists of alkylating agents such as cisplatin. There are approximately twenty five different drugs in this group, which are genotoxic to the tumor cells. The second group contains plant alkaloids such as paclitaxel. There are approximately fifteen different drugs in this group and they inhibit the mitosis of the tumor cells. The remaining groups of chemotherapy agents are not relevant to the topic of this paper and hence will not be discussed. Combination cancer chemotherapies have shown to be more effective than mono-chemical therapy, with varying degrees of success. However, the question: *"how can one best improve the outcome for a patient by changing the application frequency of a particular drug or by using a combination of drugs"* has not been extensively studied in quantitative detail and it merits significant attention. The outcome of any therapy always involves both benefits and costs. The benefits can be measured in terms of the patient survival probability or the average time to recurrence. The cost can be assessed by the severity and frequency of side effects. Unfortunately, cancers are a set of very complex diseases for which finding cures has been extremely elusive. Part of the complexity describing cancer can be attributed to the unique and complicated signalling pathways leading to cancer initiation and progression at the level of individual cancer cells. Detailed maps of the signalling pathways have been found for a number of cancers and are catalogued in the Kyoto Encyclopedia of Genes and Genomes (KEGG) database (http://www.genome.jp/kegg/). A typical KEGG pathway illustrates the molecular interactions and reaction networks for metabolism, various cellular processes, and human diseases (see [2-4]). For example, Figures 1 and 2 have been adapted from KEGG to illustrate two concrete examples how some of the available cancer chemotherapy drugs inhibit the cancer cell signaling pathways and how redundancies present in the signaling process complicate the situation.

The accumulating knowledge about the molecular biology of cancer and novel delivery tools to specifically target aberrant proteins are opening up new therapeutic possibilities. One of the most common methods of increasing cure rates using chemotherapeutic agents is to administer combination chemotherapy which most often refers to the simultaneous administration of two or more medicinal compounds or modalities to treat a single disease. This approach in cancer treatment can be traced back to 1965 when James Holland, Emil Freireich, and Emil Frei hypothesized that cancer chemotherapy should follow the strategy of antibiotic therapy for tuberculosis and use combinations of drugs, each with a different mechanism of action. Cancer cells could conceivably mutate to become resistant to a single agent, but by using different drugs concurrently it would be more difficult for the tumor to develop resistance to the combination. Holland, Freireich, and Frei simultaneously administered an antifolate, a vinca alkaloid, 6-MP and Prednisone - referred to as the POMP regimen - and induced long-term remissions in children with ALL. This approach was extended to the lymphomas and other types of cancer. Currently, nearly all successful cancer chemotherapy regimens use this paradigm of multiple drugs given simultaneously. Some types of cancers, previously

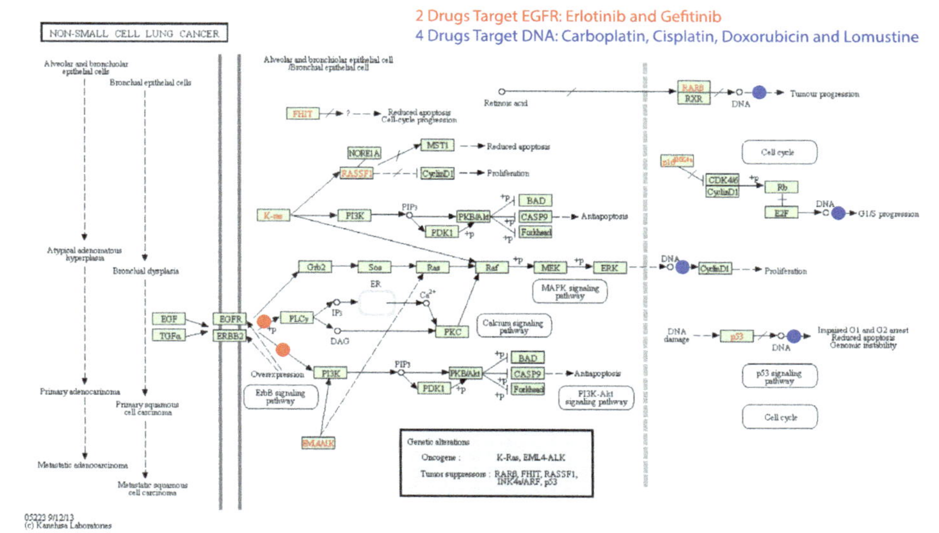

Figure 1 Lung cancer signaling pathways. The red and blue dots illustrate the action of chemotherapy agents blocking the pathways.

universally fatal, are now considered to be generally curable diseases as a result of this advance.

Current clinical trials in oncology commonly focus on three key aspects: (a) extending the scope of known drugs to new types of cancer, (b) testing new compounds, and (c) optimizing treatment by using combinations of known compounds. The latter aspect is of great interest and could benefit from a mathematical modelling approach

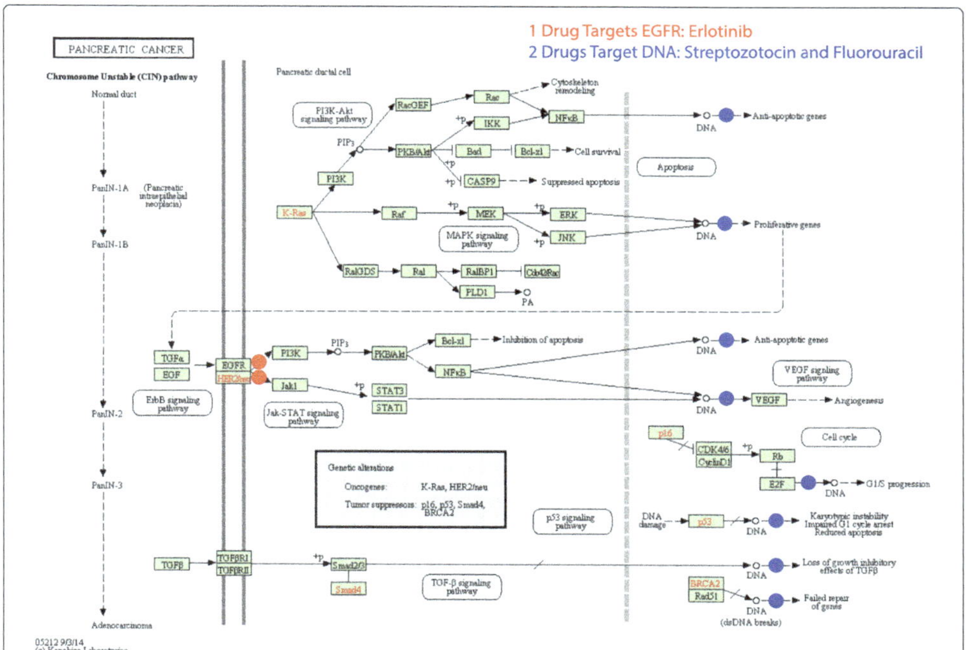

Figure 2 Pancreatic cancer signaling pathways. The red and blue dots illustrate the action of chemotherapy agents blocking the pathways.

aimed at achieving an optimized set of parameters for the dose, frequency and route of administration. This is, indeed, one of our main objectives behind the present work.

Two of the most widely used chemotherapy agents are paclitaxel and doxorubicin. Paclitaxel [5] is active against many human solid tumors related to breast cancer, ovarian cancer, non-small cell lung cancer, head and neck cancer, and advanced forms of Kaposi's sarcoma [6]. Because it is poorly water-soluble, the current formulation for Taxol® incorporates a 6 mg/ml solution in a solvent consisting of 50% polyoxyethylated castor oil (Cremophor® EL-CrEL) and 50% dehydrated alcohol (USP). Paclitaxel is typically administered by IV infusion over one or three hours but has also been administered over six and twenty four hours. Because a patient may have an anaphylactic reaction to CrEL, alternative formulations of paclitaxel have been introduced, including BMS-184476, oral paclitaxel in polysorbate 80, and ABI-007, Genexol-PM [7] and Abraxane® (nab-paclitaxel) [8]. Paclitaxel has a long residence time within the body and can stay trapped in cancer cells for over a week [9]. Paclitaxel is also highly bound to CrEL micelles, plasma proteins, platelets, and red blood cells [10].

DNA intercalators inhibit DNA polymerases and topoisomerases, resulting in the induction of apoptosis in tumor cells. DNA intercalating agents, such as amsacrine, actinomycin, mitoxantrone, and doxorubicin, have been employed as anticancer drugs and are in routine clinical use as chemotherapeutic agents [11]. It is well accepted that the antitumor activity of doxorubicin is caused by the formation of a cleavable complex of topoisomerase II, resulting in apoptosis [12,13]. Doxorubicin is indicated in the treatment of a broad spectrum of solid tumors (e.g. breast, bladder, endometrium, thyroid, lung, ovary, stomach, and sarcomas of the bone) and in the treatment of lymphoma, as well as acute lymphoblastic and myeloblastic leukemias [14]. One of the most important and clinically relevant side effects of doxorubicin is the induction of cardiomyopathy [15]. A number of mechanisms have been proposed to explain this effect of doxorubicin, including oxidative stress [16], the induction of mitochondrial damage [17], and changes in gene expression in cardiac myocytes and muscle cells in general [18,19].

Pharmacokinetics is the study of the absorption, distribution, metabolism and elimination (ADME) of drugs to, in, and from the body [20]. Pharmacological data usually consist of discrete values of the concentration of a drug in the plasma as a function of time. For drugs administered by direct intravenous (IV) infusion, a plot of these values generates a concentration-versus-time curve that rises during the infusion and then decreases after a maximum concentration value is reached. This decline may be relatively short or may last for several days, and it is mainly governed by the rate of elimination of the drug from the body. One of the key questions investigated is the functional dependence of the elimination curve and a single parameter that is often used to characterize the drug, namely its half-life. Another parameter of interest is the power law exponent of the elimination curve, which provides important information about the fate of the drug in the body and its efficacy. During clinical trials, the concentration-versus-time curves are used to determine optimum dosing regimens, potential toxicities, and drug-drug interactions.

The most common type of pharmacokinetic model is the compartmental model [20], in which a compartment is characterized by the number of drug molecules having the same probability of undergoing a set of chemical kinetic processes. The exchange of drug molecules between compartments is described by kinetic rate coefficients. Consider a one-compartment model post-infusion, such that

$$\frac{dX}{dt} = -kX, \tag{1}$$

where k is the kinetic rate coefficient describing the elimination process. If the compartment is homogeneous and instantaneously-mixed (well-stirred), the kinetics are classical and

$$k = k_o, \tag{2}$$

where k_o is a constant. The resulting solution has an exponential tail,

$$X(t) = X_o exp\left[-k(t - t_o)\right]. \tag{3}$$

Inhomogeneous compartments require a non-linear model and have been recently modeled using fractal kinetics [21,22] with a time-dependent kinetic rate coefficient:

$$k(t) = k_o t^{-\eta}, \tag{4}$$

where η is a fractal exponent, which is related to the fractal dimension of the organ in which the process takes place. The resulting solution has a stretched exponential tail,

$$X(t) = X_o \exp\left[\frac{-k_o\left(t^{1-\eta} - t_o^{1-\eta}\right)}{1-\eta}\right], \tag{5}$$

if $\eta \neq 1$. In the case $\eta = 1$, the solution is a power-law relationship,

$$X(t) = X_o\left[\frac{t_o}{t}\right]^{k_o}. \tag{6}$$

Saturable kinetics is usually modeled using Michaelis-Menten kinetics [23], where the differential equation describing a single compartment becomes

$$\frac{dX}{dt} = \frac{-V_{max}X}{K_M + X}. \tag{7}$$

The quantity V_{max} is the maximum rate of elimination, and K_M is the amount of drug required to achieve half the maximum rate of the process (i.e. when $X = K_M$ then the rate of drug elimination is; $\frac{dX}{dt} = -\frac{1}{2}V_{max}$. At very low dosage, the rate of elimination is given approximately by $\frac{dX}{dt} \approx \frac{-V_{max}X}{K_M} = -kX$. Hence the model gives the standard expression for the rate of elimination at low dosage, and as the amount of drug increases the rate of elimination becomes a constant. The corresponding amount of drug versus time curve exhibits an initial linear segment (high dosage) followed by an exponential tail at low concentrations. The implicit solution to Equation 7, assuming constant parameters, is

$$K_M \ln\left[\frac{X}{X_o}\right] + X - X_o = -V_{max}(t - t_o). \tag{8}$$

A more complex single-compartment saturable model, which produces fractal effects, is given by the kinetic equation

$$\frac{dX}{dt} = \frac{-V_{max}X^a}{K_M + X^b}, \tag{9}$$

which has the following exact solution, assuming constant parameters

$$K_M \frac{X^{1-a} - X_o^{1-a}}{1-a} + \frac{X^{1+b-a} - X_o^{1+b-a}}{1+b-a} = -V_{max}(t - t_o). \quad (10)$$

The implicit solution in Equation 10 must be modified for two special cases. The first special case is when $a = 1$, the first term must be replaced with $K_M \ln\left[\frac{X}{X_o}\right]$. The second special case is when $a - b = 1$, the second term must be replaced with $\ln\left[\frac{X}{X_o}\right]$.

Figure 3 graphically illustrates the solution given in Equation 10. In this figure; $t_o = 1$ hr, $X_o = 10$ μmol/L, $a = b$, $V_{max} = 18.80$ μmol/(L hr), and $K_M = 5.50$ (μmol/L) are taken as an illustrative example.

A single compartment model is clearly insufficient to capture the complexity of the pharmacokinetic and pharmacodynamic data exhibited by multi-drug chemotherapy. We believe that the simplest non-trivial model required for this purpose is a multi-compartment model in which the compartments are saturable due to the limitations on the number of enzymes and the molecular targets available for drug binding. The number of compartments considered for the modeling of the multi-drug pharmacokinetics of cancer chemotherapy agents should be at least four, namely: (1) the blood/plasma compartment, (2) the elimination organs compartment, (3) the healthy cells compartment, and (4) the tumor cells compartment. Compartment 1 represents the blood/plasma and is the mechanism for the transportation of the drug molecules throughout the body. Compartment 2 represents the liver and elimination organs, in which a portion of the drug is metabolized while the remaining drug is excreted from the body. Compartment 3 represents the healthy cells in the body and allows for a measure of the toxicity of the drugs. Compartment 4 represents the tumor cells. This model is clearly not as powerful as a physiologically-based approach that requires more detail as well as more model parameters describing its many compartments. However, we have decided to use the current four-compartment model to maintain simplicity in order to focus on the issue of drug interactions rather than the model's physiological relevance and accuracy. We believe this model represents a compromise between simplicity and realistic representation of the PK system. This four-compartment model is illustrated in Figure 4.

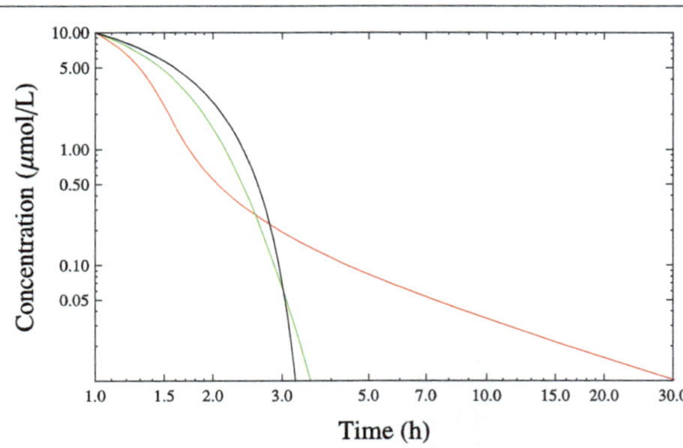

Figure 3 Nonlinear saturable kinetics. The concentration-versus-time curve of the saturable kinetics example in the text for the solution given in Equation 10 with $a = 0.75$ for the black curve, $a = 1$ for the green curve, and $a = 2$ for the red curve.

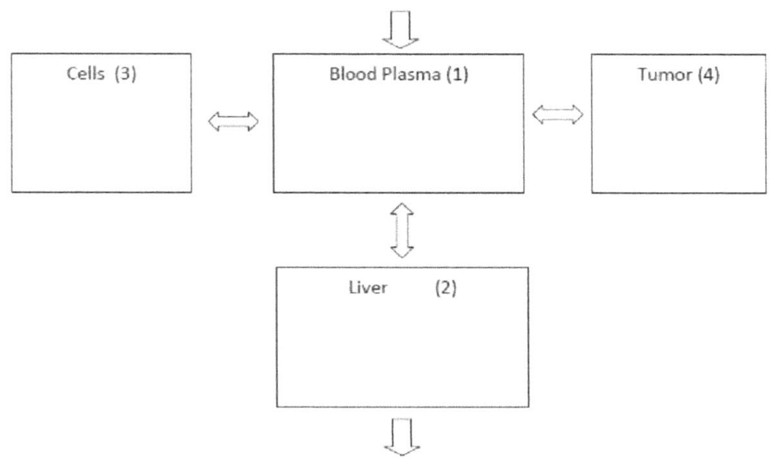

Figure 4 Four-compartment pharmacokinetic model. The four compartments in the pharmacokinetic model presented in this paper: (1) the blood/plasma compartment, (2) the elimination organs compartment, (3) the healthy cells compartment, and (4) the tumor cells compartment.

The model presented here assumes well-stirred conditions, i.e. average drug concentrations in the various compartments are the only dependent variables. The mathematical model corresponding to Figure 4 is represented by a set of first-order (nonlinear) differential equations given by

$$\frac{dX_{i,j}}{dt} = I_{i,j} + \sum_{k=0}^{4} \left[\frac{K_{i,k,j} X_{i,k}^{A(i,k,j)}}{1 + \Gamma_{i,k,j} X_{i,k}^{B(i,k,j)}} - \frac{K_{i,j,k} X_{i,j}^{A(i,j,k)}}{1 + \Gamma_{i,j,k} X_{i,j}^{B(i,j,k)}} \right], \quad (11)$$

where $X_{i,j}$ is the concentration of drug i in compartment j, $I_{i,j}$ is the rate of infusion divided by the volume of distribution of drug i in compartment j, compartment 0 ($k = 0$ in the Equation 11 sum) is the environment external to the body, $K_{i,j,k}$ is the kinetic rate coefficient for drug i moving from compartment j to compartment k, $\Gamma_{i,j,k}$ is the saturation coefficient for drug i moving from compartment j to compartment k, and $A(i, j, k)$ and $B(i, j, k)$ are the fractal exponents for drug i moving from compartment j to compartment k.

The primary motivation for this paper is to quantitatively understand the combined action of paclitaxel and doxorubicin in view of the reported interactions when these drugs are administered simultaneously. Namely, it has been demonstrated that the combination of doxorubicin followed by infusion of paclitaxel has high antitumor activity in patients with metastatic breast cancer, but is strongly limited by the associated cardiac toxicity [24]. The question that arises is how these side effects can be reduced by optimal mode of administration of the two drugs. This requires careful analysis and model development. Both paclitaxel and doxorubicin are known to be substrates for specific metabolic enzymes. Such interactions with the metabolic enzymes raise the risk of alleviating their associated toxicity and inducing unpredictable adverse side effects due to metabolites. Given this clue, here we develop a predictive model of the pharmacokinetic and pharmacodynamic interactions between paclitaxel, doxorubicin and the metabolic enzymes in the body during multi-drug chemotherapy. Our pharmacokinetic model can be used in aiding the design of optimized dosage and scheduling of the two-drug combination.

In this paper, we discuss the results obtained at both levels of computational modeling: pharmacokinetics of the two drugs under investigation and molecular dynamics of their interactions with the metabolic enzyme.

Methods

Below is a short description of the pharmacokinetic model mentioned above. The model focuses on the interactions of paclitaxel and doxorubicin with a common metabolic enzyme for the two compounds. To gain further understanding of these interactions at the molecular level, we performed a blind docking simulation protocol, which was employed to predict the binding sites, modes of binding and binding energies of the two drugs to the CYP3A4 metabolic enzyme. This was intended to explain at a molecular level why these two drugs affect each other's pharmacokinetic profiles. Below is a description of the modeling techniques used both for pharmacokinetic analysis and molecular dynamics simulations of the two drugs, paclitaxel and doxorubicin.

Pharmacokinetic modeling

The mathematical model corresponding to Figure 4 is Equation 11, which represents a system of nonlinear differential equations and in general, the kinetic rate coefficients do not have units of inverse time. Only in the special cases when all of the A's are equal to one and the drugs have the same units in all the compartments, will the kinetic rate coefficients have units of inverse time. We defined I as the rate of infusion divided by the volume of distribution for the entire body, which is commonly known for most drugs from empirical data. In our case the volume of distribution is an adjustable parameter used to optimize fits to the available data sets. The volume of distribution for specific organs would be desirable but is very hard to determine, especially for human patients. We would require empirical data for the drug concentrations in several compartments in order to generate reliable parameter fits.

The parameters in Equation 11 are determined by minimizing the *weighted percentage variance* between the solution of Equation 11 and the clinical data. The weighted variance was used because some of the clinical concentration values were for a single patient while other clinical concentration values represented the mean value of the concentration for multiple patients. A percentage ratio was used because the concentrations varied over several orders of magnitude (approximately 0.01 to 10 micromoles per liter). The *weighted percentage variance* is defined as;

$$VAR = \frac{4}{N_p} \sum_{l,k,j,i} p_{l,i,j,k} \left[\frac{X_{i,j}(t_k) - X_{i,j}^{(l)}(t_k)}{X_{i,j}(t_k) + X_{i,j}^{(l)}(t_k)} \right]^2, \tag{12}$$

where $X_{i,j}(t_k)$ is the solution of Equation 11 for the concentration of drug i in compartment j at time t_k. $X_{i,j}^{(l)}(t_k)$ is the concentration of drug i in compartment j at time t_k from clinical study l. $p_{l,i,j,k}$ is the wieghting coefficient, which is the number of patients involved in study l used to measure and calculate the mean concentration $X_{i,j}^{(l)}(t_k)$. $N_p = \sum_{l,k,j,i} p_{l,i,j,k}$. In addition, the *weighted variance* is defined as;

$$S = \frac{1}{N_p} \sum_{l,k,j,i} p_{l,i,j,k} \left[X_{i,j}(t_k) - X_{i,j}^{(l)}(t_k) \right]^2. \tag{13}$$

The initial starting values for the parameters in Equation 11 were assumed to be for a linear system. i.e. the exponents were all set to one, the interaction parameters were set to zero, and the saturation parameters were set to zero. The clinical half-life for each of the studies was used to estimate the kinetic rate coefficients. The $X_{i,j}(t_k)$ in Equation 12 were determined by solving the system of first order differential equations using standard numerical methods such as the *Runge-Kutta Method*. The maximum number of differential equations solved was fourteen for this system of drugs and their metabolites. The weighted percentage variance, *VAR*, can then be calculated for this specific set of parameters. The kinetic rate coefficients, the saturation parameters, and the exponents are then allowed to vary with the maximum number of parameters varying not exceeding the number of clinical data values for the concentrations minus one. The appropriate values for the parameters are obtained by finding the minimum value of *VAR*. The *VAR* can be minimized by varying these parameters using the *Powell's Method in Multidimensions*.

An additional aspect over and above pharmacokinetic modeling that enhances our analysis is the molecular-level understanding of the nature of interactions between each of the two drugs and the metabolic enzyme CYP3A4. The latter effort requires an entirely different modeling methodology, namely molecular dynamics, which is briefly described below.

Molecular dynamics methodology

In order to determine if both doxorubicin and paclitaxel are substrates for the same metabolic enzyme, we used blind docking combined with molecular dynamics simulations to characterize the mode of action of the two drugs at the molecular level. We used the human microsomal cytochrome P450 3A4 crystal structure (PDB: entry 1TQN) [25]. The catalytic active site was well-characterized and included a HEM group. Prior to docking simulations, protonation states of the residues constituting the CYP3A4 including the HEM group were adjusted using the software PDB2PQR [25]. The protein structure was conformationally relaxed using the NAMD molecular dynamics software with constraints on the backbone atoms. The AMBER99SB force field [26] was used for protein parameterization, while the GAFF provided parameters for the HEM group.

To carry out the blind docking protocol, the entire surface of the CYP3A4 was divided into 90 focus docking regions. For each region, the center of mass of three solvent exposed neighboring atoms was used as the center of the docking box. The dimensions of the docking cube were $90 \times 90 \times 90$ points with grid spacing of 0.03 nm. Clustering of the docking poses was performed using a 0.2 nm RMSD cut-off. To identify the most preferred binding locations, we combined all docking results and ranked them with the lowest binding energy of the largest docking cluster. As we are not only interested in the binding energies as indicators for adequate binding, we defined a hit as the docking run that includes at least 20% of the total population in its largest cluster. Finally, the solvent-exposed atoms that were used to construct the docking boxes have been used as markers for the binding locations.

All docking simulations were performed using AutoDock, version 4.0. Hydrogen atoms were added to all CYP3A4 and the two ligands followed by assigning their partial atomic charges using the Gasteiger-Marsili method. Atomic solvation parameters were assigned to the protein atoms using the AutoDock utility ADDSOL. A docking grid map with $90 \times 90 \times 90$ points and grid point spacing of 0.03 nm has been calculated using

AUTOGRID program. The grid box was centered on the active site. Rotatable bonds of each ligand were then automatically assigned using AUTOTORS utility of AutoDock. Docking was performed using the LGA method with an initial population of 400 random individuals, a maximum number of 10,000,000 energy evaluations, 100 trials, 50,000 maximum generations, a mutation rate of 0.02, a crossover rate of 0.80 and the requirement that only one individual can survive into the next generation.

Results and discussions

Pharmacokinetics of doxorubicin

The pharmacokinetics of a drug varies quite significantly between people and even in an individual person (pharmacokinetics will change with age and the health of the patient). Ideally, we would want a pharmacokinetic model with parameter values for each individual patient and the analysis would then be optimized for that specific patient; however, that is not practical. Therefore, we will calculate a set of parameters for a mean concentration versus time curve and obtain general qualitative features that will describe the people in this group as a whole.

The first stage is to determine the pharmacokinetics of doxorubicin from its concentration versus time clinical data and determine the parameters within the model in Equation 11 that fits this data without paclitaxel present. Equation 11 was fitted to several sets of experimental data for doxorubicin. The total number of data points that were digitized from the concentration-versus-time figures was 463. $N_p = 1335$, where N_p is defined in Equation 12. The total number of parameters allowed to vary was sixty five, but only forty nine parameters were used. The minimization of VAR in Equation 12 produced the forty eight parameters in Table 1 for doxorubicin. In addition, the volume of distribution for doxorubicin was found to be $V_{D,DOX} = 5.02\ L/m^2$. The Table 1 numbers give a best fit to the doxorubicin data with the weighted percentage variance $VAR = 0.24$. (The weighted variance is $S = 0.45$ micromoles squared per liter squared.) The comparison of each set of experimental data with Equation 11 using the parameters given in Table 1 is given in Appendix A.

As an example, we have presented the data from Gianni et al., Moreira et al., and Benjamin et al. [27-29] in Figure 5, and plotted the model in Equation 11 for the dosage of 60 mg/m^2 using the parameters in Table 1. The data from Gianni et al. [27] and Moreira et al. [28] stops at twenty four hours because that is when paclitaxel is introduced to the

Table 1 The PK parameters for Equation 11 modeling doxorubicin

j,k	$A_{DOX,j,k}$	$B_{DOX,j,k}$	$\Gamma_{DOX,j,k}$	$K_{DOX,j,k}$
1,2	0.90	1.99	0.0005	$8.77\left\lvert\frac{1-0.017X_{DOX,2}^{1.78}}{1+0.189X_{DOX,2}^{1.03}}\right\rvert$
1,3	1.08	1.71	0.0038	$3.55\left\lvert\frac{1+3.09X_{DOX,3}}{1+1.53X_{DOX,3}}\right\rvert$
1,4	1.97	0.96	0.0008	$0.354\left\lvert 1+0.0002X_{DOX,1}^{2.77}\right\rvert\left\lvert\frac{1+0.306X_{DOX,4}^{1.51}}{1+0.0253X_{DOX,4}^{0.40}}\right\rvert$
2,0	0.94	1.00	0.0000	0.252
2,1	0.72	0.80	0.1466	$0.132\left\lvert\frac{1+9.35X_{DOX,1}^{0.96}}{1+0.020X_{DOX,1}^{0.84}}\right\rvert$
3,1	0.97	1.16	0.0242	$0.0325\left\lvert\frac{1+2.40X_{DOX,1}^{1.34}}{1+0.165X_{DOX,1}^{1.01}}\right\rvert$
4,1	1.91	1.00	0.0000	$0.0055\left\lvert\frac{1+6.67X_{DOX,1}^{0.86}}{1+0.15X_{DOX,1}^{1.27}}\right\rvert$

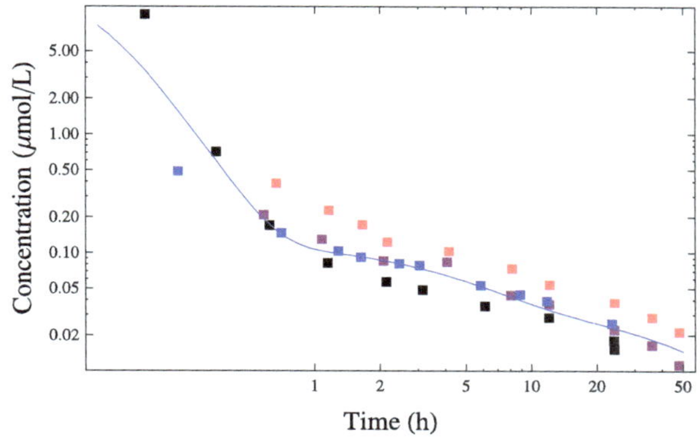

Figure 5 Comparison of theoretical and experimental doxorubicin results. The concentration-versus-time curve for doxorubicin. The black squares are the clinical data from Gianni et al. [27], the blue squares are the clinical data from Moreira et al. [28]. The purple and pink squares are the clinical data from Benjamin et al. [29]. The purple squares are for the acid alcohol data and the pink squares are for the chloroform data. The curve represents the solution of Equation 11 using the parameters in Table 1.

patients in those two studies. There are two sets of data from Benjamin et al. [29]; the pink data corresponds to chloroform being added to the extracted blood plasma and the purple data is for when an acid alcohol is added to the extracted blood plasma. In theory, these multiple patient studies should contain experimental data that overlaps along the same curve, but clearly that is not the case. The clinical data for doxorubicin is inconsistent from one study to the next with large variations in the results. The theoretical curve gives a best-fit of the data.

The metabolite, doxorubicinol, is produced from doxorubicin by the cytochrome P450 isoenzyme CYP3A4, and its pharmacokinetics can also be modeled using Equation 11. The total number of data points that were digitized from the concentration-versus-time figures was 126. $N_p = 869$, where N_p is defined in Equation 12. The total number of parameters allowed to vary was fifty one, but only forty five parameters were used. The minimization of VAR in Equation 12 for doxorubicinol produced the forty parameters in Table 2 and five parameters in Equation 14. The weighted percentage variance is $VAR = 0.24$ for doxorubicinol. (The weighted variance is $S = 0.0012$ micromoles squared per liter squared for doxorubicinol). The combined weighted percentage variance is $VAR = 0.24$ for doxorubicin and doxorubicinol. (The combined weighted variance is $S = 0.27$ micromoles squared per liter squared for doxorubicin and doxorubicinol).

Table 2 The PK parameters for Equation 11 modeling doxorubicinol

j,k	$A_{DOL,j,k}$	$B_{DOL,j,k}$	$\Gamma_{DOL,j,k}$	$K_{DOL,j,k}$
1,2	2.11	1.41	0.5836	$12.1 \left\lvert \frac{1+10.5X_{DOL,2}^{0.89}}{1+4.65X_{DOL,2}^{1.67}} \right\rvert$
1,3	1.19	1.00	0.0000	$3.41 \left\lvert 1+0.0702X_{DOL,1}^{0.90} \right\rvert \left\lvert \frac{1+0.967X_{DOL,3}^{1.92}}{1+0.0466X_{DOL,3}^{1.42}} \right\rvert$
2,0	0.79	1.36	1.8532	$8.86 \left\lvert 1-0.0228X_{DOL,2}^{1.20} \right\rvert$
2,1	1.03	1.08	8.5081	$3.30 \left\lvert 1+0.0368X_{DOL,2}^{2.09} \right\rvert \left\lvert \frac{1+0.0943X_{DOL,1}^{1.43}}{1+3.86X_{DOL,1}^{0.76}} \right\rvert$
3,1	0.64	1.00	0.0000	$0.0601 \left\lvert 1+0.0054X_{DOL,3}^{1.29} \right\rvert \left\lvert \frac{1-5.87X_{DOL,1}^{1.68}}{1+2.38X_{DOL,1}^{1.20}} \right\rvert$

The rate at which doxorubicin is being converted into doxorubicinol was found to be:

$$I_{DOL,2} = 0.252X_{DOX,2}^{0.94}\left[2.08 + \frac{0.224X_{DOL,2}^{2.02}}{1+16.1X_{DOL,2}^{0.98}}\right]. \tag{14}$$

The coefficient, 2.08, in Equation 14 represents the fraction of eliminated doxorubicin in compartment 2 being converted into doxorubicinol times the volume of distribution for doxorubicin divided by the volume of distribution for doxorubicinol. The comparison between Equation 11 using the parameters in Table 2 and each of the experimental set of data is given in Appendix A. As an example, we have presented the doxorubicin data and doxorubicinol data from Andersen et al. [30] and Greene et al. [31] in Figure 6.

Note that some of the parameters for doxorubicin in Table 1 do not play a significant roll in fitting Equation 11 to the clinical data. One of A parameters in Table 1 is close to 1, and can be set to one. Two of the saturation parameters, Γ, are very small and two are zero. In addition, the 0.0002 in row 1,4 can be set to zero. These parameters can be ignored without losing much information. This would reduce the number of doxorubicin parameters to 42. In the case of doxorubicinol, all the parameters in Table 2 have an effect with one exception. The exponent $A = 1.03$ could be set to 1 without a significant change to the results. Also, one of the three exponents in Equation 14, 0.98, is close to one and could be set to one as well. This reduces the number of doxorubicinol parameters to 43.

The best-fit parameters of doxorubicin and doxorubicinol in Tables 1 and 2 model all the data reasonably well, but not with enough accuracy for a detailed study of specific data sets because of the variations from one study to another as clearly shown in Figure 5. In order to study the interaction between doxorubicin and paclitaxel we need an accurate fit to the specific data set when there is no interaction between the two drugs. To acheive this, we begin with the best-fit parameters for all the data sets given in Tables 1 and 2 and perturb these parameters to improve the fit to the specific data set. We have adjusted and fitted the doxorubicin and doxorubicinol parameters to the clinical data obtained by Gianni et al. [27] and Moreira et al. [28] separately. The Gianni et al. [27]

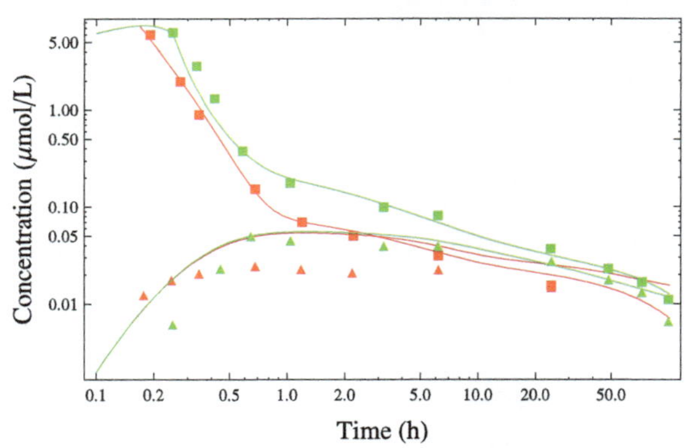

Figure 6 Comparison of theoretical and experimental doxorubicinol results. The concentration-versus-time data for doxorubicin are represented by the squares and doxorubicinol data is represented by the triangles. The red data is from Andersen et al. [30] and the green data is from Greene et al. [31]. The curves are from Equation 11 using the parameters in Tables 1 and 2.

study had 10 doxorubicin data points and 8 doxorubicinol data points, when there was no paclitaxel present. The Moreira et al. [28] study had 10 doxorubicin data points and 10 doxorubicinol data points, when there was no paclitaxel present. $N_p = 144$, for Gianni et al. [27] and 560 for Moreira et al. [28],where N_p is defined in Equation 12. We allowed all the parameters in Tables 1 and 2 to vary to determine, which parameters required the largest change. The result of fitting the theoretical model to the doxorubicin data of Gianni et al. [27] is shown in Table 3, and fitting the doxorubicinol parameters is shown in Table 4. The weighted percentage variance is $VAR = 1.7 \times 10^{-3}$ and the weighted variance is $S = 3.3 \times 10^{-4}$ micromoles squared per liter squared. The fit of the theoretical model to the data of Moreira et al. [28] is shown in Table 5 for doxorubicin, and the new doxorubicinol parameters are shown in Table 6. The weighted percentage variance is $VAR = 1.4 \times 10^{-3}$ and the weighted variance is $S = 8.5 \times 10^{-6}$ micromoles squared per liter squared. When relaxing the parameters to improve the fit with specific data, the parameters for the rate at which doxorubicin is being converted into doxorubicinol are changed in addition to the parameters in the tables. The new conversion rate for Gianni et al. [27] was found to be:

$$I_{DOL,2} = 0.264 X_{DOX,2}^{1.33} \left[2.49 + \frac{0.358 X_{DOL,2}^{1.17}}{1 + 16.5 X_{DOL,2}^{1.28}} \right]. \tag{15}$$

For Moreira et al. [28], the rate of conversion was found to be:

$$I_{DOL,2} = 0.264 X_{DOX,2}^{1.38} \left[3.98 + \frac{0.344 X_{DOL,2}^{1.87}}{1 + 9.03 X_{DOL,2}^{1.46}} \right]. \tag{16}$$

The parameters for both studies needed very little change. The result of these theoretical fits are shown in Figure 7 for Gianni et al. [27] and Figure 8 for Moreira et al. [28]. The agreement is very good as verified by the variance VAR and S defined in Equations 12 and 13, respectively.

Pharmacokinetics of paclitaxel

The pharmacokinetics of paclitaxel can be determined from its concentration-versus-time curves as well. We have digitized the data from several experimental data sets for paclitaxel and fitted to Equation 11. The number of clinical data points that were digitized from the concentration-versus-time figures and used to determine the parameters was 100. For these 100 data points $N_p = 820$, where N_p is defined in Equation 12. The

Table 3 The PK parameters for Equation 11 modeling doxorubicin for Gianni et al. [27]

j,k	$A_{DOX,j,k}$	$B_{DOX,j,k}$	$\Gamma_{DOX,j,k}$	$K_{DOX,j,k}$
1,2	0.79	2.52	0.0006	$8.14 \left\| \frac{1-0.0254 X_{DOX,2}^{2.59}}{1+0.319 X_{DOX,2}^{1.59}} \right\|$
1,3	0.83	2.43	0.0039	$3.08 \left\| \frac{1+3.12 X_{DOX,3}^{0.97}}{1+1.44 X_{DOX,3}^{0.97}} \right\|$
1,4	1.59	1.62	0.0012	$0.359 \left\| 1 + 0.0001 X_{DOX,1}^{1.55} \right\| \left\| \frac{1+0.277 X_{DOX,4}^{1.63}}{1+0.0282 X_{DOX,4}^{0.25}} \right\|$
2,0	1.33	1.00	0.0000	0.264
2,1	1.01	0.53	0.0994	$0.135 \left\| \frac{1+13.9 X_{DOX,1}^{0.87}}{1+0.1672 X_{DOX,1}^{0.66}} \right\|$
3,1	1.04	1.12	0.0287	$0.0325 \left\| \frac{1+2.02 X_{DOX,1}^{1.74}}{1+0.143 X_{DOX,1}^{0.61}} \right\|$
4,1	1.79	1.00	0.0000	$0.0046 \left\| \frac{1+7.92 X_{DOX,1}^{1.04}}{1+0.243 X_{DOX,1}^{1.95}} \right\|$

Table 4 The PK parameters for Equation 11 modeling doxorubicinol for Gianni et al. [27]

j,k	$A_{DOL,j,k}$	$B_{DOL,j,k}$	$\Gamma_{DOL,j,k}$	$K_{DOL,j,k}$
1,2	1.76	2.43	0.3371	$14.0\left\|\frac{1+10.6X_{DOL,2}^{0.72}}{1+4.54X_{DOL,2}^{1.92}}\right\|$
1,3	1.07	1.00	0.0000	$3.92\left\|1+0.1150X_{DOL,1}^{0.55}\right\|\left\|\frac{1+1.02X_{DOL,3}^{1.52}}{1+0.0482X_{DOL,3}^{2.43}}\right\|$
2,0	0.72	1.83	1.5722	$8.58\left\|1-0.0167X_{DOL,2}^{1.55}\right\|$
2,1	0.91	1.34	6.8599	$3.90\left\|1+0.0646X_{DOL,2}^{1.18}\right\|\left\|\frac{1+0.1282X_{DOL,1}^{0.95}}{1+3.97X_{DOL,1}^{1.07}}\right\|$
3,1	0.79	1.00	0.0000	$0.0594\left\|1+0.0073X_{DOL,3}^{2.26}\right\|\left\|\frac{1-6.94X_{DOL,1}^{1.17}}{1+3.19X_{DOL,1}^{0.96}}\right\|$

total number of parameters allowed to vary was sixty seven, but only fifty parameters were used. The minimization of VAR in Equation 12 produced the forty nine parameters for paclitaxel in Table 7 and the last parameter is the volume of distribution for paclitaxel, which was found to be $V_{D,PTX} = 3.53\ L/m^2$. The minimum weighted percentage variance was found to be $VAR = 0.035$. The weighted variance was determined to be $S = 0.18$ micromoles squared per liter squared. The results are shown in Figures 9 and 10. The experimental data in Figure 9 is from Gianni et al. [32]. The experimental data in Figure 10 is from Ohtsu et al. [33]. The comparison between the theoretical curve and rest of the experimental data is given in Appendix B.

Paclitaxel is metabolized by the cytochrome P450 isoenzymes CYP2C8 and CYP3A4. The primary metabolite, 6α-hydroxypaclitaxel (OHP), is formed by CYP2C8 while 3'-p-hydroxypaclitaxel and 6a,3'-p-dihydroxypaclitaxel are formed by CYP3A4. We can also model the metabolite(s) using Equation 11. We have digitized the experimental concentration-versus-time data and fitted the model in Equation 11 to the data. The total number of clinical data points that were digitized from the concentration-versus-time figures was 20. $N_p = 173$, where N_p is defined in Equation 12. We fixed the exponents of the drug concentration to 1 because of the lack of data, and the coefficients were allowed to vary. There are twenty six coefficents, so those that had the least influence were set to zero. The minimization of VAR in Equation 12 produced the fifteen parameters for 6α-hydroxypaclitaxel in Table 8 and another three parameters in Equation 17. Equation 17 is the rate at which paclitaxel is being converted into 6α-hydroxypaclitaxel:

$$I_{OHP,2} = 2.98X_{PTX,2}^{1.03}\left|\frac{1-0.0023X_{PTX,2}^{2.09}}{1+0.0876X_{PTX,2}^{1.99}}\right|\left[37.5+\frac{0.215X_{OHP,2}}{1+0.640X_{OHP,2}}\right]. \tag{17}$$

Table 5 The PK parameters for Equation 11 modeling doxorubicin for Moreira et al. [28]

j,k	$A_{DOX,j,k}$	$B_{DOX,j,k}$	$\Gamma_{DOX,j,k}$	$K_{DOX,j,k}$
1,2	0.86	2.14	0.0005	$8.86\left\|\frac{1-0.0231X_{DOX,2}^{3.12}}{1+0.157X_{DOX,2}^{0.89}}\right\|$
1,3	1.06	1.87	0.0036	$3.69\left\|\frac{1+3.09X_{DOX,3}}{1+1.61X_{DOX,3}}\right\|$
1,4	2.07	0.86	0.0007	$0.375\left\|1+0.0003X_{DOX,1}^{2.99}\right\|\left\|\frac{1+0.375X_{DOX,4}^{1.59}}{1+0.0222X_{DOX,4}^{0.31}}\right\|$
2,0	1.38	1.00	0.0000	0.264
2,1	0.91	0.51	0.2295	$0.102\left\|\frac{1+11.2X_{DOX,1}^{0.90}}{1+0.181X_{DOX,1}^{0.91}}\right\|$
3,1	1.02	0.90	0.0264	$0.0369\left\|\frac{1+1.74X_{DOX,1}^{0.96}}{1+0.298X_{DOX,1}^{1.48}}\right\|$
4,1	1.82	1.00	0.0000	$0.0052\left\|\frac{1+6.42X_{DOX,1}^{1.02}}{1+0.154X_{DOX,1}^{2.17}}\right\|$

Table 6 The PK parameters for Equation 11 modeling doxorubicinol for Moreira et al. [28]

j,k	$A_{DOL,j,k}$	$B_{DOL,j,k}$	$\Gamma_{DOL,j,k}$	$K_{DOL,j,k}$
1,2	1.57	2.51	0.3261	$15.9 \left\lvert \frac{1+14.0X_{DOL,2}^{0.50}}{1+8.04X_{DOL,2}^{1.24}} \right\rvert$
1,3	1.20	1.00	0.0000	$4.82 \left\lvert 1 + 0.0749X_{DOL,1}^{0.95} \right\rvert \left\lvert \frac{1+0.711X_{DOL,3}^{1.22}}{1+0.0433X_{DOL,3}^{1.26}} \right\rvert$
2,0	0.66	2.36	1.3731	$13.8 \left\lvert 1 - 0.0162X_{DOL,2}^{2.07} \right\rvert$
2,1	0.61	2.02	4.5141	$5.03 \left\lvert 1 + 0.0690X_{DOL,2}^{1.11} \right\rvert \left\lvert \frac{1+0.135X_{DOL,1}^{0.91}}{1+2.61X_{DOL,1}^{1.37}} \right\rvert$
3,1	0.60	1.00	0.0000	$0.0479 \left\lvert 1 + 0.0047X_{DCL,3}^{1.10} \right\rvert \left\lvert \frac{1-7.87X_{DOL,1}^{1.45}}{1+2.33X_{DOL,1}^{1.22}} \right\rvert$

The coefficient 37.5 in Equation 17 represents the fraction of eliminated paclitaxel being converted into 6α-hydroxypaclitaxel times the volume of distribution for paclitaxel divided by the volume of distribution for 6α-hydroxypaclitaxel. The weighted percentage variance for 6α-hydroxypaclitaxel was calculated to be $VAR = 0.043$ and the weighted variance was determined to be $S = 0.043$ micromoles squared per liter squared without the paclitaxel. The combined weighted percentage variance for paclitaxel and its metabolite is $VAR = 0.036$. The combined weighted variance is $S = 0.16$ micromoles squared per liter squared for paclitaxel and its metabolite.

These models describe the pharmacokinetic effects of the drugs when there is no interaction between the drugs. The clinical data shows that they do influence each other's metabolism. We intend to find out what the nature of their molecular interactions with the key metabolic enzyme are. In particular, we have hypothesized that both of the drugs compete for the same liver enzyme and the metabolic action of this enzyme, CYP3A4, can be saturated by one type of the drug molecules leaving the other unmetabolized. The drug with a higher binding free energy would have a greater probability of being metabolized. This is discussed in some detail below based on atomic level MD simulations.

The pharmacodynamic interaction of Doxorubicin and Paclitaxel with CYP3A4

The blind-docking simulations showed that both paclitaxel and doxorubicin bind to the surface of CYP3A4. Although the metabolic activity of CYP3A4 requires the binding of

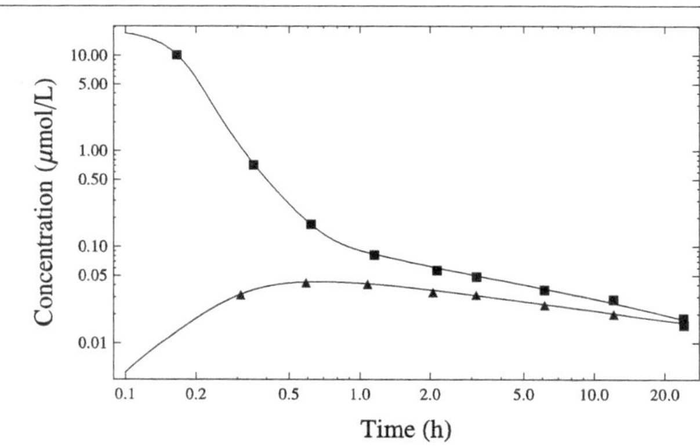

Figure 7 Comparison of our theoretical results with the experimental results of Gianni et al. [27] for doxorubicin and doxorubicinol. The concentration-versus-time data for doxorubicin is represented by the squares and the doxorubicinol data is represented by the triangles. The data is from Gianni et al. [27], with the curve from Equation 11 using the parameters in Tables 3 and 4.

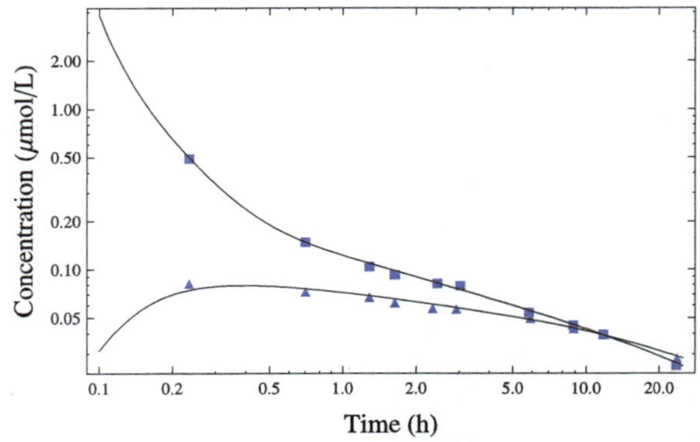

Figure 8 Comparison of theoretical and the experimental results of Moreira et al. [28] for doxorubicin and doxorubicinol. The concentration-versus-time data for doxorubicin is represented by the squares and the doxorubicinol data is represented by the triangles. The data is from Moreira et al. [28], with the curve from Equation 11 using the parameters in Tables 5 and 6.

the substrate within the active site, the binding of the ligand should occur first on the surface of the enzyme. This initial binding would provide an entry point of the substrate to the active site. Based on this assumption, we investigated all possible binding locations for both doxorubicin and paclitaxel on the surface of CYP3A4. A blind docking analysis confirmed the binding of both compounds to CYP3A4. Figure 11 shows the binding locations of paclitaxel and doxorubicin on the surface of CYP3A4. The atoms of the protein are colored using the minimal value of their binding energy to the docked ligands. This procedure allowed us to identify and rank the most probable binding locations of the two drugs. The binding energies for paclitaxel ranged from $-11\ kcal/mol$ to $-5\ kcal/mol$. On the other hand, doxorubicin bound to the surface of CYP3A4 with binding energies ranging from $-8\ kcal/mol$ to $-4\ kcal/mol$. This analysis marked the possible entry points of the two ligands to the CYP3A4 active site. It is also worth mentioning that although the adopted blind docking procedure required considerable number of docking calculations, it has two advantages compared to covering the whole protein with a single docking box. First, the spacing between grid points used by AUTOGRID can be as accurate as a normal docking simulation. Second, there is no need to increase the number of energy evolutions or the population size to proportionally high values. This blind docking protocol

Table 7 The PK parameters for Equation 11 modeling paclitaxel

j,k	$A_{PTX,j,k}$	$B_{PTX,j,k}$	$\Gamma_{PTX,j,k}$	$K_{PTX,j,k}$
1,2	1.06	1.79	0.0001	$13.8\left\|\frac{1+0.0117X_{PTX,2}^{1.05}}{1+3.37X_{PTX,2}^{0.96}}\right\|$
1,3	1.40	1.62	0.0000	$29.1\left\|1-0.0002X_{PTX,1}^{1.68}\right\|\left\|\frac{1+0.0004X_{PTX,3}^{1.24}}{1+0.551X_{PTX,3}^{1.22}}\right\|$
1,4	1.23	1.10	0.0000	$2.85\left\|\frac{1+0.0082X_{PTX,4}^{1.01}}{1+0.117X_{PTX,4}^{1.22}}\right\|$
2,0	1.03	1.99	0.0876	$2.98\left\|1-0.0023X_{PTX,2}^{2.09}\right\|$
2,1	0.94	1.09	0.0028	$0.0523\left\|\frac{1-0.0023X_{PTX,1}^{0.97}}{1+0.0015X_{PTX,1}^{1.32}}\right\|$
3,1	1.66	1.93	0.00003	$0.0230\left\|\frac{1+0.0423X_{PTX,1}^{1.75}}{1+0.0199X_{PTX,1}^{1.60}}\right\|$
4,1	1.78	1.83	0.0009	$0.0061\left\|\frac{1-0.0054X_{PTX,1}^{0.89}}{1+6.36X_{PTX,1}^{0.74}}\right\|$

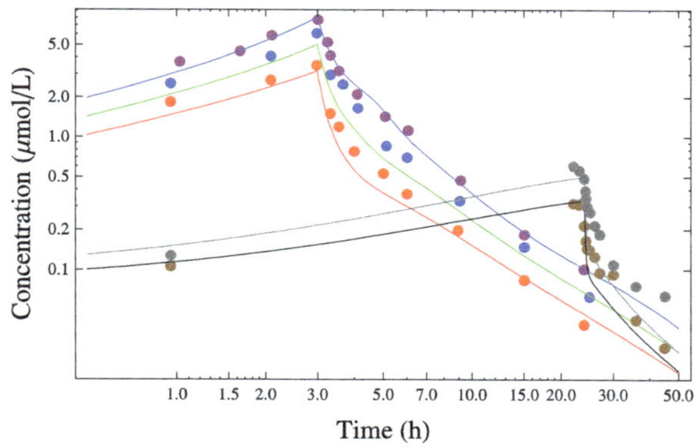

Figure 9 Comparison of theoretical curve and the experimental results of Gianni et al. [32] for paclitaxel. The concentration-versus-time data for paclitaxel is represented by the circles. The data is from Gianni et al. [32]. The curves are calculated using Equation 11 using the parameters in Table 7.

was used earlier to successfully identify the binding location of laulimalide, a microtubule stabilizing agent, on the surface of tubulin [26].

Figure 12 illustrates the binding modes of both paclitaxel and doxorubicin within the CYP3A4 active site. The protein possesses a relatively large substrate-binding cavity that accommodates the bulky paclitaxel substrate. The two compounds exhibited strong binding to the active site, although paclitaxel showed higher binding affinity. The binding energies of paclitaxel and doxorubicin were -12 *kcal*/*mol* and -10 *kcal*/*mol*, respectively. In the docked structures, paclitaxel was closer to the HEM group than doxorubicin. The two ligands formed hydrogen bonds with protein residues. For example, paclitaxel was hydrogen-bonded to residues Ile120, Leu211 and Arg372. Doxorubicin was attached to Phe108, Ile300 and Ala305. The large space available for both the ligands and protein residues allows for huge conformational changes that would accommodate the bound substrate. This correlates with similar available crystal structures of CYP3A4 bound to two distinct families of compounds [34].

Based on the above simulations, it appears that both paclitaxel and doxorubicin strongly bind to the same metabolic enzyme, CYP3A4. Their binding locations are distinct, but

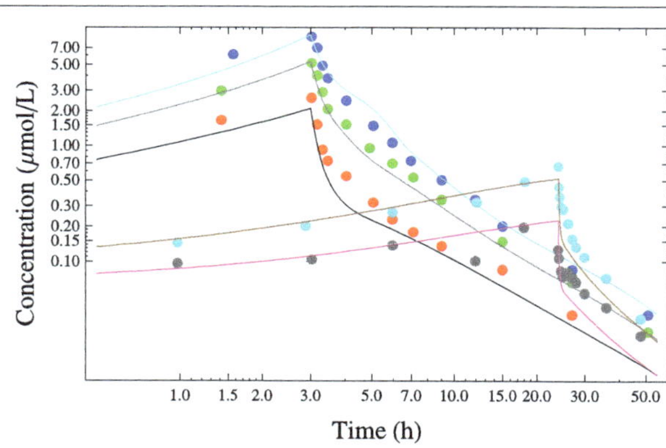

Figure 10 Comparison of theoretical curve and the experimental results of Ohtsu et al. [33] for paclitaxel. The concentration-versus-time data for paclitaxel is represented by the circles. The data is from Ohtsu et al. [33]. The curves were calculated using Equation 11 using the parameters in Table 7.

Table 8 The PK parameters for Equation 11 modeling 6α-hydroxypaclitaxel

j,k	$A_{OHP,j,k}$	$B_{OHP,j,k}$	$\Gamma_{OHP,j,k}$	$K_{OHP,j,k}$
1,2	1	1	21.3666	$12.2\left\|\frac{1-0.523X_{OHP,2}}{1+0.204X_{OHP,2}}\right\|$
1,3	1	1	0.0808	$1.67\left\|1+0.166X_{OHP,3}\right\|$
2,0	1	1	0.0000	23.9
2,1	1	1	0.0000	$0.103\left\|1+0.0182X_{OHP,2}\right\|\left\|1+2.54X_{OHP,1}\right\|$
3,1	1	1	0.0000	$0.0468\left\|1+0.0975X_{OHP,3}\right\|\left\|\frac{1-15.2X_{OHP,1}}{1+28.9X_{OHP,1}}\right\|$

conformational changes associated with the binding process may result in their competitive inhibition of each other. Furthermore, paclitaxel is predicted to bind significantly more strongly to CYP3A4 than doxorubicin giving a greater probability of being metabolized when administered simultaneously with doxorubicin. It is, therefore, reasonable to recommend the administration of doxorubicin after paclitaxel. This is further analyzed and quantified below.

The pharmacokinetic interaction between Doxorubicin and Paclitaxel

The metabolite, doxorubicinol, is produced from doxorubicin by the cytochrome P450 isoenzyme CYP3A4, while paclitaxel's metabolites are produced by the cytochrome P450 isoenzymes CYP2C8 and CYP3A4. Paclitaxel's primary metabolite is produced by CYP2C8 and hence doxorubicin is not expected to have a noticeable effect on the metabolism of paclitaxel. This is verified as illustrated in Figure 13. The three experimental studies shown in Figure 13 involve giving the patients 60 mg/m^2 of doxorubicin over five minutes and then waiting for a period of time before giving the patients 200 mg/m^2 of paclitaxel over a three-hour interval. The time for each set of paclitaxel data in Figure 13 was shifted so the time starts when the paclitaxel begins to be given to the patients. The green data corresponds to the administration of paclitaxel starting fifteen minutes after doxorubicin [27]. The black data corresponds to the administration of paclitaxel starting thirty minutes after doxorubicin [28]. The blue data corresponds to the administration of paclitaxel starting twenty four hours after doxorubicin [28]. The fact that the paclitaxel curves in Figure 13 do not change as a function of the time interval between the doxorubicin and paclitaxel administrations exemplifies that doxorubicin has no apparent effect

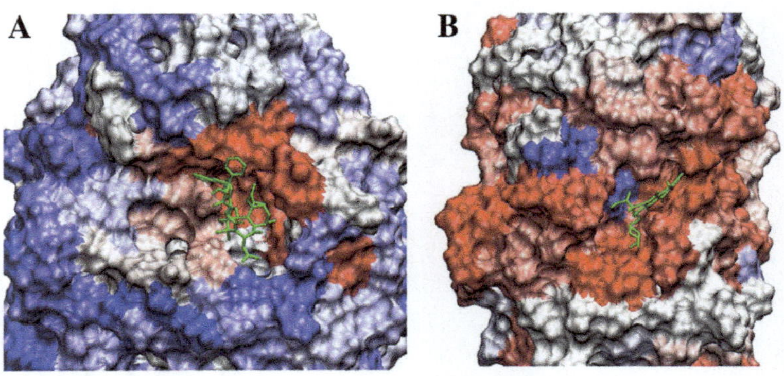

Figure 11 Blind docking of paclitaxel and doxorubicin to CYP3A4. The blind docking illustration of paclitaxel **(A)** and doxorubicin **(B)** to CYP3A4. The red areas are the favorable binding regions and the blue areas are less favorable.

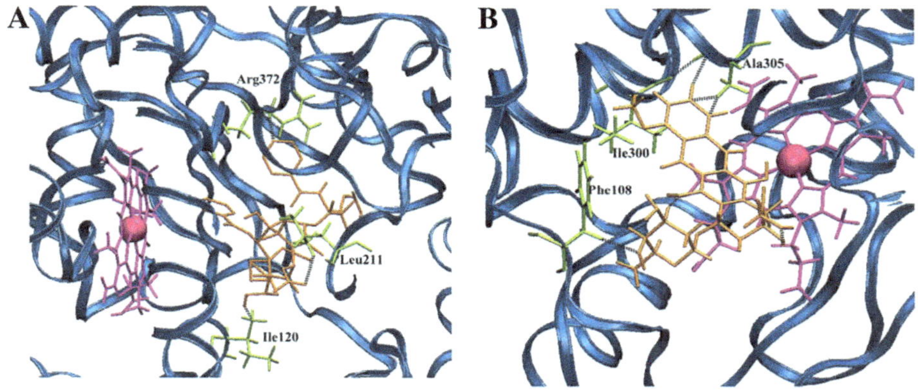

Figure 12 Binding of paclitaxel and doxorubicin within CYP3A4. The binding modes of paclitaxel **(A)** and doxorubicin **(B)** within the active site of CYP3A4.

upon the paclitaxel. The concentration-versus-time in Figure 13 also shows the theoretical curve. The theoretical curve is from Equation 11 using the parameters in Table 7 with no doxorubicin present. The three sets of clinical data and the theoretical curve agree remarkably well, which supports the assumption that doxorubicin has no influence on the paclitaxel.

In contrast, we expect the presence of paclitaxel to have a strong influence on the metabolism of doxorubicin. As predicted in the previous subsection, paclitaxel will bind to CYP3A4 more strongly than doxorubicin and if the two drugs are simultaneously present, we expect paclitaxel to replace doxorubicin that is bound to CYP3A4. The doxorubicin and doxorubicinol parameters given in Tables 3 and 4 are for the best fit to the data from Gianni et al. [27], and the doxorubicin and doxorubicinol parameters in Tables 5 and 6 are for the best fit to the data from Moreira et al. [28]. The influence of paclitaxel on doxorubicin can be included in the model given in Equation 11 by modifying the rate coefficients, $K_{i,j,k}$ given in these tables. The theoretical curve was fitted to the data by minimizing VAR in Equation 12. The number of data points in the Gianni et al. [27] study during the interaction of the drugs is 46 for doxorubicin ($N_p = 568$) and 48

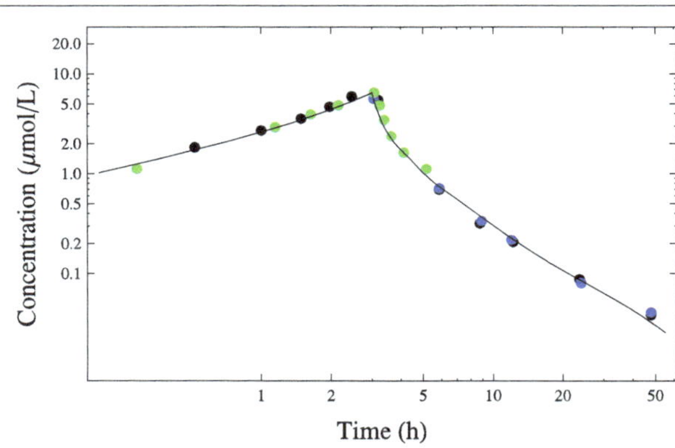

Figure 13 The administration of paclitaxel after doxorubicin. The concentration-versus-time curve for 200 mg/m^2 of paclitaxel given over a three-hour interval after the patient has recieved doxorubicin. The details about the data and the curve can be found in the discussion of the figure.

for doxorubicinol ($N_p = 494$). The weighted percentage variance is $VAR = 0.069$ and the weighted variance is $S = 1.6$ micromoles squared per liter squared for doxorubicin. The weighted percentage variance is $VAR = 0.063$ and the weighted variance is $S = 0.00010$ micromoles squared per liter squared for doxorubicinol. The combined weighted percentage variance for all the drugs (145 data values and $N_p = 1670$) is $VAR = 0.068$ and the combined weighted variance is $S = 1.1$ micromoles squared per liter squared.

The total number of parameters used to minimize VAR was 10 for doxorubicin and 39 for doxorubicinol. These parameters are shown in Equation 18, and Tables 9 (doxorubicin) and 10 (doxorubicinol) for Gianni et al. [27]. In Table 10, the row 1,2 has a term with 0.00008 and the row 3,1 has a term with 0.00003. These eight parameters can be ignored without much lost of information. In addition, the rate at which doxorubicin is being converted into doxorubicinol was found to be

$$I_{DOL,2} = 0.264X_{DOX,2}^{1.33}\left[2.49 + \frac{0.358X_{DOL,2}^{1.17}}{1 + 16.5X_{DOL,2}^{1.28}}\right]$$
$$\left[1 - \frac{0.314X_{PTX,2}}{1 + 0.0119X_{PTX,2}}\right]\left[1 + \frac{0.00611X_{PTX,2}}{1 + 0.00002X_{PTX,2}^{1.74}}\right]. \tag{18}$$

According to these parameters, the presence of the paclitaxel slows down the rate of elimination of the doxorubicin and doxorubicinol from the body. The paclitaxel also increases the kinetic rate at which doxoruciinol is moved from compartment 2 (elimination organs) to compartment 1 (blood/plasma) and reduces the kinetic rate for the reverse flow. The theoretical fit is shown in Figure 14.

The doxorubicin and doxorubicinol parameters given in Tables 5 and 6 are for the best fit to the data from Moreira et al. [28]. The influence of paclitaxel on doxorubicin and doxorubicinol can be included in the model given in Equation 11 by modifying the rate coefficients, $K_{i,j,k}$ given in these tables. The theoretical curve was fitted to the data by minimizing VAR in Equation 12. The number of data points in the Moreira et al. [28] study during the interaction of the drugs is 14 for doxorubicin ($N_p = 392$) and 16 for doxorubicinol ($N_p = 448$). The weighted percentage variance is $VAR = 0.0028$ and the weighted variance is $S = 1.2 \times 10^{-5}$ micromoles squared per liter squared for doxorubicin. The weighted percentage variance is $VAR = 0.028$ and the weighted variance is

Table 9 The PK parameters for Equation 11 modeling doxorubicin for Gianni et al. [27] with paclitaxel present

j,k	$K_{DOX,j,k}$
1,2	$8.14\left\|\frac{1-0.0254X_{DOX,2}^{2.59}}{1+0.319X_{DOX,2}^{1.59}}\right\|\left\|1+\frac{1.70X_{PTX,1}}{1+0.000323X_{PTX,1}}+\frac{0.233X_{PTX,2}}{1+0.00193X_{PTX,2}}\right\|$
1,3	$3.08\left\|\frac{1+3.12X_{DOX,3}^{0.97}}{1+1.45X_{DOX,3}^{0.97}}\right\|$
1,4	$0.359\left\|1+0.0001X_{DOX,1}^{1.55}\right\|\left\|\frac{1+0.277X_{DOX,4}^{1.63}}{1+0.0282X_{DOX,4}^{0.25}}\right\|$
2,0	$0.264\left\|1-\frac{0.314X_{PTX,2}}{1+0.0119X_{PTX,2}}\right\|$
2,1	$0.135\left\|\frac{1+13.9X_{DOX,1}^{0.87}}{1+0.167X_{DOX,1}^{0.66}}\right\|\left\|1+\frac{3.08X_{PTX,1}}{1+0.0441X_{PTX,1}}-\frac{0.0003X_{PTX,2}}{1+0.0078X_{PTX,2}}\right\|$
3,1	$0.0325\left\|\frac{1+2.02X_{DOX,1}^{1.74}}{1+0.143X_{DOX,1}^{0.61}}\right\|$
4,1	$0.0046\left\|\frac{1+7.92X_{DOX,1}^{1.04}}{1+0.243X_{DOX,1}^{1.95}}\right\|$

Table 10 The PK parameters for Equation 11 modeling doxorubicinol for Gianni et al. [27] with paclitaxel present

j,k	$K_{DOL,j,k}$
1,2	$14.0\left\lvert\frac{1+10.6X_{DOL,2}^{0.72}}{1+4.54X_{DOL,2}^{1.67}}\right\rvert\left\lvert 1-\frac{0.00008X_{PTX,1}^{0.39}}{1+0.00007X_{PTX,1}^{3.64}}-\frac{0.0368X_{PTX,2}^{3.10}}{1+0.00001X_{PTX,2}^{0.60}}\right\rvert$
1,3	$3.92\left\lvert 1+0.115X_{DOL,1}^{0.55}\right\rvert\left\lvert\frac{1+1.02X_{DOL,3}^{1.52}}{1+0.0482X_{DOL,3}^{2.43}}\right\rvert\left\lvert 1-\frac{0.00095X_{PTX,1}^{3.26}}{1+0.00006X_{PTX,1}^{5.09}}-\frac{0.0003X_{PTX,3}^{2.64}}{1+0.0001X_{PTX,3}^{2.58}}\right\rvert$
2,0	$8.58\left\lvert 1-0.0167X_{DOL,2}^{1.55}\right\rvert\left\lvert 1-\frac{0.0426X_{PTX,2}^{0.457}}{1+0.00004X_{PTX,2}^{2.28}}\right\rvert$
2,1	$3.90\left\lvert 1+0.0646X_{DOL,2}^{1.18}\right\rvert\left\lvert\frac{1+0.128X_{DOL,1}^{0.95}}{1+3.97X_{DOL,1}^{1.07}}\right\rvert\left\lvert 1+\frac{0.0223X_{PTX,1}^{2.22}}{1+0.00004X_{PTX,1}^{0.21}}+\frac{1.11X_{PTX,2}^{0.52}}{1+0.00001X_{PTX,2}^{1.12}}\right\rvert$
3,1	$0.0594\left\lvert 1+0.0073X_{DOL,3}^{2.26}\right\rvert\left\lvert\frac{1-6.94X_{DOL,1}^{1.17}}{1+3.19X_{DOL,1}^{0.96}}\right\rvert\left\lvert 1-\frac{0.00003X_{PTX,1}^{2.50}}{1+0.00002X_{PTX,1}^{0.60}}+\frac{0.0010X_{PTX,3}^{3.31}}{1+0.00004X_{PTX,3}^{5.16}}\right\rvert$

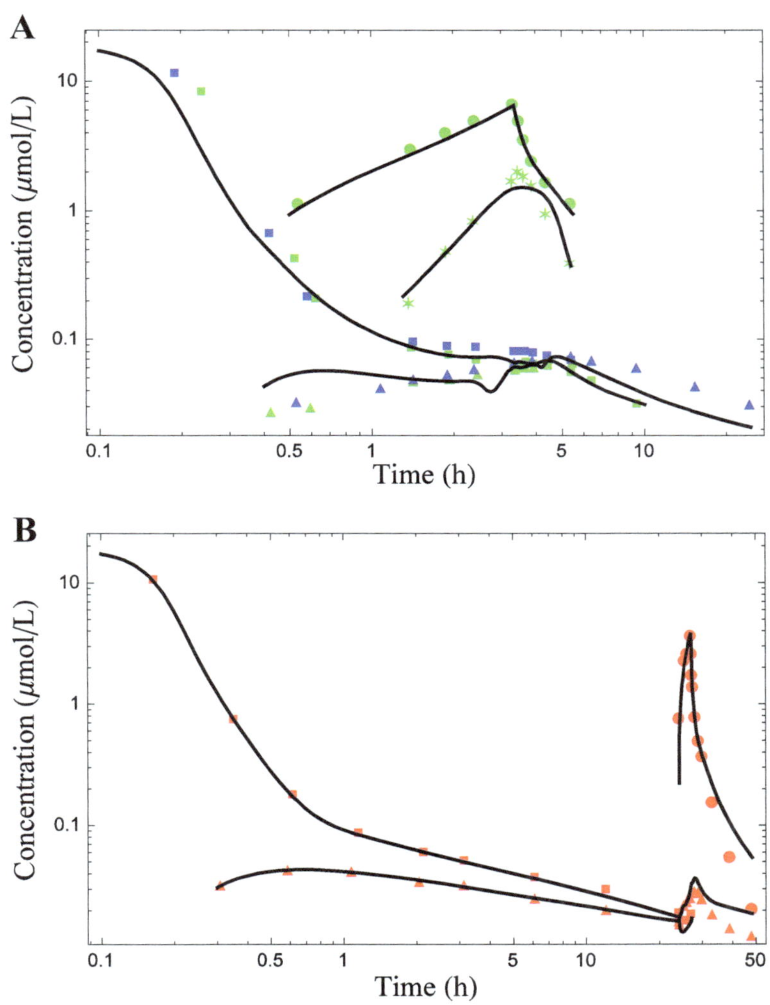

Figure 14 Comparison of our theoretical results with the experimental results of Gianni et al. [27] for doxorubicin and doxorubicinol in the presence of paclitaxel. The squares represent the doxorubicin data, the triangles represent the doxorubicinol data, the circles represent the paclitaxel data, and the stars represent the 6α-hydroxypaclitaxel data. The concentration-versus-time results for different situations are shown in parts **(A)** and **(B)** of the figure, with the black curves being our theoretical predictions. 60 mg/m^2 doxorubicin was given as a 0.083-hour bolus injection, followed by: **(A)** a 0.25-hour interval before 200 mg/m^2 paclitaxel was infused over 3 hours. The green and blue data corresponds to two different sets of patients [27]. **(B)** a 24-hour interval before 150 mg/m^2 paclitaxel was infused over 3 hours [27].

$S = 0.00019$ micromoles squared per liter squared for doxorubicinol. The combined weighted percentage variance for all the drugs (69 data values and $N_p = 1932$) is $VAR = 0.042$ and the combined weighted variance is $S = 0.082$ micromoles squared per liter squared.

The total number of parameters used to minimize VAR was 10 for doxorubicin and 24 for doxorubicinol. These parameters are given in Tables 11 (doxorubicin) and 12 (doxorubicinol) for Moreira et al. [28]. The rate of conversion of doxorubicin into doxorubicinol was found to be:

$$I_{DOL,2} = 0.264X_{DOX,2}^{1.38}\left[3.98 + \frac{0.344X_{DOL,2}^{1.87}}{1 + 9.03X_{DOL,2}^{1.46}}\right] \left[1 - \frac{0.498X_{PTX,2}}{1 + 0.101X_{PTX,2}}\right]\left[1 + \frac{0.800X_{PTX,2}^{1.28}}{1 + 0.00007X_{PTX,2}^{5.89}}\right]. \tag{19}$$

None of the parameters are insignificant. According to these parameters, the presence of the paclitaxel slows down the rate of elimination of the doxorubicin and doxorubicinol from the body. The paclitaxel repels the doxorubicin and doxorubicinol from the compartment. The paclitaxel in comparment 1 increases the rate of flow of doxorubicin and doxorubicinol into compartment 2 from 1 and slows down the flow from compartment 2 into 1. Similarly, the paclitaxel in comparment 2 increases the rate of flow of doxorubicin and doxorubicinol into compartment 1 from 2 and slows down the flow from compartment 1 into 2. The results for Moreira et al. [28] are shown in Figure 15.

The presence of paclitaxel within a compartment causes an increase in the rate at which doxorubicin flows from the compartment to the other compartment(s). In addition, the presence of paclitaxel slows the rate that doxorubicin is being eliminated from the body. The paclitaxel also causes the flow rate of the doxorubicinol to increase. The expulsion of the doxorubicin and doxorubicinol into the blood/plasma compartment is in agreement with what is observed experimentally with the spikes in the concentration-versus-time curves.

The amount of metabolite doxorubicinol being produced is also strongly affected by the presence of paclitaxel. We see from the expression for $I_{DOL,2}$ that the fraction being metabolized into doxorubicinol is increased by the presence of paclitaxel; but the

Table 11 The PK parameters for Equation 11 modeling doxorubicin for Moreira et al. [28] with paclitaxel present

j,k	$K_{DOX,j,k}$
1,2	$8.86\left\lvert\frac{1-0.0231X_{DOX,2}^{3.12}}{1+0.157X_{DOX,2}^{0.89}}\right\rvert\left\lvert 1+\frac{0.0010X_{PTX,1}}{1+0.00002X_{PTX,1}}-\frac{0.0003X_{PTX,2}}{1+0.106X_{PTX,2}}\right\rvert$
1,3	$3.69\left\lvert\frac{1+3.09X_{DOX,3}^{1.02}}{1+1.61X_{DOX,3}^{0.95}}\right\rvert$
1,4	$0.375\left\lvert 1+0.0003X_{DOX,1}^{2.99}\right\rvert\left\lvert\frac{1+0.375X_{DOX,4}^{1.59}}{1+0.0222X_{DOX,4}^{0.31}}\right\rvert$
2,0	$0.264\left\lvert 1+0.00002X_{DOX,2}^{1.90}\right\rvert\left\lvert 1-\frac{0.498X_{PTX,2}}{1+0.101X_{PTX,2}}\right\rvert$
2,1	$0.102\left\lvert\left(1+0.00001X_{DOX,2}^{1.71}\right)\frac{1+11.2X_{DOX,1}^{0.90}}{1+0.181X_{DOX,1}^{0.91}}\right\rvert\left\lvert 1-\frac{0.0010X_{PTX,1}}{1+0.00002X_{PTX,1}}+\frac{0.240X_{PTX,2}}{1+0.350X_{PTX,2}}\right\rvert$
3,1	$0.0369\left\lvert\frac{1+1.74X_{DOX,1}^{0.96}}{1+0.298X_{DOX,1}^{1.48}}\right\rvert$
4,1	$0.0052\left\lvert\left(1-0.00001X_{DOX,4}^{1.14}\right)\frac{1+6.42X_{DOX,1}^{1.02}}{1+0.154X_{DOX,1}^{2.17}}\right\rvert$

Table 12 The PK parameters for Equation 11 modeling doxorubicinol for Moreira et al. [28] with paclitaxel present

j,k	$K_{DOL,j,k}$
1,2	$15.9\left\lvert\frac{1+14.0X_{DOL,2}^{0.50}}{1+8.04X_{DOL,2}^{1.24}}\right\rvert\left\lvert 1+\frac{7.23X_{PTX,1}^{0.66}}{1+0.0070X_{PTX,1}^{1.46}}-\frac{0.701X_{PTX,2}^{0.28}}{1+0.0005X_{PTX,2}^{6.85}}\right\rvert$
1,3	$4.82\left\lvert 1+0.0749X_{DOL,1}^{0.95}\right\rvert\left\lvert\frac{1+0.711X_{DOL,3}^{1.22}}{1+0.0433X_{DOL,3}^{1.26}}\right\rvert$
2,0	$13.8\left\lvert 1-0.0162X_{DOL,2}^{2.07}\right\rvert\left\lvert 1-\frac{0.131X_{PTX,2}^{1.37}}{1+0.0004X_{PTX,2}^{6.68}}\right\rvert$
2,1	$5.03\left\lvert 1+0.0690X_{DOL,2}^{1.11}\right\rvert\left\lvert\frac{1+0.135X_{DOL,1}^{0.91}}{1+2.61X_{DOL,1}^{1.37}}\right\rvert\left\lvert 1-\frac{0.0434X_{PTX,1}^{1.10}}{1+0.0103X_{PTX,1}^{2.41}}+\frac{10.9X_{PTX,2}^{1.12}}{1+0.00007X_{PTX,2}^{6.94}}\right\rvert$
3,1	$0.0479\left\lvert 1+0.0047X_{DOL,3}^{1.10}\right\rvert\left\lvert\frac{1-7.87X_{DOL,1}^{1.45}}{1+2.33X_{DOL,1}^{1.22}}\right\rvert$

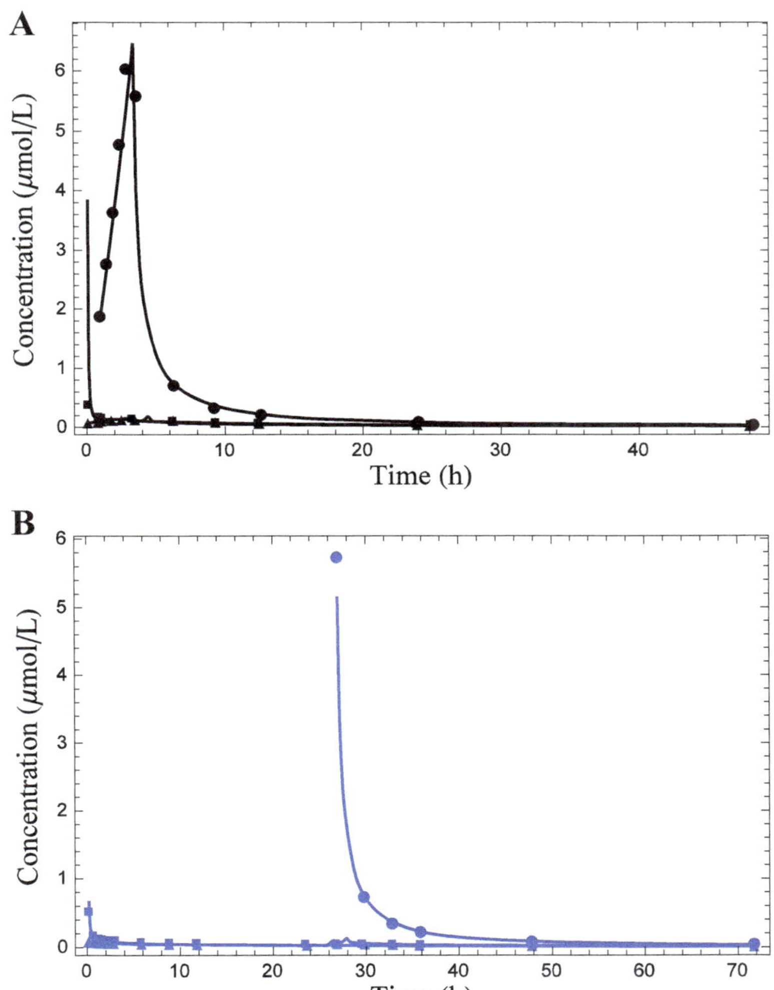

Figure 15 Comparison of theoretical and the experimental results of Moreira et al. [28] for doxorubicin and doxorubicinol in the presence of paclitaxel. The squares represent the doxorubicin data, the triangles represent the doxorubicinol data, and the circles represent the paclitaxel data. The concentration-versus-time results for different situations are shown in parts **(A)** and **(B)** of the figure with the black curves being our theoretical predictions. 60 mg/m^2 doxorubicin was given as a 0.083-hour bolus injection followed by: **(A)** 0.5-hour interval before 200 mg/m^2 paclitaxel was infused over 3 hours. **(B)** 24-hour interval before 200 mg/m^2 paclitaxel was infused over 3 hours [28].

amount of doxorubicin in compartment 2 available to be converted into doxorubicinol is reduced as well as the amount being eliminated. Doxorubicinol is a highly cardiotoxic metabolite and the reduction of its production will reduce the chances of serious side effects [35].

The parameters in Tables 9 and 10 compared to Tables 11 and 12 are similar with a few differences. These differences do not change the overall qualitative behavior and produces similar results. For example, if doxorubicin is present and then paclitaxel is introduced to the body then both predict that the flow rate of doxorubicin into the elimination organs compartment will increase. Once the paclitaxel concentration in the elimination organ has increased to a sufficient level then the flow rate out into the blood plasma has increased more than the flow rate into the organ, which causes a spike in the concentration levels seen in the experimental data.

Our findings correlate well with the experimental data and provide computational and conceptual support for a possible binding of the two drugs to CYP3A4. The question that still remains is: *what is the optimal time delay in the sequence of administration of the drugs?* To answer this question, we calculate the AUC, which is defined as;

$$AUC_{i,j} = \int_{-\infty}^{\infty} X_{i,j} dt. \tag{20}$$

To investigate the time sequence we use the dosage introduced by Gianni et al. [27] and Moreira et al. [28]. In other words, 60 mg/m^2 of doxorubicin is administered in a five-minute IV bolus and 150 mg/m^2 or 200 mg/m^2 of paclitaxel in a three-hour IV. The AUC is integrated over the period of infusion plus a one-week interval after the infusion has stopped instead of calculating the integral over all time. The portion of the curve neglected after a week has a very small contribution to the AUC, which does not change the results significantly.

The AUC for doxorubicin and doxorubicinol was calculated for the blood plasma compartment (1). In the case when doxorubicin is given first and paclitaxel is administered to the patient afterwards, the AUC of the doxorubicin is larger the closer in time the two are administered at. Unfortunately, the AUC of doxorubicinol is also very large the closer the two drugs are administered to each other reaching a maximum when both started at the same time. Therefore, we expect the toxicity of doxorubicin and doxorubicinol to be enhanced with this drug sequence. In the case when the order is switched and paclitaxel is given first there are two different situations. The first situation is when the doxorubicin is given during the first three hours while the paclitaxel is being administered. The AUC of the doxorubicin is a maximum during this interval. If the doxorubicin is given after the third hour the AUC becomes smaller the larger the interval between when the paclitaxel finishes and the doxorubicin is given. The AUC of the doxorubicinol is reduced by having the doxorubicin given after the paclitaxel. This reduces the toxicity of doxorubicinol while maximizing the effects of doxorubicin. Therefore, we maximize the effects of the combination if paclitaxel is given over three hours followed by doxorubicin given between 1 to 4 hours after the paclitaxel is started.

We conclude that doxorubicin should be administered optimally within four hours after the paclitaxel infusion is complete. These results suggest that the efficacy of doxorubicin can be increased by giving the patient paclitaxel prior to infusion of doxorubicin.

Conclusions

One of the major challenges in dose optimization is nonlinear behavior in one or more drug processes. In the current study, we investigated new ways to assess and quantify nonlinear pharmacokinetic behavior, with emphasis on their origins and therapeutic applications of the drug combination involving paclitaxel and doxorubicin. We have justified the development of a saturable compartment model with competing interactions between the two drugs by demonstrating through molecular dynamics simulations that they compete for the same metabolic enzyme with different binding affinities.

Molecular pathways are complex and the correlation of these pathways with malignancy is still an open issue, a complication that has led to systems biology approaches to cancer research [36]. While the system biology approach is promising, we focused on a well-understood and time-tested approach involving pharmacokinetic equations for a multi-compartment model leaving a more complex issue of multiple molecular pathways for future development.

Since no single drug is sufficiently efficacious to become a "silver bullet" cure for large patient populations, combinations of drugs addressing different molecular pathways appear to be capable of increasing patient survival. Optimization of combinations in terms of their sequences, schedules and dosages is a daunting task and needs a formal mathematical development. In the present paper we focused on a specific combination of two drugs and made the following two assumptions. First, saturation kinetics is in principle allowed in all compartments. Second, drug interactions are included to the extent of affecting kinetic rates of other drugs; specifically some drugs may either enhance or inhibit the transition kinetics of others depending on the built-in parameter values.

We have developed a relatively simple but robust four-compartment saturable kinetics model that has been shown to reproduce a number of empirical data sets very well. This is of practical importance in view of the toxic effects these drugs have on patients. In particular, it has been reported that doxorubicin is strongly cardio-toxic and doxorubicinol is ten times more cardio-toxic than doxorubicin [37]. Therefore, by keeping doxorubicin from being metabolized by liver enzymes, the drug will be more effective at killing the tumor as well as being less toxic to the body. This suggests that the optimal sequence is to give paclitaxel first and then almost immediately afterwards, give a dose of doxorubicin to the patient.

Appendix A: Doxorubicin experimental data

Equation 11 is the basis of our pharmacokinetic model and was fitted to several data sets that were digitized from concentration-versus-time curves. Some of the data represent mean values obtained from multiple patient sets, some of the data come from a representative patient out of a group, and some of the data describe a single patient. Due to this diversity of origin comparing different data sets is fraught with problems. However, our intention was to demonstrate the robustness of our model in fitting the data. In total, we have digitized 463 data points for doxorubicin. Multiplying the data points that represent mean values by the number of patients gave 1341 patient-data points. Similarly, for the metabolite doxorubicinol, we obtained 126 data points, which become 869 patient-data

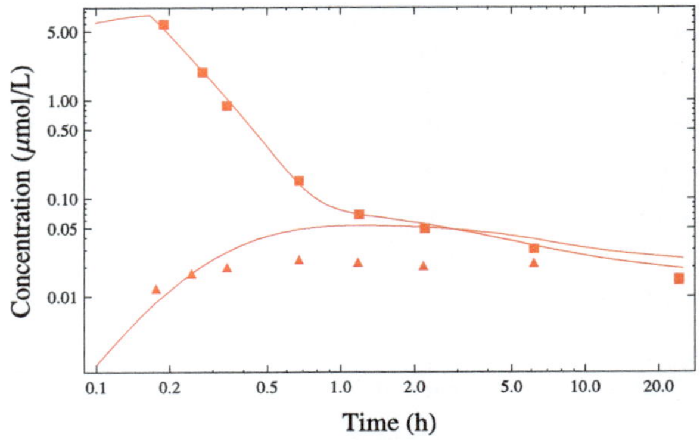

Figure 16 Doxorubicin and doxorubicinol from Andersen et al. Log-Log plot of doxorubicin and doxorubicinol. The doxorubicin data (squares) and the doxorubicinol data (triangles) are from Andersen et al. [30].

points when multiplied by the number of patients. The figures below show the individual data sets and our best fit.

The first set of data is from Andersen et al. [30]. This study involved 24 patients, each receiving 50 mg/m^2 of doxorubicin in a 10-minute IV infusion. Figure 16 is a log-log plot of their doxorubicin and doxorubicinol data compared with the theoretical curve using Equation 11 with the parameter values given in Tables 1 and 2.

The second data set is from Greene et al. [31]. This study involved 10 patients, each receiving 75 mg/m^2 of doxorubicin in a 15-minute IV infusion. Figure 17 is a log-log plot of their doxorubicin and doxorubicinol data compared with the theoretical curve using Equation 11 with the parameter values given in Tables 1 and 2.

The third data set is from Benjamin et al. [29]. This paper covered two separate studies. The first study involved 16 patients, each receiving 60 mg/m^2 of doxorubicin in a 1 to 5 minute IV infusion. The second study involved 13 patients, each receiving 60 mg/m^2 of doxorubicin in a 2 to 10 minute IV infusion. However, in the second study, doxorubicin and doxorubicinol was extracted from the plasma using chloroform: isopropyl alcohol (1:1, v/v). Figure 18 is a log-log plot of their doxorubicin and doxorubicinol data compared

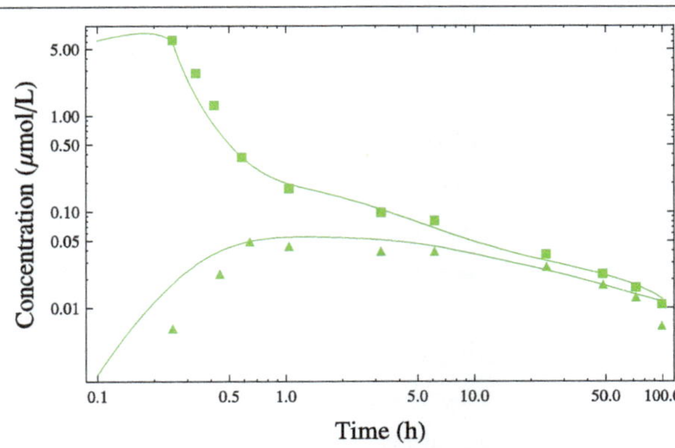

Figure 17 Doxorubicin and doxorubicinol from Greene et al. Log-Log plot of doxorubicin and doxorubicinol. The doxorubicin data (squares) and the doxorubicinol data (triangles) are from Greene et al. [31].

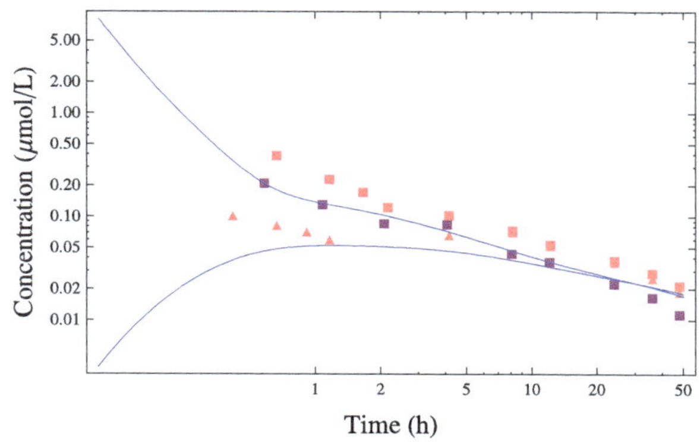

Figure 18 Doxorubicin and doxorubicinol from Benjamin et al. Log-Log plot of doxorubicin and doxorubicinol. The doxorubicin data (squares) and the doxorubicinol data (triangles) are from Benjamin et al. [29].

with the theoretical curve using Equation 11 with the parameter values given in Tables 1 and 2. The pink data correspond to the chloroform data and the purple was from the acid:alcohol extraction study.

The fourth data set is from Gianni et al. [27]. This study focused mainly on the interaction of doxorubicin with paclitaxel, but one of the published concentration-versus-time curves was for doxorubicin without any paclitaxel for the first 24 hours. This data involved 8 patients, each receiving 60 mg/m^2 of doxorubicin in a 5-minute IV infusion. Figure 19 is a log-log plot of their doxorubicin and doxorubicinol data compared with the theoretical curve using Equation 11 with the parameter values given in the text.

The fifth data set is from Moreira et al. [28]. This study focused mainly on the interaction of doxorubicin with paclitaxel, but one of the published concentration-versus-time curves was for doxorubicin without any paclitaxel for the first 24 hours. This data involved 28 patients, each receiving 60 mg/m^2 of doxorubicin in a 5-minute IV infusion. Figure 20 is a log-log plot of their doxorubicin and doxorubicinol data compared with the theoretical curve using Equation 11 with the parameter values given in Tables 1 and 2.

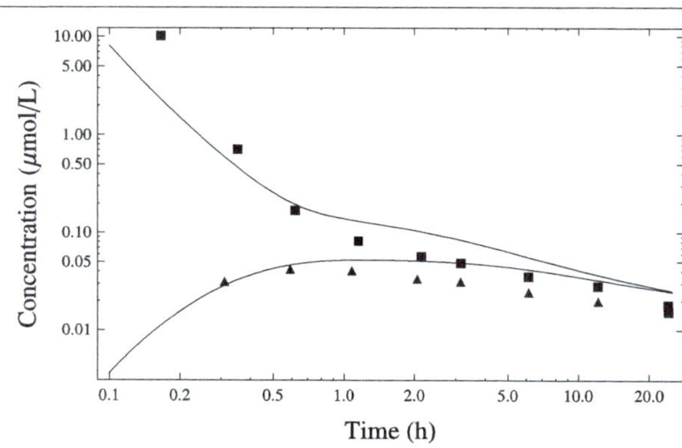

Figure 19 Doxorubicin and doxorubicinol from Gianni et al. Log-Log plot of doxorubicin and doxorubicinol. The doxorubicin data (squares) and the doxorubicinol data (triangles) are from Gianni et al. [27].

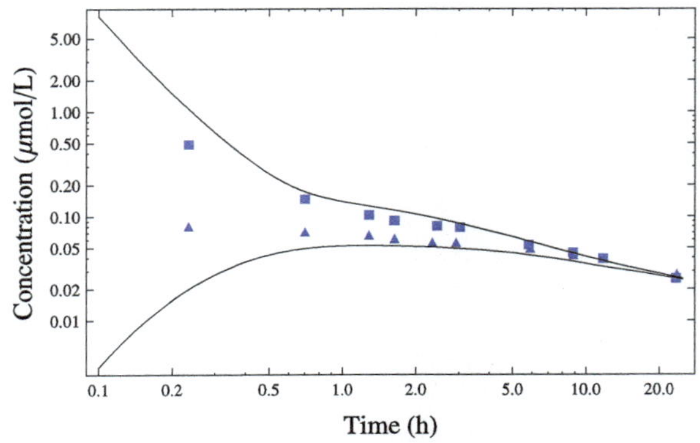

Figure 20 Doxorubicin and doxorubicinol from Moreira et al. Log-Log plot of doxorubicin and doxorubicinol. The doxorubicin data (squares) and the doxorubicinol data (triangles) are from Moreira et al. [28].

The sixth data set is from Di Fronzo et al. [38]. This study involved six patients, each receiving 0.5 mg/kg of doxorubicin. Figure 21 is a plot of their doxorubicin data compared with the theoretical curve using Equation 11 with the parameter values given in Tables 1 and 2.

The seventh, eighth, ninth, tenth and eleventh data sets are from Jacquet et al. [39]. This study involved 18 patients, each receiving 25 to 75 mg/m^2 of doxorubicin in 5 to 15 minute IV infusion. The seventh study involved one patient and shows two courses of doxorubicin, each having a dose of 51.37 mg/m^2 of doxorubicin in a 10-minute IV infusion. Figure 22 is a log-log plot of their doxorubicin data compared with the theoretical curve using Equation 11 with the parameter values given in Tables 1 and 2. The eighth study involves the same patient as the one in the seventh study and shows four courses of doxorubicin, each having a dose of 51.37 mg/m^2 of doxorubicin in a 25-minute IV infusion, shown in Figure 23. The ninth study involves a single patient and shows four courses of doxorubicin, each having a dose of 51.12 mg/m^2 of doxorubicin in a 15-minute IV infusion, which is illustrated in Figure 24. Figure 25 shows the tenth study, which involves the same patient as the ninth study and shows three courses of doxoru-

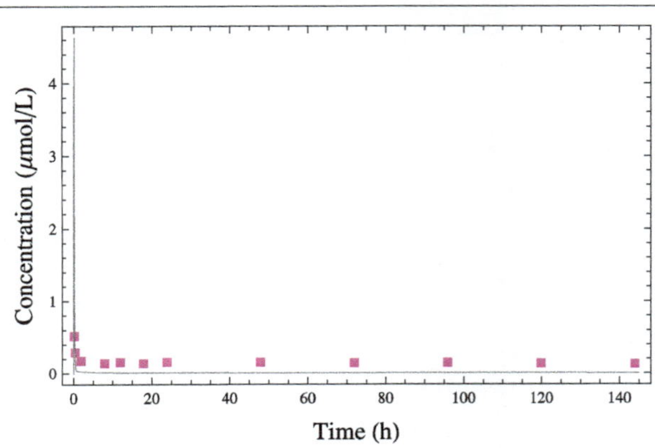

Figure 21 Doxorubicin and doxorubicinol from Di Fronzo et al. Plot of doxorubicin data from Di Fronzo et al. compared to the theoretical curve [38].

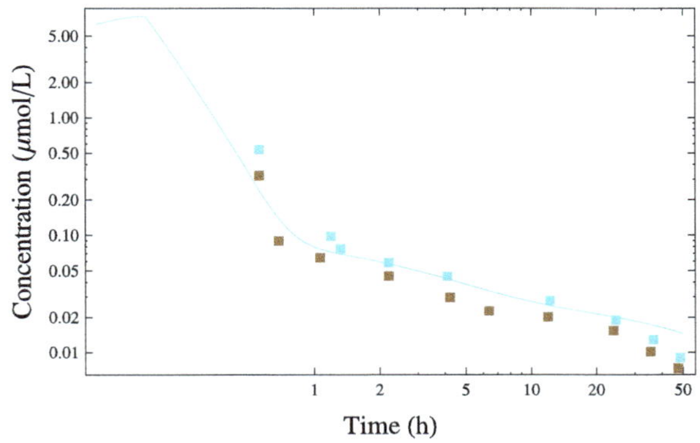

Figure 22 10-minute IV of 51.37 mg/m^2 Doxorubicin. The squares represent the doxorubicin data from Jacquet et al. for 51.37 mg/m^2 of doxorubicin during a 10-minute IV infusion [39].

bicin, each having a dose of 51.12 mg/m^2 of doxorubicin in a 10-minute IV infusion. The eleventh study involves a patient who received three courses of doxorubicin, each having a dose of 45.9 mg/m^2 of doxorubicin in a 15-minute IV infusion, which is shown in Figure 26.

The twelfth data set is from Bressole et al. [40]. This study involved a single patient receiving 80 mg of doxorubicin during each course. We assumed that the body surface area was 1.6m^2 and the infusion was over one hour. Figure 27 is a log-log plot of their doxorubicin data compared with the theoretical curve.

The thirteenth data set is from Eksborg et al. [41]. Figure 28 is a log-log plot of their doxorubicin data compared with the theoretical curves using Equation 11 with the parameter values given in Tables 1 and 2. All the data are for individual patients. The blue data are for a dose of 33.8 mg/m^2 in a 45-minute IV infusion. The pink data are for a dose of 49.3 mg/m^2 in a 2-hour IV infusion. The red data are for a dose of 49.3 mg/m^2 in a 4-hour IV infusion. The green data are for a dose of 30 mg/m^2 in an 8-hour IV infusion. We assumed a 7.5-hour infusion for the theoretical curve. The black data are for a dose of 30 mg/m^2 in a 16-hour IV infusion.

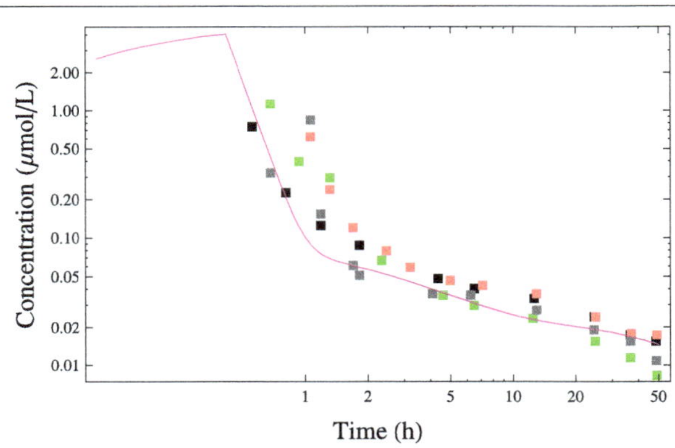

Figure 23 25-minute IV of 51.37 mg/m^2 Doxorubicin. The squares represent the doxorubicin data from Jacquet et al. for 51.37 mg/m^2 of doxorubicin during a 25-minute IV infusion [39].

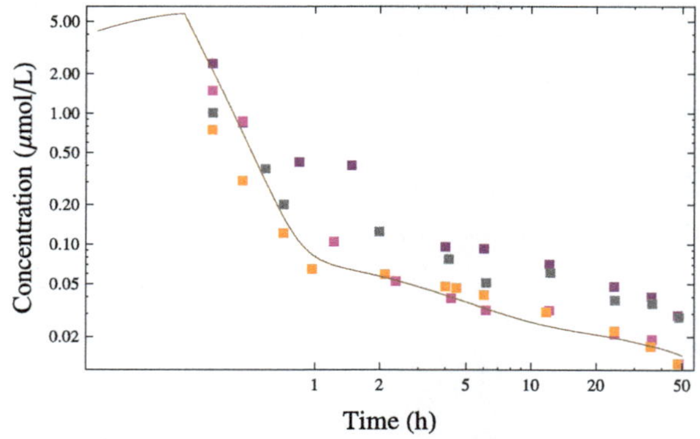

Figure 24 15-minute IV of 51.12 mg/m^2 Doxorubicin. The squares represent the doxorubicin data from Jacquet et al. for 51.12 mg/m^2 of doxorubicin during a 15-minute IV infusion [39].

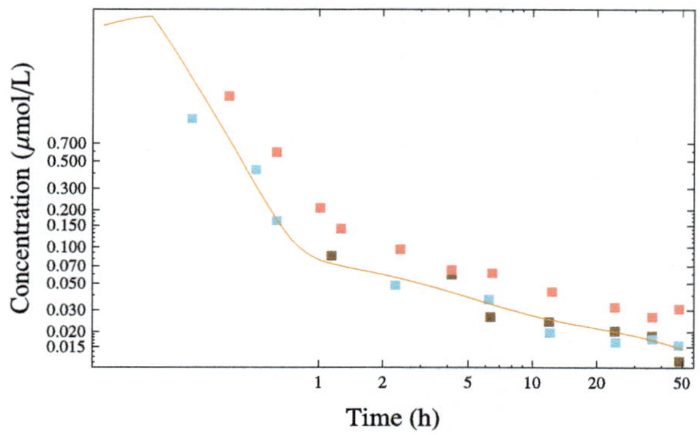

Figure 25 10-minute IV of 51.12 mg/m^2 Doxorubicin. The squares represent the doxorubicin data from Jacquet et al. for 51.12 mg/m^2 of doxorubicin during a 10-minute IV infusion [39].

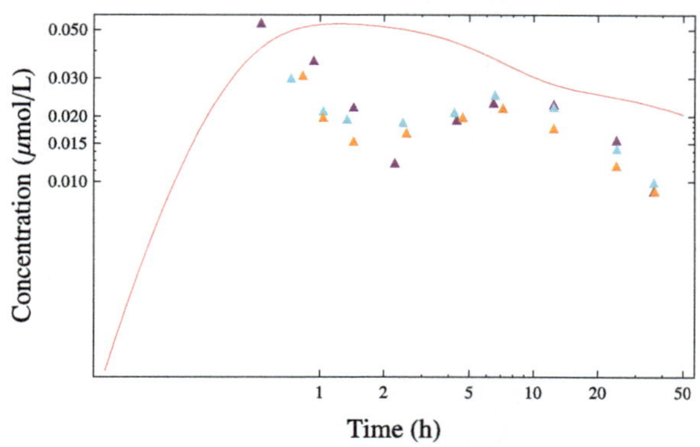

Figure 26 15-minute IV or 45.9 mg/m^2 Doxorubicinol. The triangles represent the doxorubicinol data from Jacquet et al. for 45.9 mg/m^2 of doxorubicin during a 15-minute IV infusion [39].

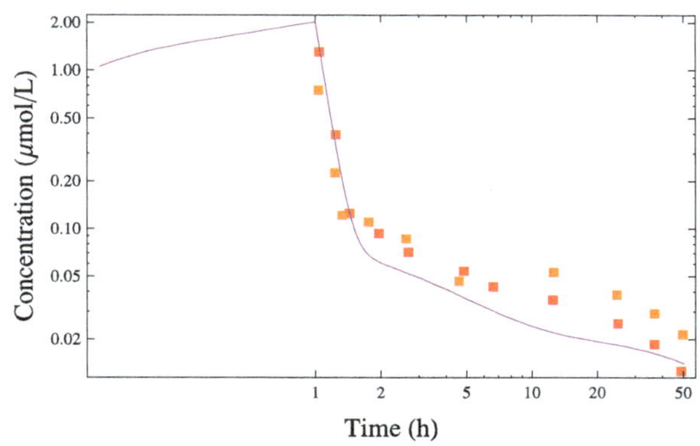

Figure 27 Doxorubicin from Bressole et al. The squares are the doxorubicin data from the Bressole et al. study [40]. The orange data is the third course and the red data is the sixth course.

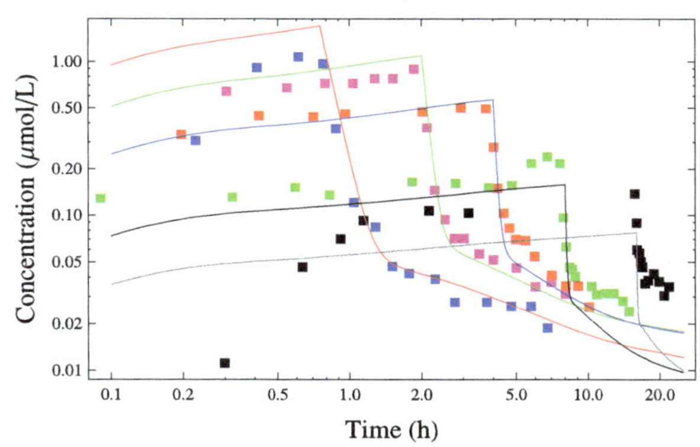

Figure 28 Doxorubicin from Eksborg et al. The doxorubicin data represented by the squares is from Eksborg et al. [41].

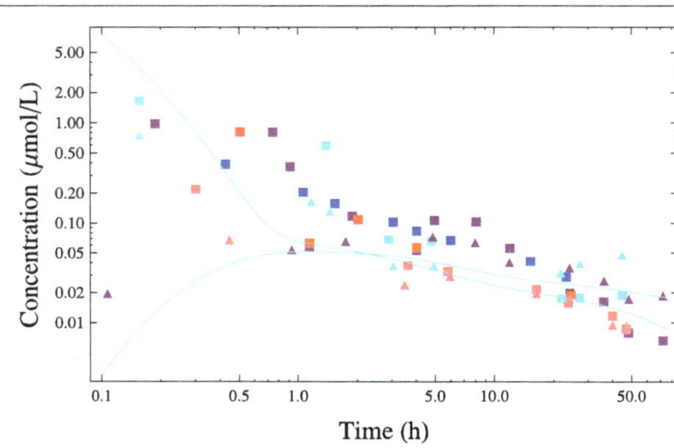

Figure 29 Doxorubicin and doxorubicinol from Chan et al. A dosage of 45 mg/m^2 of doxorubicin was administered to patients. The doxorubicin data represented by the squares and the doxorubicinol data represented by the triangles are from Chan et al. [42].

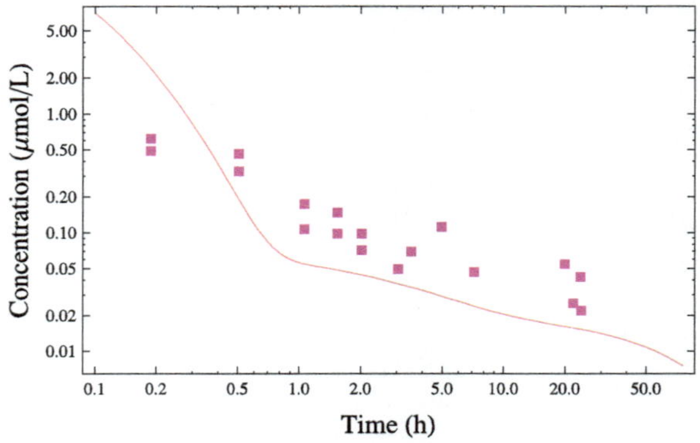

Figure 30 Doxorubicin from Chan et al. A dosage of 40 mg/m^2 of doxorubicin was administered to patients. The doxorubicin data represented by the squares and the doxorubicinol data represented by the triangles are from Chan et al. [42].

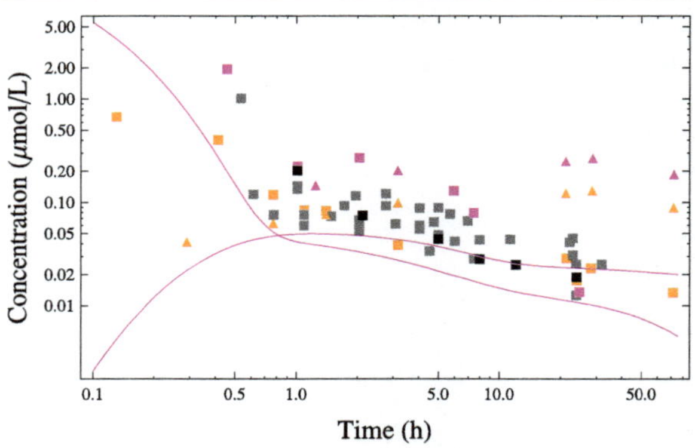

Figure 31 Doxorubicin and doxorubicinol from Chan et al. A dosage of 30 mg/m^2 of doxorubicin was administered to patients. The doxorubicin data represented by the squares and the doxorubicinol data represented by the triangles are from Chan et al. [42].

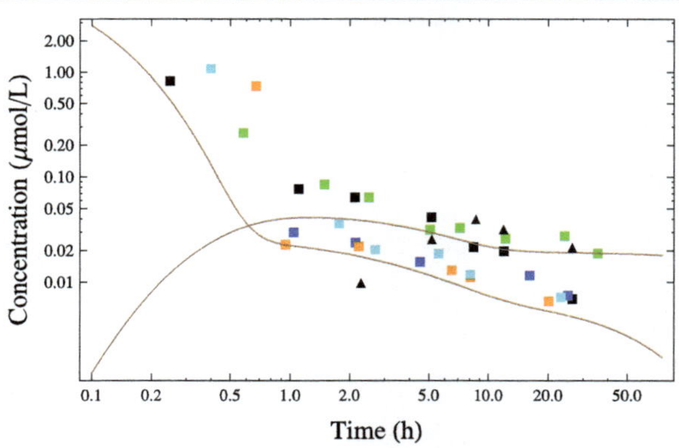

Figure 32 Doxorubicin and doxorubicinol from Chan et al. A dosage of 15 mg/m^2 of doxorubicin was administered to patients. The doxorubicin data represented by the squares and the doxorubicinol data represented by the triangles are from Chan et al. [42].

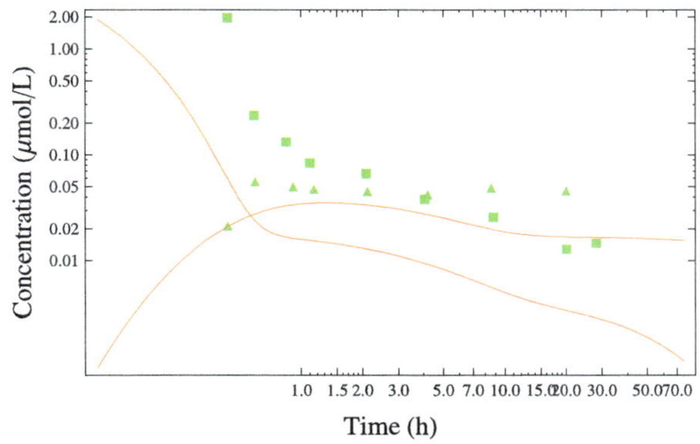

Figure 33 Doxorubicin and doxorubicinol from Chan et al. A dosage of 10 mg/m^2 of doxorubicin was administered to patients. The doxorubicin data represented by the squares and the doxorubicinol data represented by the triangles are from Chan et al. [42].

The fourteenth, fifteenth, sixteenth, seventeenth and eighteenth data sets are from Chan et al. [42]. All the theoretical curves are obtained from Equation 11 with the parameter values given in Tables 1 and 2. The fourteenth study shown in Figure 29 shows doxorubicin and doxorubicinol data for five patients, each receiving 45 mg/m^2 of doxorubicin in a bolus IV. The fifteenth study shown in Figure 30 shows several courses given to a single patient, receiving 40 mg/m^2 of doxorubicin in a bolus IV. Figure 31 shows four patients data, each receiving 30 mg/m^2 of doxorubicin in a bolus IV. The gray data represents a few courses. Figure 32 shows data for four patients, each receiving 15 mg/m^2 of doxorubicin in a bolus IV. Two courses of doxorubicin are shown for one of the patients. Figure 33 shows a patient receiving 10 mg/m^2 of doxorubicin in a bolus IV.

Appendix B: Paclitaxel experimental data

Equation 11 is the basis of our pharmacokinetic model and was fitted to several sets of paclitaxel data that were digitized from figures representing experimental concentration-versus-time curves. Some of the data represents mean values obtained from multiple patients, some of the data came from a representative patient out of a group, and some of the data describe a single patient. Due to this diversity of origin, we weighted each data set based on the number of patients it represented. We did not attempt to fit to a single set of data, but used all the data; our intention was to demonstrate the robustness of our model in fitting the data. In total, we have digitized 352 data points for paclitaxel. If we multiply the data points that represent mean values by the number of patients then we have 1104 patient-data points. Similarly, for the metabolite 6α-hydroxylpaclitaxel, we obtained 20 data points, which is still 20 patient-data points when multiplied by the number of patients. The lack of data for the metabolite means we can use only 19 parameters and the fit is not as robust as it could have been with more experimental data. The figures below show the individual data sets and our best fits using Equation 11 with the parameter values given in Tables 7 and 8.

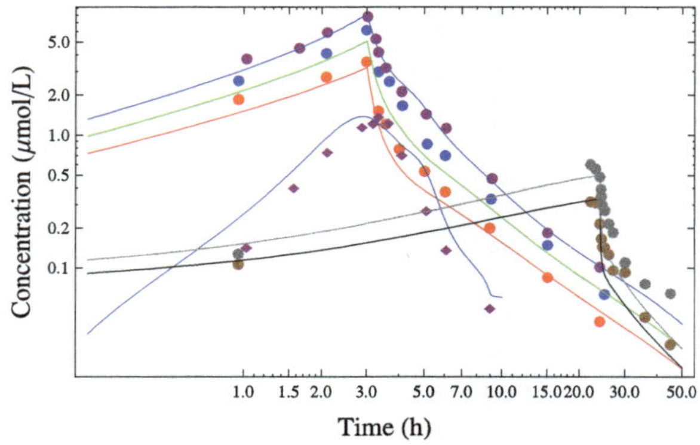

Figure 34 Paclitaxel and 6α-hydroxylpaclitaxel from Gianni et al. The paclitaxel data represented by the circles and the 6α-hydroxylpaclitaxel data represented by the diamonds are from Gianni et al. [32].

The first data set for paclitaxel is from Gianni et al. [32]. This study involved 30 patients; four patients received 135 mg/m^2 during a three-hour IV infusion, three patients received 175 mg/m^2 during a three-hour IV infusion, 15 patients received 225 mg/m^2 during a three-hour IV infusion, four patients received 135 mg/m^2 during a twenty four-hour IV infusion and 4 patients received 175 mg/m^2 during a twenty four-hour IV infusion. Figure 34 is for a single representative patient for each dosage.

The second data set is from Monsarrat et al. [43]. This study involved a female patient receiving 135 mg/m^2 during a three-hour IV infusion shown in Figure 35.

The third data set is from Mross et al. [44]. This study involved thirty patients; six patients received 150 mg/m^2 during an one-hour IV infusion, six patients received 175 mg/m^2 during an one-hour IV infusion, six patients received 200 mg/m^2 during an one-hour IV infusion, six patients receiving 225 mg/m^2 during an one-hour IV infusion and six patients receiving 250 mg/m^2 during an one-hour IV infusion shown in Figure 36. The experimental data represent the median value of the concentrations of the six patients.

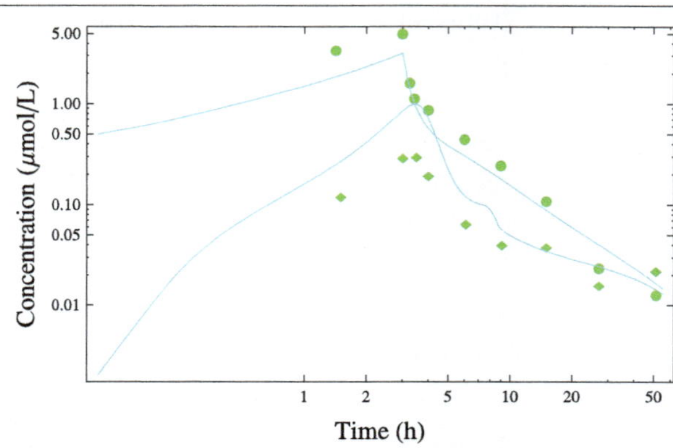

Figure 35 Paclitaxel and 6α-hydroxylpaclitaxel from Monsarrat et al. The paclitaxel data represented by the circles and the 6α-hydroxylpaclitaxel data represented by the diamonds are from Monsarrat et al. [43].

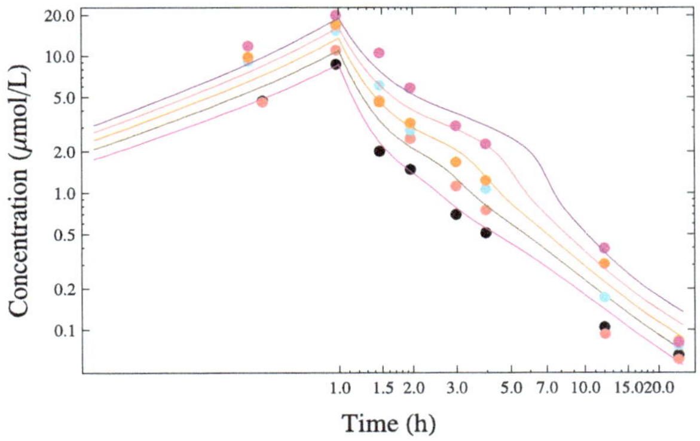

Figure 36 Paclitaxel from Mross et al. The paclitaxel data represented by the circles is from Mross et al. [44].

The fourth data set is from Brown et al. [45]. This study involved eighteen patients; seven patients receiving 175 mg/m^2 during a six-hour IV infusion, six patients receiving 250 mg/m^2 during a six-hour IV infusion, and five patients receiving 275 mg/m^2 during a six-hour IV infusion, which is shown in Figure 37. Each set of experimental data in the figure corresponds to a single representative patient from the cohort.

The fifth data set is from Ohtsu et al. [33]. This study involved twenty two patients; three patients receiving 105 mg/m^2 during a three-hour IV infusion, three patients receiving 180 mg/m^2 during a three-hour IV infusion, seven patients receiving 240 mg/m^2 during a three-hour IV infusion, three patients receiving 105 mg/m^2 during a twenty four-hour IV infusion and six patients receiving 180 mg/m^2 during a twenty four-hour IV infusion. The results are shown in Figure 38. The experimental data represent the mean values of the concentrations measured in the patients.

The sixth data set is from Longnecker et al. [46]. This study involved eight patients; two patients receiving 60 mg/m^2 during an one-hour IV infusion, three patients receiving 170 mg/m^2 during a six-hour IV infusion, and three patients receiving 265 mg/m^2 during a six-hour IV infusion, which is shown in Figure 39. The experimental data represent individual patients from the study.

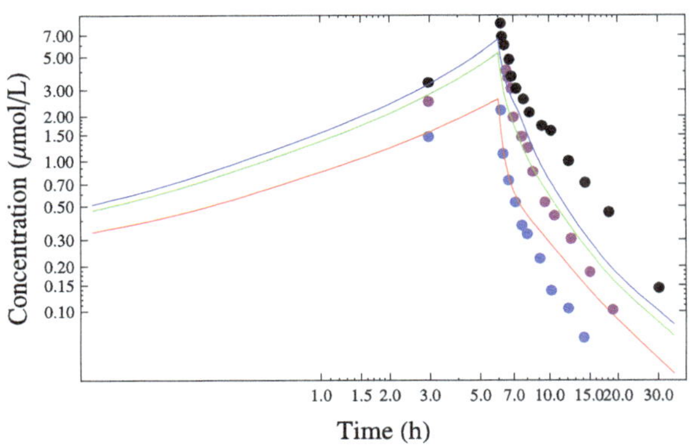

Figure 37 Paclitaxel from Brown et al. The paclitaxel data represented by the circles is from Brown et al. [45].

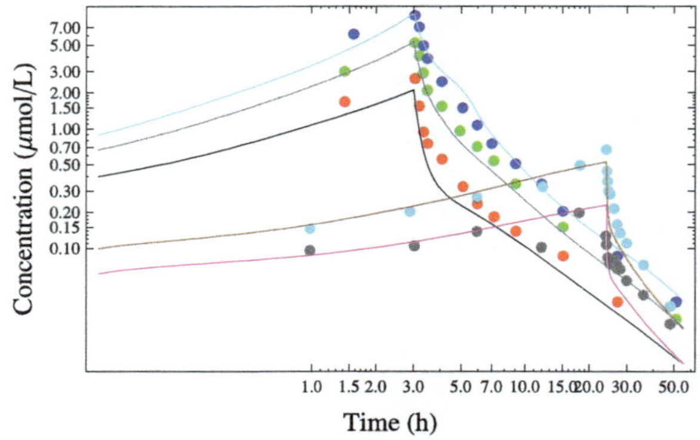

Figure 38 Paclitaxel from Ohtsu et al. The paclitaxel data represented by the circles is from Ohtsu et al. [33].

The seventh data set in Figure 40 is from Gelmon et al. [47]. This study involved five patients receiving 250 mg/m^2 during a three-hour IV infusion. The experimental data represents a single patient out of the five patients in the study.

The eighth data set shown in Figure 41 is from Wiernik et al. [48]. This study involved twelve patients; four patients receiving 175 mg/m^2 during a six-hour IV infusion, two patients receiving 200 mg/m^2 during a six-hour IV infusion, three patients receiving 230 mg/m^2 during a six-hour IV infusion and three patients receiving 275 mg/m^2 during a six-hour IV infusion. The experimental data represent the mean values of the concentrations.

The ninth data set from Maier-Lenz et al. [49] is shown in Figure 42. This study involved thirty patients; four patients receiving 150 mg/m^2 during an one-hour IV infusion, four patients receiving 175 mg/m^2 during an one-hour IV infusion, thirteen patients receiving 200 mg/m^2 during an one-hour IV infusion, eight patients receiving 225 mg/m^2 during an one-hour IV infusion and three patients receiving 250 mg/m^2 during an one-hour IV infusion. The experimental data represent the mean values of the concentrations.

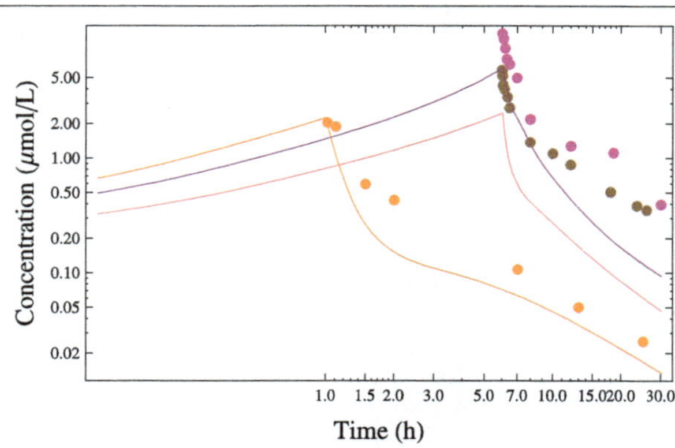

Figure 39 Paclitaxel from Longnecker et al. The paclitaxel data represented by the circles is from Longnecker et al. [46].

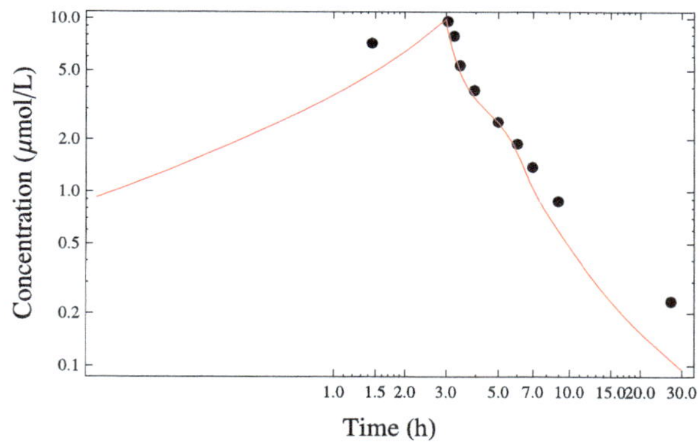

Figure 40 Paclitaxel from Gelmon et al. The paclitaxel data represented by the circles is from Gelmon et al. [47].

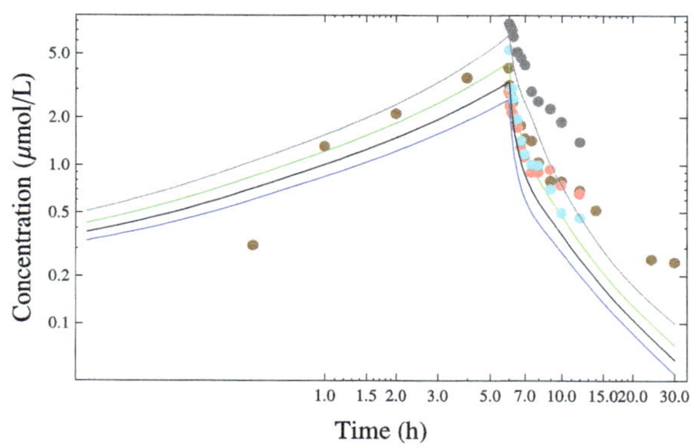

Figure 41 Paclitaxel from Wiernik et al. The paclitaxel data represented by the circles is from Wiernik et al. [48].

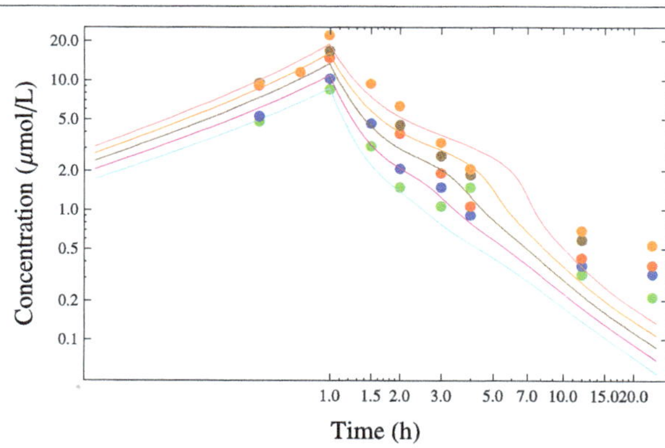

Figure 42 Paclitaxel from Maier-Lenz et al. The paclitaxel data represented by the circles is from Maier-Lenz et al. [49].

Abbreviations

ADME: Absorption, distribution, metabolism and elimination; ALA: Alanine; ALL: Acute lymphoblastic leukaemia; AMBER: Assisted model building with energy Refinement; ARG: Arginine; AUC: Area under the curve; CREL: Cremphor EL; CYP: Cytochrome P450; DNA: Deoxyribonucleic acid; DOL: Doxorubicinol; DOX: Doxorubicin; GAFF: General amber force field; HEM: Hemoglobin; ILE: Isoleucine; IV: Intravenous; KEGG: Kyoto Encyclopedia of Genes and Genomes; LEU: Leucine; LGA: Lamarckian genetic algorithm; MD: Molecular dynamics NAMD: Not (just) Another molecular dynamics program; OHP: 6a-hydroxypaclitaxel; PDB: Protein data bank; PHE: Phenylalanine; PK: Pharmacokinetics; PTX: Paclitaxel; RMSD: Root mean square deviation.

Competing interests

The authors declare that they have no competing interests.

Authors' contributions

KJEV developed the model in Equation 11, developed the numerical code for fitting the model to the experimental data, was involved in running the numerical code, and was involved in writing the manuscript. AGM was involved in collecting experimental data from the literature for both paclitaxel and doxorubicin, digitalizing the experimental concentration-versus-time data, and fitting the theoretical model to the experimental data. MGT was involved in collecting experimental data from the literature for both paclitaxel and doxorubicin, digitalizing the experimental concentration-versus-time data, and fitting the theoretical model to the experimental data. LF was involved in collecting experimental data from the literature for doxorubicin, digitalizing the experimental concentration-versus-time data, and fitting the theoretical model to the experimental data. KB was involved in all aspects of the molecular dynamics simulations. JAT was involved with the development of the model in Equation 11, development of the molecular dynamics simulations, and was involved in writing the manuscript. All authors read and approved the final manuscript.

Acknowledgements

J.A.T. acknowledges funding support for this project from NSERC, the Allard Foundation, the Canadian Breast Cancer Foundation and the Alberta Advanced Education and Technology. K.J.E.V. would like to thank J.A.T., the Cross Cancer Institute and the University of Alberta for their hospitality.

Author details

[1]Department of Physics and Astronomy, University of Lethbridge, 4401 University Drive, T1K 3M4 Lethbridge, Canada. [2]Department of Physics, University of Alberta, T6G 2J1 Edmonton, Canada. [3]Department of Engineering, Mathematics and Physics, Fayoum University, Fayoum, Egypt. [4]Department of Oncology, Faculty of Medicine and Dentistry, University of Alberta, T6G 2J1 Edmonton, Canada.

References

1. World Health Organization: **WHO global report: Noncommunicable diseases country profiles 2011.** Date: September, 2011 ISBN: 9789241502283 [http://www.who.int/nmh/publications/ncd_profiles2011/en/]
2. **KEGG PATHWAY Database - Metabolism** [http://www.genome.jp/kegg/pathway.html#metabolism]
3. **KEGG PATHWAY Database - Cellular Processes** [http://www.genome.jp/kegg/pathway.html#cellular]
4. **KEGG PATHWAY Database - Human Diseases** [http://www.genome.jp/kegg/pathway.html#disease]
5. Rowinsky EK, Wright M, Monsarrat B, Lesser GJ, Donehower RC: **Taxol: pharmacology, metabolism and clinical implications.** *Cancer Surv* 1993, **17:**283–304.
6. Huizing MT, Misser VH, Pieters RC, ten Bokkel Huinink WW, Veenhof CH, Vermorken JB, Pinedo HM, Beijnen JH: **Taxanes: a new class of antitumor agents.** *Cancer Invest* 1995, **13:**381–404.
7. Kim TY, Kim DW, Chung JY, Shin SG, Kim SC, Heo DS, Kim NK, Bang YJ: **Phase i and pharmacokinetic study of genexol-pm, a cremophor-free, polymeric micelle-formulated paclitaxel, in patients with advanced malignancies.** *Clin Cancer Res* 2004, **10:**3708.
8. Robinson DM, Keating GM: **Albumin-bound paclitaxel in metastatic breast cancer drugs.** *Drugs* 2006, **66:**941.
9. Mori T, Kinoshita Y, Watanabe A, Yamaguchi T, Hosokawa K, Honjo K: **Retention of paclitaxel in cancer cells for 1 week in vivo and in vitro.** *Cancer Chemother Pharmacol* 2006, **58:**665–675.
10. Henningsson A, Karlsson MO, Vigano L, Gianni L, Verweij J, Sparreboom AA: **Mechanism-based pharmacokinetic model for paclitaxel.** *J Clin Oncol* 2001, **19:**4065.
11. Brana MF, Cacho M, Gradillas A, Pascual-Teresa B, Ramos A: **Intercalators as anticancer drugs.** *Curr Pharm Des* 2001, **7:**1745–1780.
12. Hickman JA, Beere HM, Wood AC, Waters CM, Parmar R: **Mechanisms of cytotoxicity caused by antitumor drugs, 6th international congress of toxicology: toxicology from discovery and experimentation to the human perspective.** *Toxicol Lett* 1992, **64–65:**553–561.
13. Kiechle FL, Zhang XB: **Apoptosis: biochemical aspects and clinical implications.** *Clinica Chimica Acta* 2002, **326:**27–45.
14. Carter SK, Comis RL: **Integration of chemotherapy into a combined modality approach for cancer treatment for pancreatic adenocarcinoma.** *Cancer Treat Rev* 1975, **2:**193–214.
15. Lenaz L, Page JA: **Cardiotoxicity of adriamycin and related anthracyclines.** *Cancer Treat Rev* 1975, **3:**111–120.
16. Locker GY, Doroshow JH, Myers CE: **Glutathione peroxidase - its role in adriamycin cardiotoxicity.** In *Proceedings of the American Association for Cancer Research: Volume 18.* Philadelphia: American Association for Cancer Research; 1977:87.

17. Wallace KB: **Doxorubicin-induced cardiac mitochondrionopathy.** *Pharmacol Toxicol* 2003, **93:**105–115.
18. Kurabayashi M, Jeyaseelan R, Kedes L: **Doxorubicin represses the function of the myogenic helix-loop-helix transcription factor myod - involvement of id gene induction.** *J Biol Chem* 1994, **269:**6031–6039.
19. Boucek RJ, Miracle A, Anderson M, Engelman R, Atkinson J, Dodd DA: **Persistent effects of doxorubicin on cardiac gene expression.** *J Mol Cell Cardiol* 1999, **31:**1435–1446.
20. Gibaldi M, Perrier D: *Pharmacokinetics.* 2nd edition. New York: M. Dekker; 1982.
21. Macheras P: **A fractal approach to heterogeneous drug distribution: calcium pharmacokinetics.** *Pharm Res* 1996, **13:**663–670.
22. Fuite J, Marsh RE, Tuszynski JA: **Fractal pharmacokinetics of the drug mibefradil in the liver.** *Phys Rev E* 2002, **66:**021904.
23. Marsh RE, Tuszynski JA, Sawyer MB, Vos KJE: **Emergence of power laws in the pharmacokientics of paclitaxel due to competing saturable processes.** *J Pharmacy Pharmaceutical Sci* 2008, **11:**77–96.
24. Hornak V, Abel R, Okur A, Strockbine B, Roitberg A, Simmerling C: **Comparison of multiple amber force fields and development of improved protein backbone parameters.** *Proteins* 2006, **65:**712–25.
25. Yano JK, Wester MR, Schoch GA, Griffin KJ, Stout CD, Johnson EF: **The structure of human microsomal cytochrome p450 3a4 determined by x-ray crystallography to 2.05-a resolution.** *J Biol Chem* 2004, **279:**38091–38094.
26. Bennet MJ, Barakat K, Huzil JT, Tuszynski JA, Schriemer DC: **Discovery and characterization of the laulimalide-microtubule binding mode by mass shift perturbation mapping.** *Chem Biol* 2010, **17:**725–734.
27. Gianni L, Vigano L, Locatelli A, Capri G, Giani A, Tarenzi E, Bonadonna G: **Human pharmacokinetic characterization and in vitro study of the interaction between doxorubicin and paclitaxel in patients with breast cancer.** *J Clin Oncol* 1997, **15:**1906–1915.
28. Moreira A, Lobato R, Morais J, Silva S, Ribeiro J, Figueira A, Vale D, Sousa C, Araujo F, Fernandes A, Oliveira J, Passos-Coelho JL: **Influence of the interval between the administration of doxorubicin and paclitaxel on the pharmacokinetics of these drugs in patients with locally advanced breast cancer.** *Cancer Chemotherapy Pharmacol* 2001, **48:**333–337.
29. Benjamin RS, Riggs CE, Bachur NR: **Influence of the interval between the administration of doxorubicin and paclitaxel on the pharmacokinetics of these drugs in patients with locally advanced breast cancer.** *Cancer Chemotherapy Pharmacol* 2001, **48:**333–337.
30. Andersen A, Holte H, Slordal L: **Pharmacokinetics and metabolism of doxorubicin after short-term infusions in lymphoma patients.** *Cancer Chemotherapy Pharmacol* 1999, **44:**422–426.
31. Greene RF, Collins JM, Jenkins JF, Speyer JL, Myers CE: **Plasma pharmacokinetics of adriamycin and adriamycinol: Implications for the design of in vitro experiments and treatment protocols.** *Cancer Res* 1983, **43:**3417–3421.
32. Gianni L, Kearns CM, Giani A, Capri G, Vigano L, Locatelli A, Bonadonna G, Egorin MJ: **Nonlinear pharmacokinetics and metabolism of paclitaxel and its pharmacokinetic/pharmacodynamic relationships in humans.** *J Clin Oncol* 1995, **13:**180–190.
33. Ohtsu T, Sasaki Y, Tamura T, Miyata Y, Nakanomyo H, Nishiwaki Y, Saijo N: **Clinical pharmacokinetics and pharmacodynamics of paclitaxel: a 3-hour infusion versus a 24-hour infusion.** *Clin Cancer Res* 1995, **1:**599–606.
34. Ekroos M, Sjogren T: **Structural basis for ligand promiscuity in cytochrome p450 3a4.** *Proc Natl Acad Sci U S A* 2006, **103:**13682–13687.
35. Boucek RJ, Olson RD, Brenner DE, Ogunbunmi EM, Inui M, Fleischer S: **The major metabolite of doxorubicin is a potent inhibitor of membrane-associated ion pumps: A correlative study of cardiac muscle with isolated membrane fractions.** *J Biol Chem* 1987, **262:**15851–15856.
36. Hornberg JJ, Bruggeman FJ, Westerhoff HV, Lankelma J: **Cancer: a systems biology disease.** *Biosystems* 2006, **83:**81–90.
37. Mross K, Niemann B, Massing U, Drevs J, Unger C, Bhamra R, Swenson C: **Pharmacokinetics of liposomal doxorubicin (tlc-d99; myocet) in patients with solid tumors: an open-label, single-dose study.** *Cancer Chemother Pharmacol* 2004, **54:**514–524.
38. Di Fronzo G, Lenaz L, Bonadonna G: **Distribution and excretion of adriamycin in man.** *Biomedicine* 1973, **19:**169–171.
39. Jacquet JM, Bressolle F, Galtier M, Bourrier M, Donadio D, Jourdan J, Rossi JF: **Doxorubicin and doxorubicinol: intra- and inter-individual variations of pharmacokinetic parameters.** *Cancer Chemother Pharmacol* 1990, **27:**219–225.
40. Bressolle F, Ray P, Jacquet JM, Bres J, Galitier M, Donadio D, Jourdan J, Rossi JF: **Bayesian estimation of doxorubicin pharmacokinetic parameters.** *Cancer Chemother Pharmacol* 1991, **29:**53–60.
41. Eksborg S, Strandler HS, Edsmyr F, Naslund I, Tahvanainen P: **Pharmacokinetic study of iv infusions of adriamycin.** *Eur J Pharmacol* 1985, **28:**205–212.
42. Chan KK, Chlebowski RT, Tong M, Chen HSG, Gross JF, Bateman JR: **Clinical pharmacokinetics of adriamycin in hepatoma patients with cirrhosis.** *Cancer Res* 1980, **40:**1263–1268.
43. Monsarrat B, Chatelut E, Royer I, Alvineirie P, Dubois J, Dezeuse A, Roche H, Cros S, Wright M, Canal P: **Modification of paclitaxel metabolism in a cancer patient by induction of cytochrome p450 3a4.** *Drug Metab Dispos* 1998, **26:**229–233.
44. Mross K, Hollander N, Hauns B, Schumacher M, Maier-Lenz H: **The pharmacokinetics of a 1-h paclitaxel infusion.** *Cancer Chemother Pharmacol* 2000, **45:**463–470.
45. Brown T, Havlin K, Weiss G, Cagnola J, Koeller J, Kuhn J, Rizzo J, Craig J, Phillips J, Von Hoff D: **A phase 1 trial of taxol given by a 6-hour intravenous infusion.** *J Clin Oncol* 1991, **9:**1261–1267.
46. Longnecker SM, Donehower RC, Cates AE, Chen TL, Brundrett RB, Grochow LB, Ettinger DS, Colvin M: **High-performance liquid chromatography assay for taxol in human plasma and urine and pharmacokinetics in a phase i trial.** *Cancer Treat Rep* 1987, **71:**53–59.

47. Gelmon K, Eisenhauer E, Bryce C, Tolcher A, Mayer L, Tomlinson E, Zee B, Blackstein M, Tomiak E, Yau J, Batist G, Fisher B, Iglesias J: **Randomized phase ii study of high-dose paclitaxel with or without amifostine in patients with metastatic breast cancer.** *J Clin Oncol* 1999, **17:**3038–3047.
48. Wiernik PH, Schwartz EL, Strauman JJ, Dutcher JP, Lipton RB, Paietta E: **Phase I clinical and pharmacokinetic study of taxol.** *Cancer Res* 1987, **47:**2486–2493.
49. Maier-Lenz H, Hauns B, Haering B, Koetting J, Mross K, Unger C, Bauknecht T, duBois A, Meerpohl HG, Hollaender N, Diergarten K: **Phase i study of paclitaxel administration as a 1-hr infusion: Toxicity and pharmacokinetics.** *Semin Oncol* 1997, **24:**19–161919.

7

Realistic human muscle pressure for driving a mechanical lung

Emeline Fresnel[1,2*], Jean-François Muir[3] and Christophe Letellier[1,2]

*Correspondence:
emeline.fresnel@coria.fr
[1]CORIA-UMR 6614 Normandie Université, CNRS-Université et INSA de Rouen, Campus Universitaire du Madrillet, F-76800 Saint-Etienne du Rouvray, France
[2]ADIR Association, Hôpital de Bois-Guillaume, F-76031 Rouen, France
Full list of author information is available at the end of the article

Abstract

Background: An important issue in noninvasive mechanical ventilation consists in understanding the origins of patient-ventilator asynchrony for reducing their incidence by adjusting ventilator settings to the intrinsic ventilatory dynamics of each patient. One of the possible ways for doing this is to evaluate the performances of the domiciliary mechanical ventilators using a test bench. Such a procedure requires to model the evolution of the pressure imposed by respiratory muscles, but for which there is no standard recommendations.

Methods: In this paper we propose a mathematical model for simulating the muscular pressure developed by the inspiratory muscles and corresponding to different patient ventilatory dynamics to drive the ASL 5000 mechanical lung. Our model is based on the charge and discharge of a capacitor through a resistor, simulating contraction and relaxation phases of the inspiratory muscles.

Results: Our resulting equations were used to produce 420 time series of the muscle pressure with various contraction velocities, amplitudes and shapes, in order to represent the inter-patient variability clinically observed. All these dynamics depend on two parameters, the ventilatory frequency and the mouth occlusion pressure.

Conclusion: Based on the equation of the respiratory movement and its electrical analogy, the respiratory muscle pressure was simulated with more consistency in regards of physiological evidences than those provided by the ASL 5000 software. The great variability in the so-produced inspiratory efforts can cover most of realistic patho-physiological conditions.

Keywords: Respiratory muscle pressure; Mechanical lung; Mechanical ventilation

Background

Issues in noninvasive mechanical ventilation

The main goal of mechanical ventilation is to assist the spontaneous breathing of a patient with acute or chronic respiratory failure. The ventilator improves the blood oxygenation and unloads the respiratory muscles by supplying a suitable level of pressure support [1]. Typically, a ventilator delivers a high pressure (often named "Inspiratory Positive Airway Pressure") during the inspiration and a low pressure (named "Expiratory Positive Airway Pressure") during the expiration. Noninvasive mechanical ventilation is thus quite tricky since the pressure cycle delivered by the ventilator must be synchronized with the patient breathing cycle, that is, the pressure rise to reach the upper pressure level must be triggered at the onset of the patient inspiratory effort and the pressure release must be triggered when the patient ends his inspiration for breathing out. Such a good synchronization between the patient breathing cycle and the pressurization cycle delivered by the ventilator is important to ensure a better comfort and to reduce the patient work of breathing, although these points have not yet been clearly validated [2]. Patient-ventilator asynchrony is defined as a phase shift between the patient ventilatory cycle and the pressurization cycle delivered by the ventilator. There are various types of asynchrony events [3,4] which can be detected using noninvasive measurements [5]: reducing their incidence allows to decrease the duration for which a patient needs mechanical ventilatory assistance [3]. The quality of the synchronization between the patient ventilatory cycle and the pressurization cycle delivered by the ventilator depends on patient physiological characteristics and on ventilator settings. Some parameters are considered as being critical for synchronizing the ventilator pressure cycle to the patient breathing dynamics: the sensitivity of the high pressure trigger [3], the low pressure value [6], the level of pressure support (the difference between the high and low pressure values) [7,8] or the time during which the high pressure is delivered [9]. However, the way according which the pressure cycle is governed by patient inspiratory demand depends on the considered ventilators. Moreover, the terminology as well as the units of the settings present strong heterogeneities among the available ventilators [10].

Nowadays, bench test studies appear as the most appropriate and most effective way to compare performances of ventilators. Unfortunately, these studies reveal that there is a serious need for defining standardized protocols in order to obtain reliable and reproducible results [11]. We choose to focus on the problem of simulating the inspiratory effort delivered by the patient, that is, the dynamics of the breathing patterns as it can be produced with the active mechanical lung ASL 5000 (IngMar Medical, Pittsburgh, USA), one of the most flexible active mechanical lungs available in the market. The choice for this device results from the fact that it becomes very often used for testing performances of domiciliary ventilators [12-17]. It consists of a piston chamber driven by a software and whose dynamics can be adjusted by realistic parameters such as the resistance of the airways and the lung compliance. With such a device, it is quite easy to simulate the inter-patient variability clinically observed since the muscular pressure responsible for the inspiratory effort can be parametrized using predefined waveforms or designed by the operator (in the latter case, the curve is downloaded to the device). There is no consensus for designing the inspiratory effort [11] and all parameter values required to reproduce muscular pressure are very rarely fully reported, with the exception of Chatburn's studies [18]. Most often a semi-sinusoidal muscle pressure — as predefined in the ASL 5000 — is used (Figure 1), but this was never compared to the (rare) physiological data available. This is mainly due to the fact that measuring a muscular pressure is not a trivial task and necessarily requires invasive measurements. Indeed, in common studies only the airflow and the pressure within the ventilatory circuit (as measured by the ventilator) are available. Airflow thus results from the interactions between the patient and his ventilator; it corresponds to the output of the system, the muscle pressure being the source signal. The ASL 5000 was thus designed for producing the airflow from a preset muscular pressure. In order to conduct a parametric study for investigating the quality of the synchronization between patient breathing cycles and the pressure cycles delivered by the ventilator, it is therefore convenient to produce muscular pressure time series depending on a few

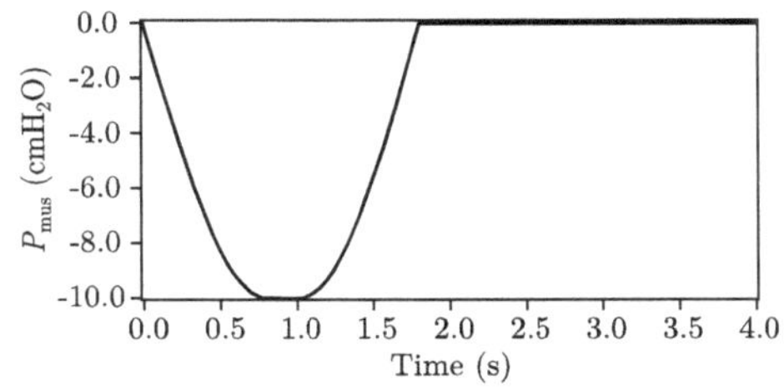

Figure 1 Muscle pressure curve. Respiratory muscular pressure used to drive the ASL 5000 mechanical lung. Sinusoidal half-wave model as proposed by the ASL 5000 software. Parameter values: Increase = 20%, Hold = 5%, Release = 20%, T_{tot} = 4 s, Amp=10 cmH_2O. Increase, hold and release are expressed as a fraction of T_{tot}.

parameters varied (with an increment as small as possible). Each parameter setting can thus be investigated for hundreds of patient ventilatory dynamics, that is, as it was tested on a cohort of hundreds of patients. In practice, the latter is nearly impossible to conduct because it is not possible to measure muscular pressure in a huge cohort of patients with the expected parameters, that is, according to our objective, varying the ventilatory frequency and the occlusion pressure within the target intervals. It is therefore convenient to simulate the corresponding muscular pressure time series for driving a mechanical lung as the ASL 5000: ventilators are thus tested as if they were connected to a large number of different patients. Such a procedure allows to make the study reproducible by any group having an ASL 5000 (something which is impossible to achieve with a database of measurements in patients).

Our objective is therefore to construct time series of the respiratory muscle pressure in agreement with physiological evidences for investigating ventilator performances under quite realistic conditions. For that purpose, we started from the equation of motion for the respiratory system, used to determine the muscle pressure P_{mus}. Based on an electrical analogy, the resulting mathematical model is then presented and some examples of inspiratory efforts are discussed.

Mechanics of the respiratory system: historical aspects

In the early 20th century, Armand Imbert investigated the relative velocities of spontaneous and triggered muscle contractions [19]. Using an electrical excitation produced by the discharge of a capacitor connected to a power source, contractions of the explored muscle were recorded through a drum maintained with a bandage. Muscle contraction can be divided into two phases (Figure 2): a relatively stiff contraction followed by a relaxation phase. Consequently, there is an obvious asymmetry between the contraction and the relaxation phases. The rate of contraction increases as the contraction becomes stronger. Similar results were found with spontaneous muscle contractions [19]. Regarding to the respiratory muscles, stronger contractions allow to increase the volume of the thoracic cage, thus inducing a larger negative pressure and then producing a stronger inspiratory flow.

Equation of motion for the respiratory system

The respiratory mechanics is mainly governed by the equation expressing the muscle pressure required for producing the airflow Q_v. There are three contributions to this

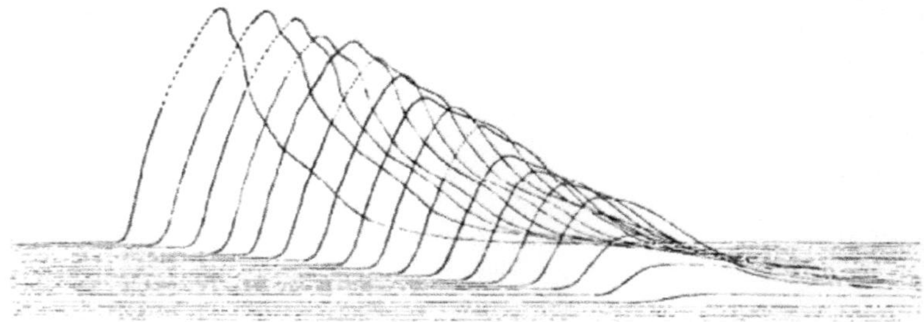

Figure 2 Imbert's muscle contractions. Triggered by capacitor discharges, muscle contractions are represented from bottom to top with an increasing intensity (from Imbert, 1911).

equation. First, the static property of the respiratory system implies that the energy per unit volume (physically equivalent to a pressure), P_{el}, used to deform the respiratory system (lungs and thoracic cage) is proportional to the inspired volume V and to the elastance E of the respiratory system, that is,

$$P_{\text{el}}(t) = E\,V(t)\,. \tag{1}$$

Due to the relative precision of the data measured in patients, the elastance may be here considered as constant, at least during inspiration and expiration. In fact, the difference of elastance between inspiration and expiration will be implicitely taken into account in the different time constant τ_c and τ_r considered for modelling the muscle pressure as explained in the subsequent part of this paper. Second, the volume energy P_{res} dissipated by the resistance R of the respiratory system to airflow is given by

$$P_{\text{res}}(t) = R\,Q_v = R\,\frac{\mathrm{d}V(t)}{\mathrm{d}t}\,. \tag{2}$$

This component results from the air viscosity, that is, the loss of energy only due to interactions between the air and the airway walls: this term corresponds to the so-called Hagen-Poiseuille law. Third, the volume energy P_{in} required to inflate the lung depends on the inertance I of the respiratory system according to

$$P_{\text{in}}(t) = I\,\frac{\mathrm{d}Q_v(t)}{\mathrm{d}t} = I\,\frac{\mathrm{d}^2V(t)}{\mathrm{d}t^2}\,. \tag{3}$$

Elastance E characterizes the static property of the lung mechanics and, resistance R and inertance I correspond to the dynamical properties of the respiratory system. The total pressure required to move the lungs is thus expressed as

$$P_{\text{tot}}(t) = P_{\text{el}}(t) + P_{\text{res}}(t) + P_{\text{in}}(t) = E\,V(t) + R\,\frac{\mathrm{d}V(t)}{\mathrm{d}t} + I\,\frac{\mathrm{d}^2V(t)}{\mathrm{d}t^2}\,. \tag{4}$$

At a normal respiratory rate, the inertial pressure P_{in} can be neglected when compared to the elastic and resistive pressures; thus we consider that

$$P_{\text{tot}}(t) \approx P_{\text{el}}(t) + P_{\text{res}}(t) = E\,V(t) + R\,\frac{\mathrm{d}V(t)}{\mathrm{d}t}\,. \tag{5}$$

For a patient breathing spontaneously, the total pressure P_{tot} applied to the respiratory system results from the (negative) muscle pressure P_{mus} and from the mouth pressure P_m, as

$$P_{tot}(t) = P_m(t) - P_{mus}(t) = E\,V(t) + R\,\frac{dV(t)}{dt}\,. \tag{6}$$

This equation is the dual of the equation of motion

$$P_{mus}(t) - P_m(t) = E\,V'(t) + R\,\frac{dV'(t)}{dt}\,, \tag{7}$$

proposed by Mead & Agostini [20], where V' is the lung volume related to the equilibrium volume and $\frac{dV'}{dt}$ the airflow. A mechanical model corresponding to Eq. (7) was proposed by Mecklenburgh & Mapleson [21] (Figure 3).

A fitted quadratic model for P_{mus}

The action of respiratory muscles is equivalent to a (negative) pressure applied to the chest wall. It has been shown that when compliance C and resistance R of a given subject are known, it is possible to reconstruct the times series of P_{mus} from continuous measurements of the airflow Q_v and from the mouth pressure P_m [22]. In this latter work, these measures were performed in 12 healthy subjects i) spontaneously breathing and ii) under ventilatory assistance. According to these results, the pressure P_{mus} resulting from the respiratory muscles takes the form of a so-called relaxation oscillation. A breathing cycle

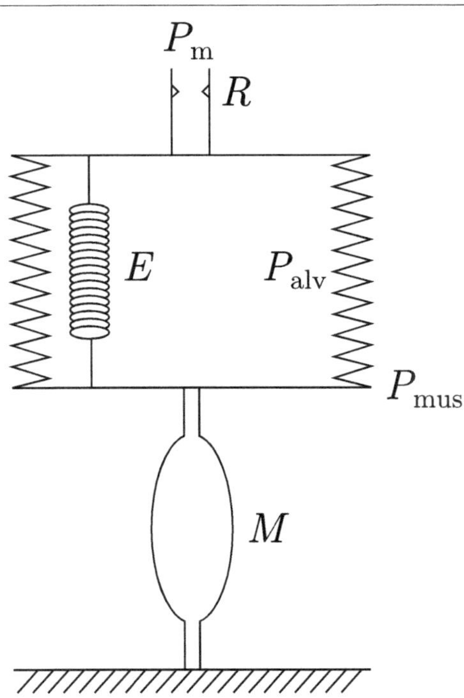

Figure 3 Mechanical analogy of the respiratory system. This mechanical model is equivalent to the ventilatory mechanics when inertance is neglected. The alveolar compartment is separated from the mouth by a resistance R. P_{alv} represents the pressure in the alveoli and P_m the pressure at the mouth. The respiratory muscles (both inspiratory and expiratory) are combined into a single muscle component M which produces a negative pressure P_{mus}. $E = \frac{1}{C}$ is the elastance (inverse of compliance C) of the respiratory system (redrawn from [21]).

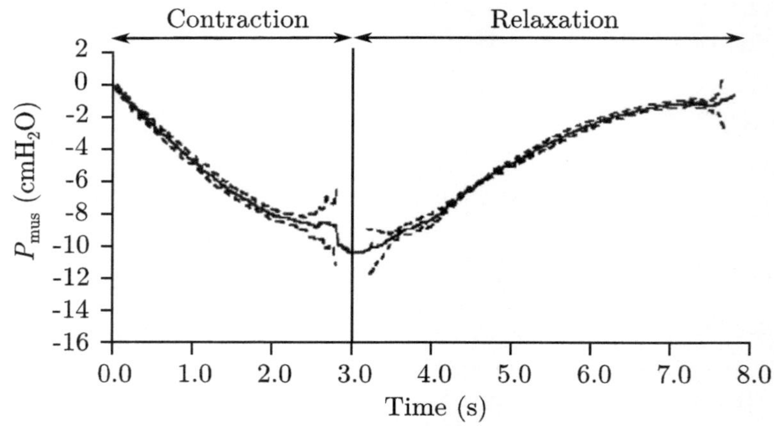

Figure 4 Measured pressure developed by respiratory muscles. Measured P_{mus} (solid line) averaged over 12 subjects and 95% confidence limits (dotted lines) for the contraction and relaxation phases of a ventilatory cycle (adapted from [22]).

is thus split into two phases, inspiration and expiration, corresponding to contraction and relaxation of the respiratory muscles, respectively (Figure 4).

By fitting quadratic polynomials on the mean measured P_{mus}, Mecklenburg & Mapleson described the relaxation oscillation of P_{mus} by four quadratic polynomials in time according to

$$P_{mus}(t) = \left| \begin{array}{ll} -0.8445t + 0.12848t^2 & 0 < t \leq 1.71\text{s} \\ 4.4408 - 6.0281t + 1.6975t^2 & 1.71 \leq t \leq 1.97\text{s} \\ -2.5980 + 1.1317t - 0.1232t^2 & 1.97 \leq t \leq 4.59\text{s} \\ 0 & 4.59 \leq t \leq 4.74\text{s} \end{array} \right. \tag{8}$$

where one could consider that ventilatory cycle is split into i) a main contraction, ii) a kind of end-inspiration pause, iii) a main relaxation and iv) an end-expiratory pause. The resulting P_{mus} time series (corresponding to a breathing cycle with a duration of 4.74 s) is shown in Figure 5. The shape of the P_{mus} time series is not significantly affected by a ventilatory assistance or a resistance loading, and only the duration of the contraction and relaxation phases is changed [23]. Such a model for P_{mus} is therefore quite general but only works for a ventilatory frequency set at 12.7 cycles per minute (cpm). Unfortunately, the

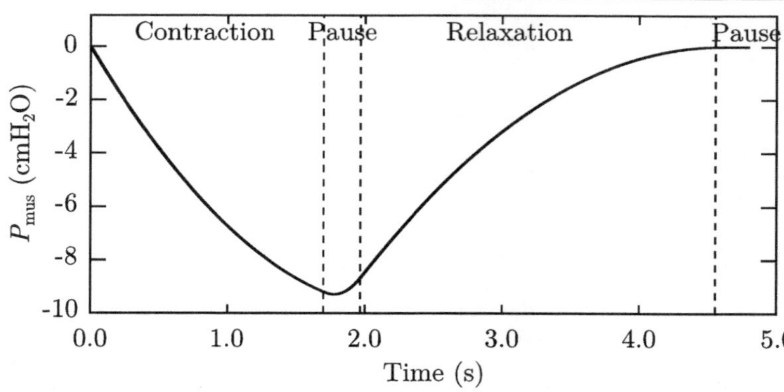

Figure 5 P_{mus} produced by the quadratic model. Time series of the muscle pressure P_{mus} solution to the set of equations (8) corresponding to a mean breathing cycle.

ventilatory frequency f_v among patients is commonly observed in the range [10; 30] cpm and, consequently, the P_{mus} model (8) is not appropriate to simulate P_{mus} time series for various breathing dynamics.

Methods

A flexible dynamical model from an electrical analogy

To be able to reproduce the inter-patient variability, it is thus desirable to construct a flexible model allowing to vary the amplitude as well as the duration of the contraction and relaxation phases. Since the time series shown in Figure 5 suggests charge and discharge of a conductance through a resistance, we developed a model dedicated to the muscle pressure and based on a RC-circuit as shown in Figure 6. We considered the inspiratory muscles as a powered system (through nerve impulses) producing a contraction followed by a relaxation as ventilatory muscles do during a breathing cycle. The contraction corresponds to the charge and the relaxation to the discharge of a capacitance C_m through a resistance R_m. The capacitance C_m characterizes the distension property of respiratory muscles and the resistance R_m corresponds to the resistance produced via tendons and intramuscular structures such as myofibrils and connective tissue.

In this configuration, we consider that the input of the system is the voltage U produced by the central nervous system and the output is the voltage U_{rm} across the capacitance C_{m} which corresponds to the voltage delivered by the respiratory muscles. We assumed that the capacitor is discharged at time $t = 0$, that is $U_{\text{rm}}(0) = 0$. The characteristic equation of a capacitance is

$$i_C = C_m \frac{\text{d}U_{\text{rm}}}{\text{d}t} \tag{9}$$

where i_C is the current flowing through the resistance R_m and U_{rm} the voltage across the capacitance C_m. By applying Kirchhoff's second rule to the circuit shown in Figure 6, we obtained the voltage U_{rm} corresponding to the muscle contraction as

$$U = U_R + U_{\text{rm}} \tag{10}$$

where the voltage U_R across the resistance is expressed as

$$U_R = R_m\, i_C = R_m C_m \frac{\text{d}U_{\text{rm}}}{\text{d}t}\,. \tag{11}$$

During inspiration, $U = U_{\text{max}}$ and a solution to the differential equation

$$U = U_{\text{rm}} + R_m C_m \frac{\text{d}U_{\text{rm}}}{\text{d}t} \tag{12}$$

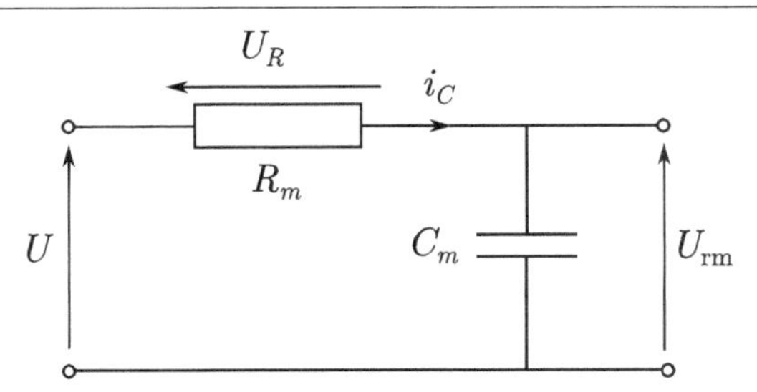

Figure 6 *RC*-circuit equivalent to the respiratory muscles. Electrical circuit used to simulate the respiratory muscle activity.

is

$$U_{\rm rm}(t) = U_{\rm max}\left(1 - e^{-\frac{t}{\tau}}\right), \tag{13}$$

where $U_{\rm max}$ corresponds to the maximum pressure delivered by the muscle contraction and $\tau = R_m C_m$ is the time constant of the circuit. During expiration, it can reasonably be assumed that $U = 0$: the end of the nervous impulse induces the muscle relaxation. The solution to the resulting differential equation

$$0 = U_C + R_m C_m \frac{dU_{\rm rm}}{dt} \tag{14}$$

is thus

$$U_{\rm rm}(t) = U_{\rm max}\, e^{-\frac{t}{\tau}}, \tag{15}$$

where $U_{\rm rm}(0) = U_{\rm max}$ as expected at the end of the inspiration (and so at the beginning of the expiration). From the mechanical point of view, the positive muscular contraction induces a negative pressure $P_{\rm mus}$ in the respiratory system. Thus, $P_{\rm mus} = -U_{\rm rm}$.

When simulating an inspiratory effort, the charge corresponds to the inspiratory phase whose duration must be chosen, and the discharge is associated with the expiratory phase. The duration of the breathing cycle is directly related to the patient ventilatory frequency f_v that we choose to vary from 10 to 30 cpm (cycles per minute) in order to simulate "normal" ventilatory dynamics corresponding to stable health conditions and faster dynamics observed in acute situations or during physical efforts.

Parameter $U_{\rm max}$ appearing in solutions (13) and (15) corresponds to the amplitude of the pressure delivered by the respiratory muscles. Solutions (13) and (15) are negative, that is, they are consistent with Mecklenburg and co-workers as well as with the definitions used in the ASL 5000.

In an electronic circuit, the time constant τ is the same for the charge and the discharge. In the respiratory system, the contraction (inspiration) is active and the expiration (relaxation) is passive. As a consequence, there are two distinct time constants τ_c and τ_r (this is equivalent to consider that the elastance is slightly different between inspiration and expiration) as discussed below. An oscillation of the pressure $P_{\rm mus}$ corresponding to a breathing cycle is made of the two solutions (13) and (15) but with two different time constants as exemplified in Figure 7 for a breathing cycle whose duration is $T_{\rm tot} = 4.29$ s and the inspiration duration is $T_{\rm I} = 1.29$ s. The global morphology of $P_{\rm mus}$ is in a good agreement with the mean curve provided by measurements (compare Figure 7 to Figure 4). Comparing the first two quadratic polynomials used by Mecklenburg and coworkers for the contraction with the single exponential function (13) shows a transition between contraction and relaxation stiffer than those observed in measurements. Nevertheless, triggering mechanisms in home ventilators are not sensitive to such a departure and this approximation will have (if any) a marginal effect on the synchronization between the pressure cycle delivered by the machine and the breathing cycle supplied by the mechanical lung.

Inspiratory and expiratory durations

The transition between inspiration and expiration is assumed to correspond to the onset of the relaxation of the ventilatory muscles. Solution (13) retained for the contraction

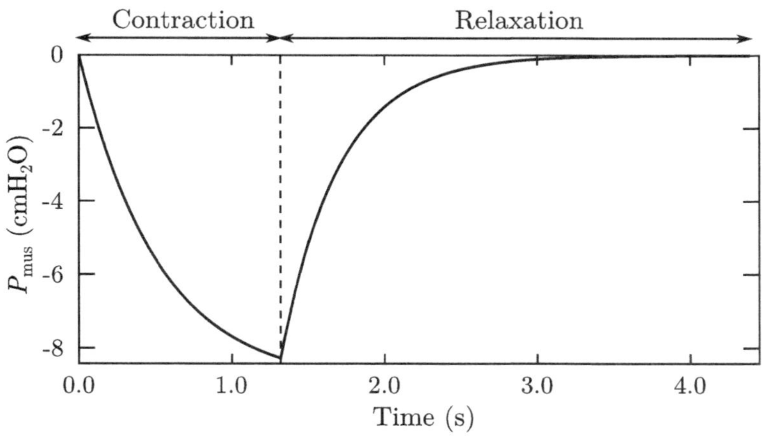

Figure 7 Simulated muscular pressure produced by the electric circuit. Solutions during inspiration (13) and expiration (15) were concatenated.

is valid between $t = 0$ and $t = T_I$, and solution (15) for relaxation is used between $t = T_I$ and $t = T_{tot}$, where T_{tot} is the duration of the breathing cycle. It has been shown that $\frac{T_I}{T_{tot}}$ is not constant when the ventilatory frequency $f_v \left(= \frac{1}{T_{tot}}\right)$ is varied [24]. When f_v is increased, the expiratory time decreases faster than the inspiratory duration and, consequently, $\frac{T_I}{T_{tot}}$ increases. Physiologically, the inspiratory duration cannot be shorter than 0.5 s and longer than 1.5 s. From the data provided in [24], $\frac{T_I}{T_{tot}}$ is measured in healthy subjects within the range [0.4; 0.6] when f_v is varied within [16; 51] cpm using various levels of exercise. The ratio $\frac{T_I}{T_{tot}}$ for resting subjects is commonly between 0.3 and 0.4 [25].

From the data collected by Scheuermann and Kowalchuk, it appears that the ratio $\frac{T_I}{T_{tot}}$ is significantly correlated to the ventilatory frequency ($r = 0.71$, $p < 0.01$). Nevertheless, the linear regression obtained from their data cannot be used since their measures were performed in subjects under exercise, contrary to patients who are commonly ventilated during their sleep. We thus retained a linear dependence roughly matching with $T_I = 1.5$ s for $f_v = 10$ cpm and $T_I = 1.0$ s for $f_v = 30$ cpm, as observed in previous studies [26,27]. The ratio $\frac{T_I}{T_{tot}}$ thus depends on the ventilatory frequency according to

$$\frac{T_I}{T_{tot}} = 0.0125 f_v + 0.125 \,. \qquad (16)$$

The computed $\frac{T_I}{T_{tot}}$ ratios and the inspiratory and expiratory durations are reported in Table 1. The obtained $\frac{T_I}{T_{tot}}$ are between 0.25 ($f_v = 10$ cpm) and 0.50 ($f_v = 30$ cpm) and thus cover most of the range found in the published data.

Amplitude and stiffness of the muscular pressure

The respiratory muscle function is commonly assessed in clinical works by using the so-called mouth occlusion pressure $P_{0.1}$, measured 100 ms after the onset of inspiration during quiet breathing. $P_{0.1}$ is representative of the central unconscious control of breathing [28] and of the strength of the inspiratory demand. In healthy adults at rest, $P_{0.1}$ equals about 1 cmH$_2$O [29] with an unavoidable inter-subject variability [30,31]. The value for quiet breathing in healthy adults is $P_{0.1} = 0.93 \pm 0.48$ cmH$_2$O

Table 1 Time parameters defining the phases of the breathing cycles: total duration T_{tot}, $\frac{T_I}{T_{tot}}$ ratio calculated from equation (16), and the associated inspiratory and expiratory times

f_V (cpm)	10	11	12	13	14	15	16
T_{tot} (s)	6.0	5.455	5.0	4.615	4.286	4.0	3.75
$\frac{T_I}{T_{tot}}$	0.250	0.263	0.275	0.288	0.300	0.313	0.325
T_I (s)	1.5	1.432	1.375	1.327	1.286	1.25	1.219
T_E (s)	4.5	4.023	3.625	3.288	3.0	2.75	2.531
f_V (cpm)	17	18	19	20	21	22	23
T_{tot} (s)	3.529	3.333	3.158	3.0	2.857	2.727	2.609
$\frac{T_I}{T_{tot}}$	0.338	0.350	0.363	0.375	0.388	0.400	0.413
T_I (s)	1.191	1.167	1.145	1.125	1.107	1.091	1.076
T_E (s)	2.338	2.166	2.013	1.875	1.75	1.636	1.533
f_V (cpm)	24	25	26	27	28	29	30
T_{tot} (s)	2.5	2.4	2.308	2.222	2.143	2.069	2.0
$\frac{T_I}{T_{tot}}$	0.425	0.438	0.450	0.463	0.475	0.488	0.500
T_I (s)	1.063	1.05	1.039	1.028	1.018	1.009	1.0
T_E (s)	1.437	1.35	1.269	1.194	1.125	1.060	1.0

in [32] and, 0.75 ± 0.32 cmH_2O in [33]. In this latter work, the occlusion pressure is found to be 2.83 ± 1.27 cmH_2O in patients with chronic obstructive pulmonary diseases (COPD) and 2.41 ± 1.01 cmH_2O in patients with restrictive lung disease. In a large cohort of 464 patients with chronic hypercapnic respiratory failure [34], the occlusion pressure is $P_{0.1} = 4.69$ (3.57; 6.63) cmH_2O for patients with COPD, $P_{0.1} = 3.67$ (2.45; 5.51) cmH_2O for patients with an obesity hypoventilation syndrome (OHS) and, $P_{0.1} = 2.55$ (1.43; 3.77) cmH_2O for patients with other various diseases. The main value is the median and values in parenthesis are quantile values. The occlusion pressure in patients with acute respiratory failure under mechanical ventilation assistance can be found within the range [6; 10] cmH_2O [35-37] when the pressure support level is quite low.

According to these studies, most of clinical cases are considered when $P_{0.1}$ is varied from 0.5 to 10 cmH_2O (with an increment equal to 0.5 cmH_2O). Combined with the ventilatory frequency f_V which is varied between 10 and 30 cpm with an increment equal to 1 cpm, the ventilatory dynamics is studied for

$$\left(\frac{30-10}{1}+1\right) \times \left(\frac{10-0.5}{0.5}+1\right) = 21 \times 20\,, \tag{17}$$

that is, 420 different ventilatory dynamics, each pair $(f_V, P_{0.1})$ corresponding to a given lung model (or a patient in a clinical equivalent). Our simulations then represent 420 lung models, thus taking into account the inter-patient variability.

Results and discussion

Parameters defining the inspiratory effort

Time constants τ appearing in equations (13) for describing the contraction and (15) for the relaxation of the respiratory muscles characterize the stiffness of the muscle dynamics. If this time constant is unique in a RC-circuit, this is not the case in the respiratory system because the muscular contraction is active and the relaxation is passive. Consequently, contraction is a shorter phenomenon than relaxation. These time constants depend on two parameters: the ventilatory frequency f_v and the mouth occlusion pressure $P_{0.1}$. The first dependence is identical for both contraction and relaxation: we empirically choose to set the time constant τ at $\frac{10}{f_v}$. For contraction, such a relation leads to $\tau_c = 1$ s (0.33 s) for $f_v = 10$ cpm (30 cpm), thus corresponding to the inspiratory duration $T_\mathrm{I} = 1.5$ s (1 s). The asymmetry between contraction and relaxation is introduced via the second dependence, that is the way the occlusion pressure $P_{0.1}$ affects such a constant. In both cases, increasing $P_{0.1}$ leads to an increase of the time constant. Since contraction is shorter than relaxation, the time constant τ_c for contraction must be more sensitive to $P_{0.1}$ than the time constant τ_r for relaxation. After trials and errors, we found that

$$\tau_c = \frac{10}{f_v + 4\,\mathrm{P}_{0.1}} \tag{18}$$

and

$$\tau_r = \frac{10}{f_v + \frac{P_{0.1}}{2}}\,. \tag{19}$$

induce P_mus time series leading to airflow and pressure time series which look like those measured in patients (as we will show in the last section). In these two equations, τ is expressed in second, f_v in cpm and $P_{0.1}$ in $\mathrm{cmH_2O}$ (implicitly per 0.1 s). Some values for these two time constants are reported in Table 2 for few values of f_v and $P_{0.1}$. Typically, these time constants are in the range [0.14; 0.98] s. They must be of the same order of magnitude as those associated with respiratory systems given by $\tau = RC$ as for any RC-circuit: the characteristic time constant is equal to 0.10 s for a restrictive patient ($R = 5\ \mathrm{cmH_2O.l^{-1}.s^{-1}}$ and $C = 20\ \mathrm{ml.cmH_2O^{-1}}$) and to 1.25 s for an obstructive patient ($R = 25\ \mathrm{cmH_2O.l^{-1}.s^{-1}}$ and $C = 50\ \mathrm{ml.cmH_2O^{-1}}$). Time constants for the respiratory muscles are thus in a range corresponding to physiological values for the respiratory system.

The respiratory muscle pressure is thus explicitly governed by

$$P_\mathrm{mus}(t) = \left| \begin{array}{ll} P_\mathrm{max}\left(1 - e^{-\frac{f_v + 4P_{0.1}}{10}t}\right) & 0 < t \leq T_\mathrm{I} \\ P_\mathrm{max}\left(e^{-\frac{f_v + \frac{P_{0.1}}{2}}{10}t}\right. & T_\mathrm{I} < t \leq T_\mathrm{tot}\,. \end{array} \right. \tag{20}$$

The ventilatory dynamics is clinically characterized by the ventilatory frequency f_v (in cpm) and the occlusion pressure $P_{0.1}$ ($\mathrm{cmH_2O}$ per 0.1 s). Since the mechanical lung is set using the maximum pressure P_mus reached during the ventilatory cycle, it is thus useful

Table 2 Examples of the time constant τ_c for contraction and τ_c for relaxation for few values of the ventilatory frequency f_v and of the occlusion pressure $P_{0.1}$

	$f_v = 10$ cpm		$f_v = 20$ cpm		$f_v = 30$ cpm	
$P_{0.1}$ ($\mathrm{cmH_2O}$)	τ_c	τ_r	τ_c	τ_r	τ_c	τ_r
0.5	0.83	0.98	0.45	0.49	0.31	0.33
5	0.33	0.80	0.25	0.44	0.20	0.31
10	0.20	0.67	0.17	0.40	0.14	0.29

Time constants are in second.

to express P_{max} in terms of $P_{0.1}$ and f_v. By definition,

$$P_{0.1} = P_{mus}(0.1) = P_{max}\left(1 - e^{-\frac{f_v + 4P_{0.1}}{10}0.1}\right) \tag{21}$$

thus leading to

$$P_{max} = \frac{P_{0.1}}{1 - e^{-\frac{0.1\cdot(f_v + 4P_{0.1})}{10}}}, \tag{22}$$

which can be used to define P_{max}. Values of P_{max} are reported in Figure 8 for the 420 respiratory dynamics we are using for designing our "cohort of mechanical patients". Using Eq. (20), the P_{mus} time series can thus be computed from the knowledge of f_v and $P_{0.1}$, P_{max} and T_I being determined using Eqs. (22) and (16), respectively. These time series thus computed are then loaded in the mechanical lung software to drive the piston. Examples of the resulting P_{mus} time series are shown in Figure 9 for four different ventilatory dynamics.

Some simulations using the mechanical lung

To validate our model, we chose to compare the airflow time series obtained from some simulations with the ASL 5000 to time series measured during a protocol which was conducted by Rabarimanantsoa-Jamous at the Rouen University Hospital during her Ph.D. thesis [38], in which written approval was obtained from the patients. In order to investigate how asynchrony events could affect the quality of noninvasive mechanical ventilation and sleep quality, airflow and airway pressure were measured during one night. The database consists of a cohort of 35 patients with respiratory diseases and daily assisted using a mechanical ventilator (VPAP III STA, ResMed, Australia). Some patients ($n = 20$)

Obstructive diseases
Healthy subjects
Restrictive conditions

Stable situation ← $P_{0.1}$ (cmH_2O) → Acute situation

Ventilatory frequency f_v (cpm)

	0.5	1	1.5	2	2.5	3	3.5	4	4.5	5	5.5	6	6.5	7	7.5	8	8.5	9	9.5	10
10	4.4	7.7	10.1	12.1	13.8	15.2	16.4	17.5	18.4	19.3	20.1	20.8	21.5	22.1	22.7	23.3	23.9	24.4	24.9	25.4
11	4.1	7.2	9.6	11.6	13.2	14.6	15.8	16.9	17.9	18.8	19.6	20.3	21.0	21.7	22.3	22.9	23.5	24.0	24.5	25.0
12	3.8	6.8	9.1	11.0	12.7	14.1	15.3	16.4	17.4	18.3	19.1	19.8	10.6	21.2	21.9	22.5	23.1	23.6	24.1	24.7
13	3.6	6.4	8.7	10.6	12.2	13.6	14.8	15.9	16.9	17.8	18.6	19.4	20.1	20.8	21.5	22.1	22.7	23.2	23.8	24.3
14	3.4	6.1	8.3	10.1	11.7	13.1	14.3	15.4	16.4	17.3	18.2	19.0	19.7	20.4	21.1	21.7	22.3	22.9	23.4	24.0
15	3.2	5.8	7.9	9.7	11.3	12.7	13.9	15.0	16.0	16.9	17.8	18.6	19.3	20.0	20.7	21.3	21.9	22.5	23.1	23.6
16	3.0	5.5	7.6	9.4	10.9	12.3	13.5	14.6	15.6	16.5	17.4	18.2	19.0	19.7	20.3	21.0	21.6	22.2	22.8	23.3
17	2.9	5.3	7.3	9.0	10.6	11.9	13.1	14.2	15.2	16.2	17.0	17.8	18.6	19.3	20.0	20.7	21.3	21.9	22.5	23.0
18	2.8	5.1	7.0	8.7	10.2	11.6	12.8	13.9	14.9	15.8	16.7	17.5	18.3	19.0	19.7	20.3	21.0	21.6	22.2	22.7
19	2.6	4.9	6.8	8.5	9.9	11.3	12.5	13.5	14.6	15.5	16.4	17.2	17.9	18.7	19.4	20.0	20.7	21.3	21.9	22.4
20	2.5	4.7	6.6	8.2	9.6	11.0	12.1	13.2	14.2	15.2	16.0	16.9	17.6	18.4	19.1	19.7	20.4	21.0	21.6	22.2
21	2.4	4.5	6.3	7.9	9.4	10.7	11.9	12.9	13.9	14.9	15.7	16.6	17.3	18.1	18.8	19.4	20.1	20.7	21.3	21.9
22	2.3	4.4	6.1	7.7	9.1	10.4	11.6	12.7	13.6	14.6	15.5	16.3	17.1	17.8	18.5	19.2	19.8	20.4	21.1	21.6
23	2.3	4.2	6.0	7.5	8.9	10.2	11.3	12.4	13.4	14.3	15.2	16.0	16.8	17.5	18.2	18.9	19.6	20.2	20.8	21.4
24	2.2	4.1	5.8	7.3	8.7	9.9	11.1	12.1	13.1	14.0	14.9	15.7	16.5	17.3	18.0	18.7	19.3	19.9	20.6	21.2
25	2.1	4.0	5.6	7.1	8.5	9.7	10.8	11.9	12.9	13.8	14.7	15.5	16.3	17.0	17.7	18.4	19.1	19.7	20.3	20.9
26	2.0	3.9	5.5	6.9	8.3	9.5	10.6	11.7	12.6	13.6	14.4	15.2	16.0	16.8	17.5	18.2	18.8	19.5	20.1	20.7
27	2.0	3.8	5.3	6.8	8.1	9.3	10.4	11.4	12.4	13.3	14.2	15.0	15.8	16.5	17.3	17.7	18.6	19.3	19.9	20.5
28	1.9	3.7	5.2	6.6	7.9	9.1	10.2	11.2	12.2	13.1	14.0	14.8	15.6	16.3	17.0	17.7	18.4	19.0	19.7	20.3
29	1.9	3.6	5.1	6.5	7.7	8.9	10.0	11.0	12.0	12.9	13.8	14.6	15.4	16.1	16.8	17.5	18.2	18.8	19.5	20.1
30	1.8	3.5	5.0	6.3	7.6	8.7	9.8	10.8	11.8	12.7	13.6	14.4	15.2	15.9	16.6	17.3	18.0	18.6	19.3	19.9

Figure 8 Maximum pressure P_{max} (in cmH_2O). of the inspiratory muscular efforts as a function of $P_{0.1}$ (in cmH_2O) and of the ventilatory frequency f_v (in cpm). The shaded areas are identified as corresponding to different groups of patients.

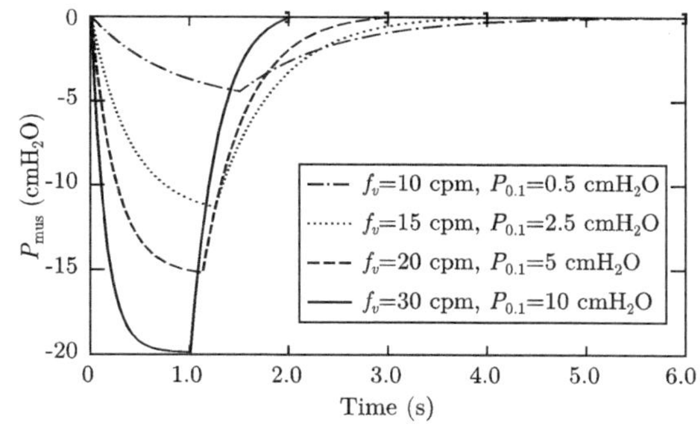

Figure 9 Simulated muscle pressures. Time series of the muscle pressure P_{mus} simulated for four different ventilatory dynamics. The ratio $\frac{T_I}{T_{tot}}$ is determined according to Eq. (16).

were affected by obesity hypoventilation syndrome (OHS) and some other ($n = 15$) by chronic obstructive pulmonary disease (COPD), associated or not with sleep apnea. Mean body index (BMI) for these patients is 42 ± 10.5 and their mean age is 62 ± 11.7. We selected three airflow time series in this database, our aim being, first, to exemplify how different the dynamics underlying airflow time series can be different and, second, to show that we are able to reproduce these different dynamics with our model for the muscle pressure.

The first case corresponds to a 45 years old patient affected by OHS and sleep apnea; the ventilator is set with a high pressure $P_h = 22$ cmH$_2$O, a low pressure $P_l = 5$ cmH$_2$O and the shortest pressure rise time. The measured airflow

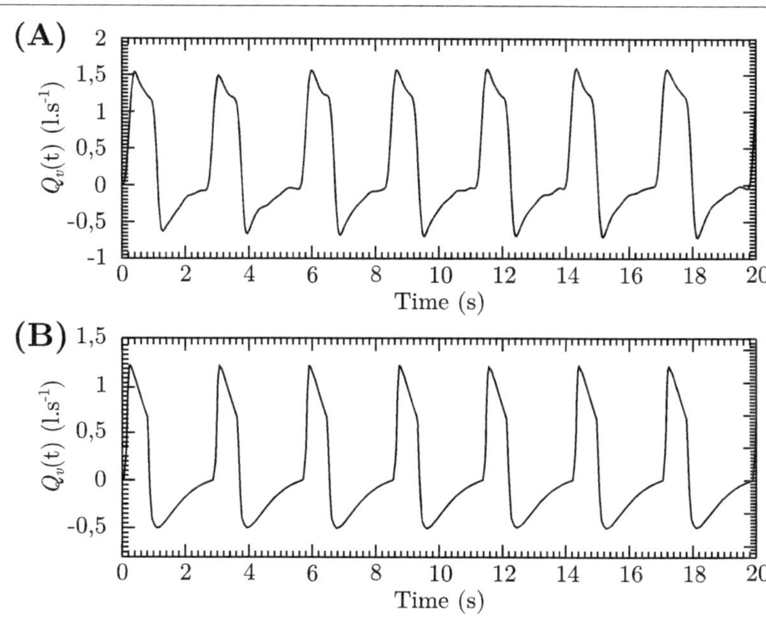

Figure 10 Patient airflow (A) and simulated curve (B). Case of a 45 years old male. (BMI= 35.4 kg.m^{-1}) with an OHS and a sleep apnea syndrome, ventilated with a VPAP III ventilator (ResMed) in a pressure support mode. Pressure values: $P_h = 22$ cmH$_2$O and $P_l = 5$ cmH$_2$O. These values were also used with the mechanical lung. We set $P_{0.1} = 1$ cmH$_2$O, $R = 5$ cmH$_2$O/l/s, and $C = 20$ ml/cmH$_2$O, as commonly used for a restrictive patient (the actual values were not measured).

(Figure 10A) presents a stiff rise up to nearly 1.5 $l.s^{-1}$ which is followed by a fast decrease. Expiration is characterized by a fast return to a quasi-null airflow, with a plateau up to the end of the breathing cycle. The main characteristics of the airflow were reproduced in the time series (Figure 10B) simulated with a restrictive lung model (patient with OHS are considered as being restrictive). The ventilatory frequency was set at the measured value (14.3 cpm). The ventilator was a VPAP III as used in the protocol with patients, and pressure levels were set as for this 45 years old male. We just adjusted the occlusion pressure (which was not measured in the the protocol) to optimize the airflow time series: we used $P_{0.1} = 1$ cmH_2O.

The second case is a 74 years old woman also affected by OHS and sleep apnea; the ventilator pressure levels were $P_h = 23$ cmH_2O and $P_l = 9$ cmH_2O. A long pressure rise time was used. The corresponding airflow time series (Figure 11A) presents a stiff rise, then a plateau which results from the long pressure rise time, and finally a fast decrease of the airflow up to a minimum negative value, followed by a slow return to a null airflow. Most of the characteristics of the measured airflow were reproduced with our restrictive lung model driven with a ventilatory frequency f_v equal to the measured value (17 cpm). The VPAP III ventilator was set as for this patient. The most corresponding airflow was obtained with the occlusion pressure $P_{0.1} = 3$ cmH_2O: this value greater than the one used for the first case can be justified by the BMI of the patient which is roughly twice the BMI of the first patient. The small plateau at nearly null airflow at the end of expiration was observed for the longest breathing cycles. Reducing the ventilatory frequency would have led to such a plateau with our simulation.

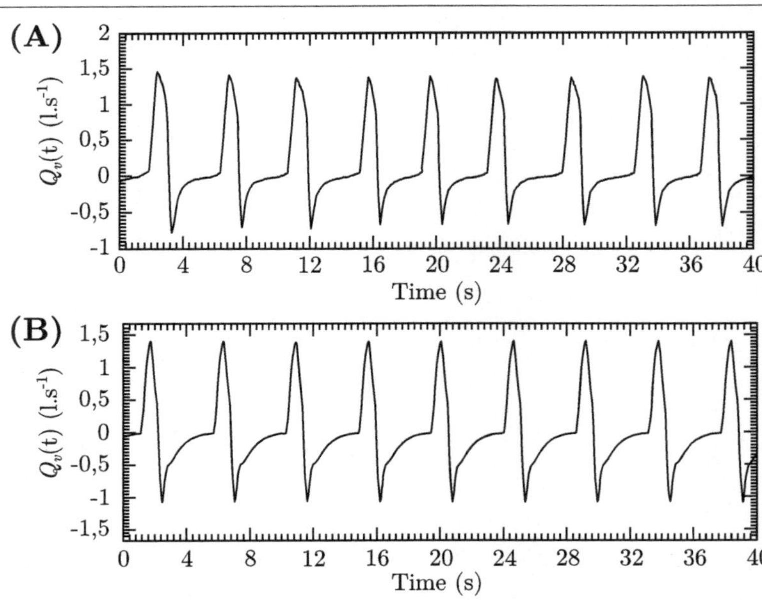

Figure 11 Patient airflow (A) and simulated curve (B). Case of a 74 years old female. (BMI $= 66.5$ $kg.m^{-1}$) with an OHS and a sleep apnea syndrome,0 ventilated with a VPAP III ventilator (ResMed) in a pressure support mode. Pressure values: $P_h = 23$ cmH_2O and $P_l = 9$ cmH_2O. These values were also used with the mechanical lung. We set $P_{0.1} = 3$ cmH_2O. R and C values were those used for the first case shown in Figure 10.

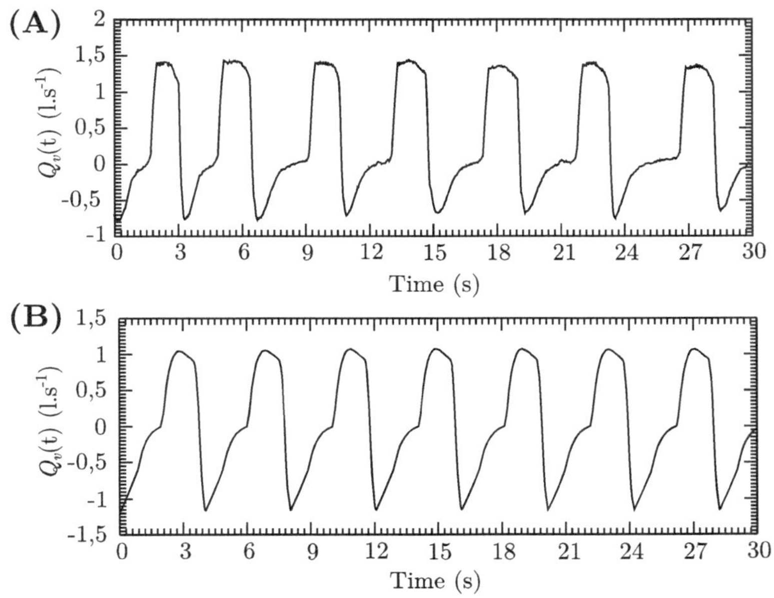

Figure 12 Patient airflow (A) and simulated curve (B). Case of a 62 years old male. (BMI= 50.2 kg.m^{-1}) with a COPD, ventilated with a VPAP III ventilator (ResMed) in a pressure support mode. Pressure values: $P_h = 30$ cmH$_2$O and $P_l = 6$ cmH$_2$O. These values were also used with the mechanical lung. We set $P_{0.1} = 4$, $R = 25$ cmH$_2$O/l/s, and $C = 50$ ml/cmH$_2$O, as commonly used for an obstructive patient (the actual values were not measured).

The third case is a 62 years old male affected by COPD ; the ventilator was set with pressure levels $P_h = 30$ cmH$_2$O and $P_l = 6$ cmH$_2$O. The shortest pressure rise time was chosen. The associated airflow (Figure 12A) presents a stiff rise: once the maximum airflow is reached, there is a first slow decrease followed by an abrupt decrease of the airflow up to the minimum negative airflow. The second part of expiration is characterized by a rather slow increase of the airflow up to the nearly null value. The airflow time series simulated with an obstructive lung model (as required for a patient with a COPD) well reproduces the main features observed in the measured time series. The ventilatory frequency was set to the measured value ($f_v = 23.1$ cpm). The best time series was obtained with $P_{0.1} = 4$ cmH$_2$O.

For the three simulated time series, the amplitude of the expiratory airflow is always less than the expiratory airflow measured with patients. This is due to the fact that the airflow in ventilatory circuit with true patients was measured after the mask which contains intentional leaks to send the carbon dioxide out of the ventilatory circuit, thus avoiding carbon dioxide rebreathing. Contrary to this, the airflow in ventilatory circuit with the mechanical lung is measured within the piston chamber (the artificial lung), before any intentional leak. The presence of intentional leak has also an effect on the inspiratory airflow: the measured airflow with patient is overestimated because it takes into account the part flowing through the intentional leak and, consequently, not provided to the patient.

The transitions between inspiration and expiration in our simulations are stiffer than in the measured airflow. This could be also an effect induced by the location of the airflow sensor which differs in the two circuits. Such a stiffness is perhaps also induced by the fact that in our model the parameters used to simulate the pathophysiological conditions (the airway resistance R and the thoracopulmonary compliance C) are kept constant during the simulations whereas they could change during inspiration and expiration [39]. As exemplified with the simulated time series shown in Figures 10, 11 and 12, such approximation did not prevent us to obtain airflow time series with characteristics close to those measured in ventilatory circuit with patient. We have therefore now a realistic mechanical lung model to assess performances of ventilators.

Conclusion

In this paper we designed a realistic respiratory muscle pressure to drive a mechanical lung such as the ASL 5000. In order to do that, we started from the scarce data available in the literature to define the main properties of the dynamics underlying the pressure driven by the respiratory muscles. Using an electrical analogy, we used two exponential functions, one for inspiration and one for expiration. In order to have a muscular dynamics only depending on the ventilatory frequency and the mouth occlusion pressure — two clinical parameters commonly measured —, we introduced a linear dependency between the ratio of the inspiratory duration to the breathing cycle duration and the ventilatory frequency as suggested by clinical evidences. The resulting model was validated by comparing simulated airflow time series measured in the mechanical lung driven by our model with those measured in ventilatory circuits with patients. By varying the ventilatory frequency and the mouth occlusion pressure we are now able to reproduce the inter-patient variability and, consequently, to investigate performances of ventilators on a "cohort" of realistic lung models.

Consent

Written informed consent was obtained from the patients for the publication of this report and any accompanying images.

Competing interests

The authors declare that they have no competing interests.

Authors' contributions

EF and CL have developed the approach of the work and have written the paper, while JFM has provided medical expertise. All authors read and approved the final manuscript.

Acknowledgements

We would like to thank Jean-Christophe Richard, Jean-Paul Janssens and Antoine Cuvelier for stimulating discussions at the origin of this work.

Author details

[1]CORIA-UMR 6614 Normandie Université, CNRS-Université et INSA de Rouen, Campus Universitaire du Madrillet, F-76800 Saint-Etienne du Rouvray, France. [2]ADIR Association, Hôpital de Bois-Guillaume, F-76031 Rouen, France. [3]Service de pneumologie – CHU de Rouen, Hôpital de Bois-Guillaume, F-76230 Bois-Guillaume, France.

References

1. Brochard L, Harf A, Lorino H, Lemaire F: **Inspiratory pressure support prevents diaphragmatic fatigue during weaning from mechanical ventilation.** *Am Rev Respir Dis* 1989, **139:**513–521.
2. Sassoon CS, Foster GT: **Patient-ventilator asynchrony.** *Curr Opin Crit Care* 2001, **7:**28–33.

3. Thille AW, Rodriguez P, Cabell B, Lellouche F, Brochard L: **Patient-ventilator asynchrony during assisted mechanical ventilation.** *Intensive Care Med* 2006, **32:**1515–1522.
4. Naeck R, Bounoiare D, Freitas US, Rabarimanantsoa H, Portmann A, Portier F, Cuvelier A, Muir J-F, Letellier C: **Dynamics underlying patient-ventilator interactions during nocturnal noninvasive ventilation.** *Int J Bifurcation Chaos* 2012, **22**(2):1250030.
5. Cuvelier A, Achour L, Rabarimanantsoa H, Letellier C, Muir J-F, Fauroux B: **A noninvasive method to identify ineffective triggering in patients with noninvasive pressure support ventilation.** *Respiration* 2010, **80:**198–206.
6. Chao DC, Scheinhorn DJ, Stearn-Hassenpflug M: **Patient-ventilator trigger asynchrony in prolonged mechanical ventilation.** *Chest* 1997, **112:**1592–1599.
7. Leung P, Jubran A, Tobin MJ: **Comparison of assisted ventilator modes on triggering, patient effort, and dyspnea.** *Am J Respir Crit Care Med* 1997, **155:**1940–1948.
8. Nava S, Bruschi C, Rubini F, Palo A, Iotti G, Braschi A: **Respiratory response and inspiratory effort during pressure support ventilation in COPD patients.** *Intensive Care Med* 1995, **21:**871–879.
9. Tassaux D, Gainnier M, Battisti A, Jolliet P: **Impact of expiratory trigger setting on delayed cycling and inspiratory muscle workload.** *Am J Respir Crit Care Med* 2005, **15:**1283–1289.
10. Chopin C, Chambrin M: **Essai de classification des modes actuels de ventilation mécanique en pression positive.** *Réanimation Urgence* 1998, **7:**87–99.
11. Olivieri C, Costa R, Conti G, Navalesi P: **Bench studies evaluating devices for non-invasive ventilation: critical analysis and future perspectives.** *Intensive Care Med* 2012, **38:**160–167.
12. Stell IM, Paul G, Lee KC, Ponte J, Moxham J: **Non-invasive ventilator triggering in chronic obstructive pulmonary disease. A test lung comparison.** *Am J Respir Crit Care Med* 2001, **164:**2092–2097.
13. Costa R, Navalesi P, Spinazzola G, Rossi M, Cavaliere F, Antonelli M, Proietti R, Conti G: **Comparative evaluation of different helmets on patient-ventilator interaction during non-invasive ventilation.** *Intensive Care Med* 2008, **34:**1102–1108.
14. Ferreira JC, Chipman DW, Kacmarek RM: **Trigger performance of mid-level ICU mechanical ventilators during assisted ventilation: a bench study.** *Intensive Care Med* 2008, **34:**1669–1675.
15. Borel JC, Sabil A, Janssens JP, Couteau M, Boulon L, Lévy P, Pépin JL: **Intentional leaks in industrial masks have a significant impact on efficacy of bilevel non-invasive ventilation: a bench test study.** *Chest* 2009, **135:**669–677.
16. Ferreira JC, Chipman DW, Hill NS, Kacmarek RM: **Bilevel vs ICU ventilators providing non-invasive ventilation: effect of system leaks: a COPD lung model comparison.** *Chest* 2009, **136:**448–456.
17. Costa R, Navalesi P, Spinazzola G, Ferrone G, Pellegrini A, Cavaliere F, Proietti R, Antonelli M, Conti G: **Influence of ventilator settings on patient-ventilator synchrony during pressure support ventilation with different interfaces.** *Intensive Care Med* 2010, **36:**1363–1370.
18. Chatburn RL: **Which ventilators and modes can be used to deliver noninvasive ventilation?** *Respir Care* 2009, **54**(1):85–101.
19. Imbert A: **Vitesses relatives des contractions musculaires volontaires et provoquées.** *L'année psychologique* 1911, **18:**47–54.
20. Mead J, Agostoni E: *Dynamics of breathing. Handbook of Physiology.* Washington, DC: American Physiological Society; 1964.
21. Mecklenburgh JS, Al-Obaidi TAA, Mapleson WW: **A model lung with direct representation of respiratory muscle activity.** *Br J Anaesth* 1992, **68:**603–612.
22. Mecklenburgh JS, Mapleson WW: **Ventilatory assistance and respiratory muscle activity. 1: Interaction in healthy volunteers.** *Br J Anaesth* 1998, **80:**422–433.
23. Mecklenburgh JS, Mapleson WW: **Ventilatory assistance and respiratory muscle activity. 2: Simulation with an adaptive active ("aa"İ or "a-squared") model lung.** *Br J Anaesth* 1998, **80:**434–439.
24. Scheuermann BW, Kowalchuk JM: **Breathing patterns during slow and fast ramp exercise in man.** *Exp Physiol* 1999, **84**(1):109–120.
25. Whipp BJ, Pardy RL: *Breathing during exercise. Handbook of Physiology.* Baltimore: Williams & Wilkins for APS; 1986.
26. Rabarimanantsoa H, Achour L, Letellier C, Cuvelier A, Muir J-F: **Recurrence plots and Shannon entropy for a dynamical analysis of asynchronisms in mechanical non-invasive ventilation.** *Chaos* 2007, **17:**013115.
27. Achour L, Letellier C, Cuvelier A, Vérin E, Muir J-F: **Asynchrony and cyclic variability in pressure support noninvasive ventilation.** *Comput Biol Med* 2007, **37:**1308–1320.
28. American Thoracic Society: **European Respiratory Society: ATS/ERS Statement on respiratory muscle testing.** *Am J Respir Crit Care Med* 2002, **166:**518–624.
29. Burki NK, Mitchell LK, Chaudhary BA, Zechman FW: **Measurement of mouth occlusion pressure as an index of respiratory centre output in man.** *Clin Sci Mol Med* 1977, **53:**117–123.
30. Whitelaw WA, Derenne JP: **Airway occlusion pressure.** *J Appl Physiol* 1993, **74:**1475–1483.
31. Derenne JP, Couture J, Iscoe S, Whitelaw WA, Milic-Emili J: **Occlusion pressure in man rebreathing CO_2 under methoxyflurane anesthesia.** *J Appl Physiol* 1976, **40:**805–814.
32. Tobin MJ, Laghi F, Walsh JM: *Monitoring of respiratory neuromuscular function., Principles and Practice of Mechanical Ventilation.* New York: McGraw-Hill; 1994.
33. Scott GC, Burki NK: **The relationship of resting ventilation to mouth occlusion pressure. An index of resting respiratory function.** *Chest* 1990, **98**(4):900–906.
34. Budweiser S, Jörres RA, Criée C-P, Langer V, Heinemann F, Hitzl AP, Schmidbauer K, Windisch W, Pfeifer M: **Prognostic value of mouth occlusion pressure in patients with chronic ventilatory failure.** *Respir Med* 2007, **101:**2343–2351.
35. Herrera M, Blasco J, Venegas J, Barba R, Doblas A, Marquez E: **Mouth occlusion pressure (P0.1) in acute respiratory failure.** *Intensive Care Med* 1985, **11:**134–139.
36. Berger KI, Sorkin IB, Norman RG, Goldring RM: **Mechanism of relief of tachypnea during pressure support ventilation.** *Chest* 1996, **109:**1320–1327.

37. Perrigault P-F-O, Pouzeratte YH, Jaber S, Capdevila XJ, Hayot M, Boccara G, Ramonatxo M, Colson P: **Changes in occlusion pressure (P0.1) and breathing pattern during pressure support ventilation.** *Thorax* 1999, **54:**119–123.
38. Rabarimanantsoa-Jamous H: **Qualité des interactions patient-ventilateur en ventilation non invasive nocturne.** *PhD Thesis*. Université de Rouen; 2008.
39. Nikischin W, Gerhardt T, Everett R, Bancalari E: **A new method to analyze lung compliance when pressure-volume relationship is nonlinear.** *Am J Respir Crit Care Med* 1998, **158:**1052–1060.

8

Lumping Izhikevich neurons

Sid Visser[1,2,3*] and Stephan A Van Gils[2,3]

*Correspondence:
sid.visser@nottingham.ac.uk
[1] School of Mathematical Sciences, University of Nottingham, Nottingham, UK
[2] Department of Applied Mathematics, University of Twente, Enschede, The Netherlands
Full list of author information is available at the end of the article

Abstract

We present the construction of a planar vector field that yields the firing rate of a bursting Izhikevich neuron can be read out, while leaving the sub-threshold behavior intact. This planar vector field is used to derive lumped formulations of two complex heterogeneous networks of bursting Izhikevich neurons. In both cases, the lumped model is compared with the spiking network. There is excellent agreement in terms of duration and number of action potentials within the bursts, but there is a slight mismatch of the burst frequency. The lumped model accurately accounts for both intrinsic bursting and post inhibitory rebound potentials in the neuron model, features which are absent in prevalent neural mass models.

Keywords: Neural mass; Izhikevich neuron; Bursting; Subthreshold behavior

Background

The Izhikevich neuron is a computationally simple, yet biologically realistic neuron model [1,2], which belongs to the class of nonlinear integrate-and-fire neurons. Bifurcation analysis of a general class of nonlinear integrate-and-fire neurons is given in [3,4], whereas a more detailed analysis of the quadratic form is given in [5].

When coupled in a complex network, knowledge of the individual neurons does not suffice to predict the network behavior. As large networks are not suitable for bifurcation analysis, one seeks to formulate low dimensional reductions that capture the space-time-averaged behavior of the network. An example is the Wilson-Cowan reduction [6,7] where the mean population activities are governed by a low dimensional system of ordinary differential equations (ODEs) or integro-differential equations respectively. Just as many other lumped formulations of neural activity, these models pivot on the relation between the individual cells' firing rates and aggregate network activity [8-10].

Indeed, the most traditional form of a lumped model for a single population of neurons has the form

$$Qx = f(x, I) \tag{1}$$

with x representing either the mean membrane potential or the postsynaptic conductances/currents.

Q is a linear differential operator which characterizes the synaptic dynamics. Here, f depicts the firing rate as function of the population's own activity x, due to connections within the population, and some external input I. This basic framework is easily extended to include multiple populations or other features, such as spike frequency adaptation and spatial structures (e.g. neural fields). Yet, it should be noted that models of this form only describe the dynamic evolution of populations averaged quantities.

Population density methods, on the other hand, describe the evolution of the distribution of neurons over the state space, which depends on the quantities that are used in the underlying neuron model [11-14]. In [14] application of the moment closure method is investigated, whose steady-state solution, in its simplest form (closure at the first moment), is the so-called mean-field firing-rate model. Next, in [15] this method is applied to arrive at a planar system of switching ODEs for all-to-all coupled networks of Izhikevich neurons with different topologies. This system is subsequently analyzed by means of numerical bifurcation methods [16] and direct simulation. Its variables are the mean adaptation and the synaptic input (here the adjective 'mean' can be omitted due to the all-to-all coupling). One assumption is that the mean adaptation follows the mean firing rate, rather than the individual firing rates. Furthermore a separation of time scales is assumed: the adaptation time scale is assumed to be much larger than the membrane time-constant.

The analogy with [15] is that we also consider networks of Izhikevich neurons. The novelty is *the construction of the Izhikevich Planar Vector Field (IPVF) from which the firing rate function can be read out.* The resulting reduced vector field is three dimensional in contrast to the two-dimensional vector field obtained by [15]; we keep the mean membrane potential as a state variables. This state variable follows naturally from the construction from the associated planar vector field. We claim that it is necessary to retain the membrane potential, particularly to deal with post-inhibitory rebound spikes/bursts.

This paper is structured as follows: Izhikevich' spiking neuron model is studied with parameters corresponding to an intrinsically bursting neuron. Using phase plane analysis, the discontinuous resets are approximated with a smooth branch, which relates to the firing rate of the cell. Most importantly, the proposed approximation leaves all of subthreshold dynamics intact. The resulting model is shown to match quantitatively with the original spiking model, with respect to both spike count and spike timing. Upon making several assumptions relating to network connectivity and synchrony, a random network of Izhikevich neurons can be collapsed into a system of ODEs. Finally, we give two examples of random networks, not necessarily all-to-all connected, of heterogeneous Izhikevich neurons where we compare the true network behavior with the properties of the corresponding reduced vector field.

Methods

Izhikevich neuron in the phase plane

The Izhikevich neuron is given by the following equations [1]

$$\begin{cases} \dot{V} = 0.04V^2 + 5V + 140 - u + I_{\text{ext}}(t) \\ \dot{u} = a(bV - u) \end{cases}, \qquad V(t^-) = 30 \Rightarrow \begin{cases} V(t^+) = c \\ u(t^+) = u(t^-) + d \end{cases} \tag{2}$$

Indeed, with four parameters the model is able to reproduce many of the key features observed in real neural tissue, e.g. intrinsic spiking/bursting, spike-frequency adaptation, and rebound spikes/bursts. Due to a reset process, present in all other integrate-and-fire models, the model does not require a fast recovery variable relating to a resonating current. Hence, the variable u is often seen as a slow variable within the neuron, e.g. Calcium concentration.

Since the Izhikevich model has two variables, phase plane analysis is a natural starting point for characterizing the model. Figure 1 depicts the phase portrait of an intrinsically bursting neuron: Starting on the V-nullcline at point A, the dynamics are exclusively governed by the u component which are, in this case, directed downwards. As u decreases slowly, the fast dynamics in the V direction push the system towards the V-nullcline, causing the orbit to closely follow the quadratic V-nullcline. After passing point B, V increases fast until the threshold $V = 30$ is reached and V is reset onto the green line with increased u. As long as the system's reset point is right of the unstable branch of the V-nullcline, another spike is generated. Note that this condition is equivalent to the resets points lying below point C: the intersection point of the V-nullcline and the reset line $V = c$ (in green). Upon being reset above C, the dynamics in the V-direction are pointing towards the left and the system evolves in that direction until the limit cycle is completed when hitting A.

The main objective is to replace the fast spiking dynamics by a slowly changing variable which represents the neurons firing rate, whilst leaving the subthreshold dynamics intact. In particular:

- The right branch of the V-nullcline corresponds with the firing threshold.
- The left branch of the V-nullcline characterizes the fast dynamics in the subthreshold regime.
- The u-nullcline left of the firing threshold determines the slow dynamics in the subthreshold regime.
- Point B corresponds to the entrance of the bursting regime.
- Point C corresponds to the exit of the bursting regime.
- Point A corresponds to the undershoot after burst termination.

An interpretation of these criteria, expressed as a set of ODEs, is as follows.

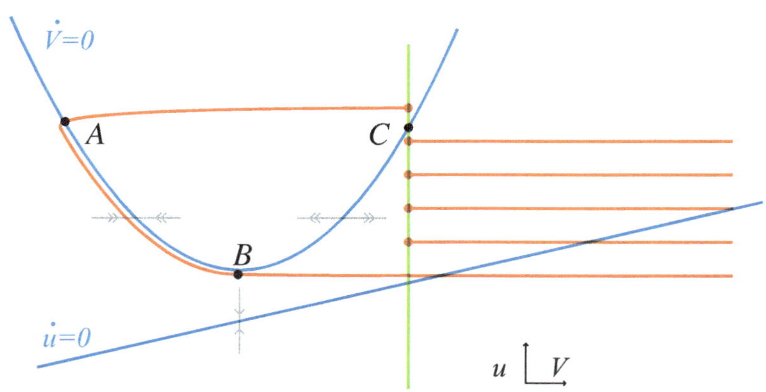

Figure 1 $V - u$ phase plane of the Izhikevich neuron. (Blue) nullclines of V and u, (green) reset value c, and (red) periodic solution of the system. The red dots correspond with resets. See text for description of the points A, B, and C.

Reduction

First, the model's ability to produce spikes, i.e. V quickly reaches 30mV, is taken away by adding another branch to the V-nullcline, see Figure 2. Orbits close to this additional branch correspond with a neuron being in burst-state and, since the condition (non-) bursting is binary, the precise shape and position of this artificial branch are of little importance.

In the bursting regime, the time between consecutive spikes is approximated as follows. Assuming that $J = -u + I_{\text{ext}}(t)$ varies sufficiently slow to be considered as a constant parameter for V, the reduced membrane potential v satisfies:

$$\dot{v} = 0.04v^2 + 5v + 140 + J, \quad v(0) = c \tag{3}$$

where the fact is used that v is reset to c after each spike. Albeit nonlinear, this equation has an analytic solution $v(t; J, c)$ which is used to find the time-to-spike T_s for which $v(T_s; J, c) = 30$:

$$T_s(J; c) = T(J; 30) - T(J; c),$$

with function T given by:

$$T(J; v) = \frac{10}{\sqrt{4J - 65}} \arctan\left(\frac{2v + 125}{5\sqrt{4J - 65}}\right)$$

In order to have $T_s(J; c)$ positive, it is required that (3) has a positive right hand side at the initial value:

$$0.04c^2 + 5c + 140 + J > 0. \tag{4}$$

Inside the bursting regime the instantaneous firing rate is now readily determined as the reciprocal of the inter spike times.

For reasons explained later, the firing rate inside the bursting regime is capped at a minimal value:

$$f_b(J; c) = \begin{cases} \max\left(\frac{1}{T_s(J; c)}, f_{\min}\right) & 0.04c^2 + 5c + 140 + J > 0, \\ f_{\min} & \text{otherwise} \end{cases}$$

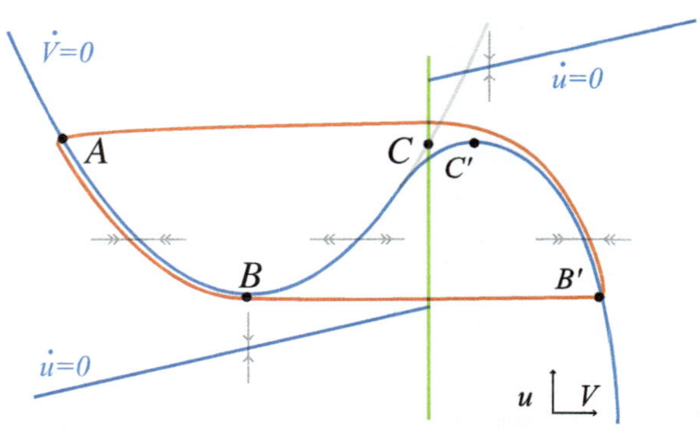

Figure 2 $V - u$ phase plane of the reduced model. (Blue) nullclines of V and u, (green) reset value c, and (red) periodic solution of the system. See text for description of the points A, B, C, and C'.

In order to distinguish between bursting or non-bursting regime, the actual firing rate (where J is substituted) is set to

$$f(V, u, I_{\text{ext}}; c) = H_{\omega}(V - c) f_b(-u + I_{\text{ext}}; c) \tag{5}$$

with H_{ω} a smooth approximation to the Heaviside step function with steepness-parameter ω. The step function aligned at the reset value now distinguishes between the bursting and non-bursting regimes of the neuron. In the remainder of this text, the following is used:

$$H_{\omega}(V) = \frac{1}{1 + \exp(-\omega V)}$$

with $\omega = 20$.

This firing rate is not only the main output variable of the neuron, for it also finds its way back into in the dynamical system. Namely, every time a spike is generated, the recovery variable is increased by d, which yields an time-averaged increase rate of fd in the reduced model. Hence, the recovery variable in the reduced model takes the form:

$$\dot{u}_r = a(bV_r - u_r) + d \cdot f(V_r, u_r, I_{\text{ext}}; c).$$

As a final step, the V-nullcline has to be adjusted such that another branch appears in the phase portrait, c.f. Figure 2. The most natural choice would be to have this branch follow the time-averaged value of V during a spike, i.e.

$$\bar{v} = \frac{1}{T_s(J; c)} \int_0^{T_s(J;c)} v(t) dt$$

Although a closed expression can be found for this integral, this approach is not pursued here. The primary reason is that the precise position and shape of the additional branch are merely insignificant. Instead, a much simpler expression is chosen which does not obscure the resulting equations as much as the expression for $\bar{v}$ would.

Despite the fact that the shape of the new branch is of little importance, the local maximum at C' is not. Since at this point the stability of the slow manifold, i.e. V-nullcline, changes due to a fold bifurcation, it corresponds with the termination of the burst. For that reason it is important that this maximum is in line with the original termination point C, such that the undershoot at A is preserved accurately. Taking into account the requirement that subthreshold behavior should not be modified too much, the new branch is conveniently introduced by an exponential. In this case, the full reduced system takes the form:

$$\begin{aligned} \dot{V}_r &= g(V_r, u_r, I_{\text{ext}}) := 0.04V_r^2 + 5V_r + 140 - u_r + I_{\text{ext}} - \xi \exp(\eta(V_r - c)) \\ \dot{u}_r &= h(V_r, u_r, I_{\text{ext}}) := a(bV_r - u_r) + d \cdot f(V_r, u_r, I_{\text{ext}}; c) \end{aligned} \tag{6}$$

with η and ξ parameters characterizing the additional branch. Appropriate values for these parameters can be determined as follows: Have $u = n_v(V)$ depict the V-nullcline of the original Izhikevich model, then the level of point C in Figure 2 is given by $n_v(c)$.

For $u = n_{vr}(V_r) = n_v(V_r) - \xi \exp(\eta(V_r - c))$ the V_r-nullcline of the reduced model, the following equations should hold to have point C' at the same level as C:

$$n_{vr}(c+\kappa) = n_v(c)$$
$$n'_{vr}(c+\kappa) = 0$$

with κ representing the horizontal separation between C and C', and n'_{vr} denoting the derivative of n_{vr}. Solving the equations results in:

$$\eta = \frac{n'_v(c+\kappa)}{n_v(c+\kappa) - n_v(c)} \qquad \xi = \frac{n_v(c+\kappa) - n_v(c)}{e^{\eta\kappa}}.$$

where n'_v is the derivative of n_v. It is undesirable to choose low values of κ since the additional branch will be very steep and the stiffness of the dynamical system increases unnecessarily. Large values, on the contrary, will result in a branch which is too gradual, such that both the timing and the subthreshold behavior are affected. Values of κ in the interval [0.5, 2] give merely identical results; here is chosen for $\kappa = 0.8$.

Remark

Since the appearance $f_{\min}$ in the reduction is unexpected, it is worthwhile to discuss both the necessity and interpretation for this parameter. In the reduced vector field, an equilibrium on the right branch of the V_r-nullcline corresponds to a tonically firing neuron, while absence of such an equilibrium corresponds to an intrinsically bursting neuron. In the case of the bursting neuron, passage from the right branch of the V_r-nullcline to the left is guaruanteed by raising the u_r-nullcline such that possible equilibria are removed. This raise is achieved by the introduction of $f_{\min}$.

Physiologically, $f_{\min}$ represents the minimum the firing rate for a neuron at which it can still be considered 'bursting'. While this depends on the type of cell at hand, it is generally expected that large cells will have lower firing rates (also within bursts) than small cells.

Single cell comparison

In order to assess the quality of the proposed reduction (6), simulations are performed for different parameters of the model. Not only the membrane potentials and recovery variables of both models, also the output firing rate of the reduced model is matched with the discrete spikes of the original model. For that, define:

$$N(t) := \text{number of spikes in } [0, t] \tag{7a}$$
$$N_r(t) := \int_0^t f(V_r(\tau), u_r(\tau), I_{\text{ext}}(\tau))\, d\tau \tag{7b}$$

for the Izhikevich and the reduced model respectively.

Figure 3 shows a typical comparison of a tonically bursting neuron. The general impression is that the reduction is very successful in mimicking the original Izhikevich model. Although in the latter half of the simulation the inter-burst frequency of the reduced model is higher, which results in a phase difference between both models, the average spiking behavior, on the other hand, is accurately preserved by the reduction.

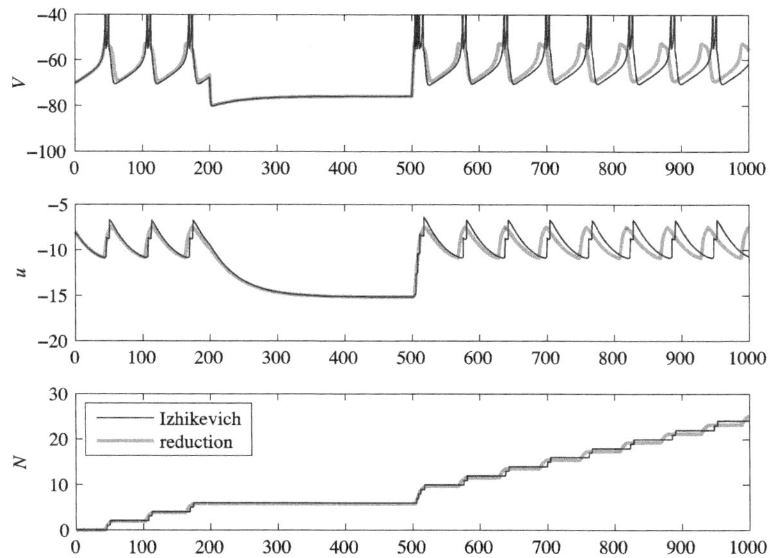

Figure 3 Tonic bursting neuron, which is suppressed with a negative current. Top, middle, and bottom diagrams correspond with V, u and N (c.f. (7)) respectively. Parameters as follows: $a = 0.02$, $b = 0.2$, $c = -55$, $d = 2$, $\kappa = 0.8$, and $f_{\min} = 0.1$. Furthermore, a background current I_0 is given of magnitude 6, which is set to -6 between 200–500 ms, resulting in a suppression of the bursting pattern.

A comparison for the generation of post-inhibotory rebound potentials of a single cell in given in Appendix A.

A population of Izhikevich neurons

With the reduction of a single neuron into place, the next natural step is to consider the global activity patterns of a network of neurons. A single population of merely identical cells is considered first, which can be generalized to multiple populations afterwards.

For now, consider a network of N neurons, which are randomly connected. It is assumed that each spike arriving at a synaptic terminal at time $\hat{t}$ results in an identical response of the synapse, say $q(t - \hat{t})$, yielding the postsynaptic current

$$I_{\text{post}}(t) = \bar{g}q(t - \hat{t})(V_{\text{post}}(t) - E_{\text{syn}})$$

with $\bar{g}$ representing the maximal conductance, E_{syn} the associated reversal potential and V_{post} the membrane potential of the postsynaptic neuron. Furthermore, let q be the impulse response of a linear operator Q such that

$$Qq = \delta$$

where δ is the Dirc delta mass centered at 0 and equality holds in the sense of distributions. For any neuron i, let $T_i(t) := \{\hat{t}_j, j = 1, \ldots, n | 0 < \hat{t}_1 < \ldots < \hat{t}_n \leq t\}$ be the (possibly empty) set of spike times of neuron i up to and including time t. Define the synaptic activation s_i of postsynaptic channels as the convolution of q with the spike times T_i, then the following two formulations are equivalent

$$s_i(t) := \sum_{\hat{t} \in T_i(t)} q(t - \hat{t}) \Leftrightarrow Qs_i(t) = \sum_{\hat{t} \in T_i(t)} \delta(t - \hat{t})$$

Next, have w_{ij} represent the synaptic weight of the connection from neuron j onto neuron i and note that $w_{ij} = 0$ implies that neuron j does not connect with i. The postsynaptic current arriving at cell i is then given by

$$I_{\text{syn},i}(t) = \bar{g}\sum_{j=1}^{N} w_{ij}s_j(t)(V_i(t) - E_{\text{syn}}) \tag{8}$$

with V_i the membrane potential of neuron i.

In order to deduce the average network activity, a number of assumptions has to be made. Since all of these assumptions are common practice in neural mass modeling (c.f. [17]), we only state them briefly in the following part. The reader, however, is encouraged to challenge these assumptions and reflect on their implications.

- **Temporal averaging.** First the individual spike trains are averaged by representing them with the firing rate. That is, every neuron is described by the reduced system (6) as described in the previous section. The corresponding synaptic variable satisfies:

  $$Qs_i = f(V_i, u_i, I_{\text{ext}} - I_{\text{syn},i})$$

- **Randomly connected network.** Let W be a random variable representing the synaptic weight of connections, such that all synaptic weights w_{ij} are assumed to be independently identically distributed as W. Note that if $P(W = 0) > 0$, the network is not necessarily all-to-all connected. Define

  $$\bar{w} := \mathbb{E}W, \qquad S(t) := \sum_{i=1}^{N} s_i(t)$$

  which gives rise to the expected postsynaptic current at cell i

  $$I_{\text{syn},i}(t) = \bar{g}\bar{w}S(t)\,(V_i(t) - E_{\text{syn}}) \tag{9}$$

- **Synchrony.** Neurons receiving a similar input S are assumed to be reside closely together in phase space. Hence, the averages

  $$V(t) := \frac{1}{N}\sum_{i=1}^{N} V_i(t), \qquad u(t) := \frac{1}{N}\sum_{i=1}^{N} u_i(t) \tag{10}$$

  suffice to approximate, c.f. (6)

  $$\dot{V} = \frac{1}{N}\sum_{i=1}^{N} g(V_i, u_i, I_{\text{ext}} - I_{\text{syn},i}) \approx g(V, u, I_{\text{ext}} - I_{\text{syn}}) \tag{11a}$$

  $$\dot{u} = \frac{1}{N}\sum_{i=1}^{N} h(V_i, u_i, I_{\text{ext}} - I_{\text{syn},i}) \approx h(V, u, I_{\text{ext}} - I_{\text{syn}}) \tag{11b}$$

  with

  $$I_{\text{syn}} = I_{\text{syn}}(t) = \bar{g}\bar{w}S(t)\,(V(t) - E_{\text{syn}})$$

  Finally, due to both synchrony and linearity of Q, one finds

  $$QS = Nf(V, u, I_{\text{ext}} - I_{\text{syn}}) \tag{11c}$$

The reduced model (11) now represents the lumped network activity.

Extensions to multiple populations are easily incorporated in a similar fashion. An inhibitory population, for instance, can be included by introducing another differential operator, say Q_2, which corresponds with a slower postsynaptic current, combined with an appropriate reversal potential.

Results

Comparison

Now that the general derivation of a neural mass model based on Izhikevich neurons is discussed, the correspondence with the network of spiking neurons they represent is studied next. Therefore, simulations of both the neural mass and the detailed model are performed and the relation between them is investigated.

First, only the dynamics of a single population of merely identical excitatory neurons are considered, since we are primarily interested in the quantitative and qualitative comparison of both formulations. A model with multiple populations is studied afterwards.

Setup

To start, exponential synapses with a decay rate α are chosen for all connections in all network, hence

$$Q := \frac{d}{dt} + \alpha$$

Furthermore, in the spiking neuron network we set p_{con} as the probability for one neuron to make a connection to another. The synaptic weights of these connections will all be equal and set to w_0. The expected weight for a connection, is now given by:

$$\bar{w} = w_0 p_{\text{con}} \tag{12}$$

The random generation of connections between cells will lead to slight inhomogeneities in the network. Additional inhomogeneities are introduced by randomizing the parameters of the individual cells. This type of perturbation will ensure that the synchronization patterns we observe are due to the network interactions, rather than the intrinsic properties of the neurons. Within the reduced model, the expected value of each of these parameters is used.

Simulations

Simulations of the spiking neuron network are performed with Norns — Neural Network Studio[a]: a dedicated C++ simulation tool with an intuitive Matlab interface. Networks are generated with their connections as described above. The spike trains obtained from Norns are post-processed within Matlab, which we discuss in detail below.

The formulated neural mass model can, depending on the parameters, become a stiff system of ODEs. In that case, Matlab's `ode23s` is used to for numerical integration, while `ode45` is used otherwise.

Comparison

To assess the similarity between the models, it is desirable to compare the same physical quantities within both models. In this light, the synaptic activation S appears to be the best choice: it can be determined from individual spike trains and it is the key variable in the lumped model.

For the lumped model, S is readily available from the simulations. In the network of spiking neurons, we reconstruct the synaptic activation of all neurons by convolving their spike trains with the shape of a postsynaptic response.

One excitatory population

Simulations of a population of excitatory intrinsically firing neurons are shown in Figure 4. The top diagram shows the behavior of a single isolated neuron in the the network and its corresponding firing rate reduction, c.f. Figure 3. The raster plot of a 1000 randomly connected such neurons is shown in the middle and the total activity of both the network and the lumped model are shown in the bottom diagram. It can be seen that, although all connections are excitatory, the frequency of bursting of the network is lower than that of the single cell.

This is explained as follows: as all connections in the network are excitatory, the V-nullcline is effectively lifted during the network bursts due to the external input from synapses. When the network burst terminates, the V-nullcline drops to its initial position. This creates the larger range of u which must be traversed for both the spiking and quiescent phase of the bursting, resulting in more spikes and longer interburst intervals respectively.

These network bursts are, apart from a small difference in interburst-frequency, accurately reproduced by the network reduction (11). Although the interburst-frequency is off, the last network burst (starting at $t = 900$ ms) reveals that the amplitude, shape, and duration are each mimicked in great detail by the neural mass model.

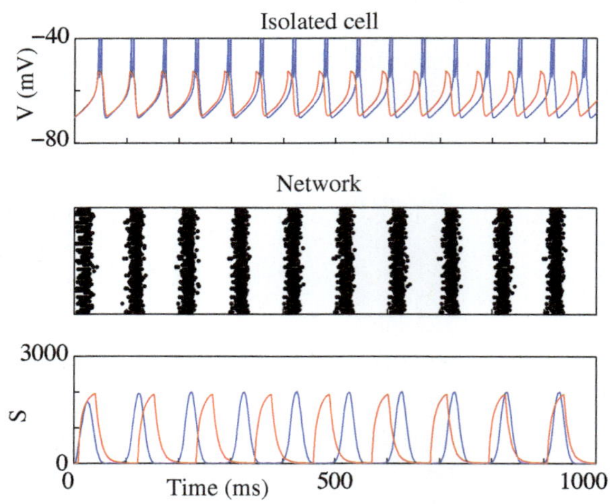

Figure 4 Simulations for an isolated cell and a 1000 cell network of excitatory intrinsically bursting neurons. Top: membrane potential of isolated Izhikevich neuron with average parameters showing the original spiking neuron (blue) and the approximation by the reduced model (red). A random network of 1000 such spiking neurons is created, of which spike trains of 50 neurons are depicted in the raster plot (middle). Bottom diagram depicts the convoluted spike trains of all neurons in the network (blue), c.f. Section 'Comparison'. The synaptic variable S of the corresponding lumped model of this network is also shown (red). Neuron parameters: $a \sim N(0.02, 0.001)$, $b \sim N(0.2, 0.01)$, $c \sim N(-55, 1)$, $d \sim N(2, 0.1)$, $\kappa = 0.8$, $f_{\min} = 0.2$, and $I_{\text{ext}} \sim N(6, 0.2)$. Network parameters: $\alpha = 10^{-1}$, $\bar{g} = 0.002$, $E_{\text{syn}} = 0$, $p_{\text{con}} = 0.03$, and $w_0 = 1$.

Network with rebound bursts

One of the key features of the reduction is that the subthreshold dynamics of the Izhikevich neuron are left intact, and, as a consequence, it is able to generate post-inhibitory rebound potentials, c.f. Appendix A. To illustrate this distinct feature of the model, simulations are performed with a network whose rhythms relies heavily on rebound activity.

A network consisting of three populations of neurons, one inhibitory (I) of 100 neurons and two excitatory (E_1 and E_2) of 300 neurons each, is set up with connections as in Figure 5A. All parameters of the network are given in Appendix B, but here we give a global outline of its properties. Due to inhomogeneities, neurons in E_1 fire either sporadically or are quiescent otherwise and, furthermore, these neurons are capable of firing rebound bursts in response to an inhibitory barrage by neurons in population I. Population E_2 contains intrinsically bursting neurons, which, similar to the example in Section 'One excitatory population', produce regular network bursts. The slow parameter a, however, is decreaed such that the burst-frequency is lower. These bursts will activate the inhibitory neurons in population I.

A simulation of this particular network is shown in Figure 5B by means of raster plots. The pattern arises because a population bursts in E_2 gives rise to activation of I, which, in turn, causes a synchronized post-inhibitory burst in E_1. This burst in E_1 is large enough to initiate the next population burst in E_2 before intrinsic mechanisms would. This completes the cycle. Figure 5C shows the summed synaptic variables of each population in blue.

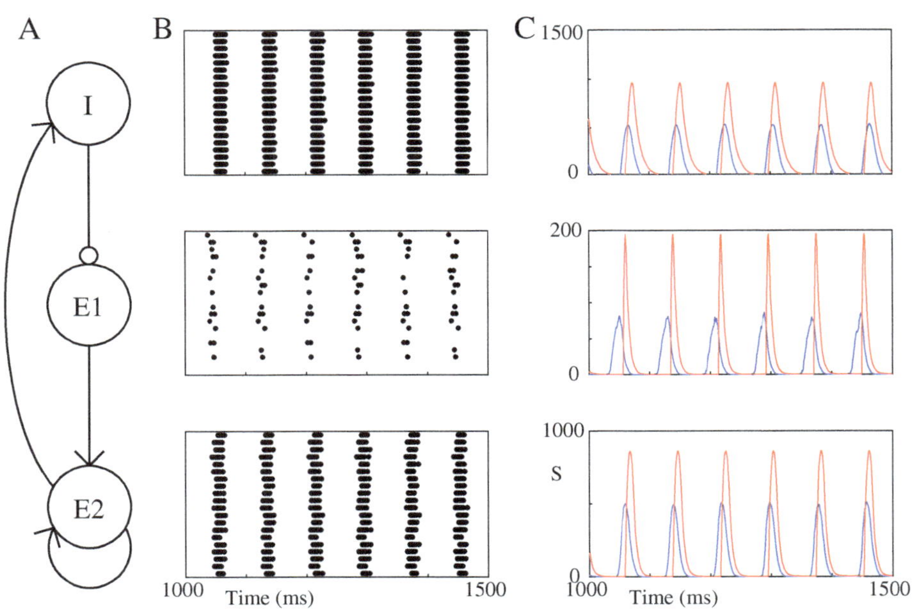

Figure 5 Simulations for a network with post-inhibitory rebound potentials. (A) depicts the network structure of the three populations: two excitatory populations, E_1 and E_2, each consisting of 300 cells and one inhibitory population I of 100 cells. **(B)** shows raster plots for 20 neurons within the populations I, E_1 and E_2 respectively. Note that some cells in E_1 are quiescent. The activation of synapses S for each population is shown in **(C)**, where (blue) corresponds to the spiking neuron network and (red) represents the reduced model. See text for a description of network activity. $E_{inh} = -80$ and other parameters in Appendix B.

The population reduction from this model follows readily from the single population discussed before; for clarity though, the model and its parameters are stated in Appendix B. The time evolution of the resulting reduced model is shown in red in Figure 5C. Apart from a change in amplitude, the frequency of the reduction matches closely with the original spiking neuron network. The fact that the period of this solution matches better than in the single population example, see previous section, could be explained by the following: In the single population model, c.f. Figure 4, the network bursts are initiated when the slow variable u reaches a certain threshold value and are, therefore, dependent on the value of the slow variable at burst termination. Since the network of Izhikevich neurons increases the slow variable u with discrete steps d, a discrepancy is to be expected with respect to the reduced model, which increases u in a time-continuous manner. Although the slow variable plays a role in the rebound bursts in the network at hand, the timing of the (rebound-)bursts depends more so on the synaptic time constants — which are identical in both models.

Next, we note that, apart from a difference in amplitudes, the shapes of the bursts in both populations I and E_2 are accurately mimicked. The discrepancy is largest in population E_1, with respect to both amplitude and burst duration. It appears that it is this gain in E_1 in amplitude which gives rise to an increased excitatory input onto E_2, resulting in a burst with increased magnitude. Subsequently, the stronger burst in E_2 explains the amplitude gain of the inhibitory population I.

The natural question to ask now is why the post-inhibitory rebound bursts in E_1 are poorly reproduced by the reduced model. The fact that some neurons are quiescent during the bursts, as follows from the raster plot in Figure 5, suggests that inhomogeneities play a role in this; either due to variations in single cell parameters, or to differences in number of network connections. Indeed, further investigation has revealed that, at the level of a single cell, the generation of rebound bursts is particularly sensitive to variations in parameter b. Simulations (not shown) of the spiking neuron network in which for all cells b is set to a constant — rather than drawn from a normal distribution, c.f. Appendix B — show that the bursts of neurons in E_1 are more alike than in Figure 5. As a consequence, the result lies closer to the response of the reduced model (in which all inhomogeneities of the original network are averaged); not only for population E_1, also for I and E_2.

Finally, we illustrate the models capability to distinguish between different types of inhibition. Figure 6 shows simulation results for an identical network structure as before, with the exception that the reversal potential of the inhibitory synapses is set to $E_{\text{inh}} = -65$. For this value is close to the resting membrane potentials of the neurons in E_1, the inhibition becomes effectively of the shunting type. Because shunting inhibition does not alter the membrane potential of a cell, it is unable to activate the slow currents responsible for the post-inhibitory rebound potentials and, hence, the neuron is unable to fire such action potentials. This matches with the results in Figure 6B: the raster plot of E_1 shows not only less activity, but it appears more irregular too. This is better seen by the synaptic variable in Figure 6C, which is 3–4 times smaller than in 5C. Here it is also visible that this background activity of E_1 is absent during inhibitory activity of population I, hence the shunting inhibition successfully suppresses this activity.

Without the rebound potentials of E_1, neurons in E_2 receive less excitatory input and the frequency of network bursts is merely determined by the intrinsic dynamics of population

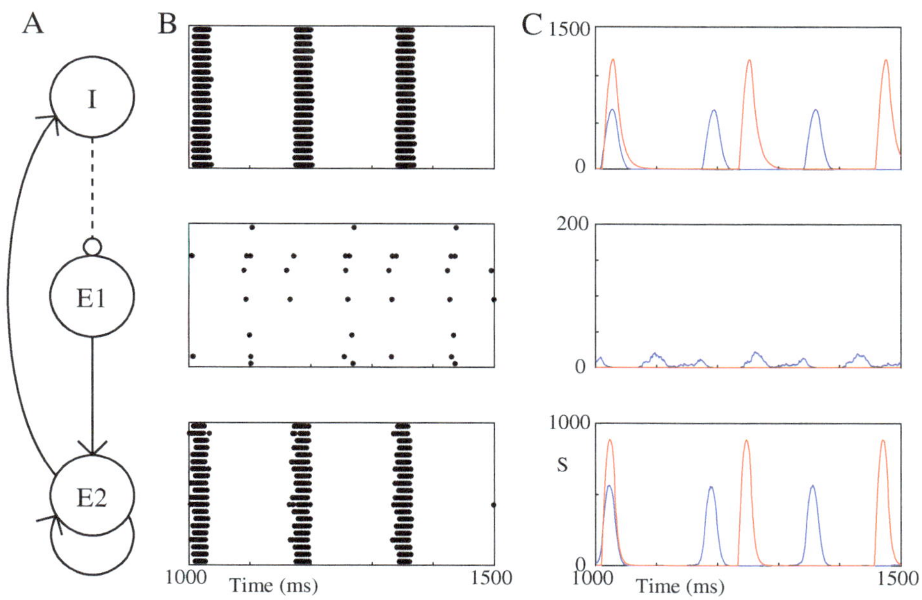

Figure 6 Similar to Figure 5, except that the post-inhibitory rebound potentials are taken away by raising the reversal potential of the inhibitory synapses to $E_{inh} = -65$. (A) depicts the network structure, with the reduced inhibition denoted by the dashed line. Raster plots of the three populations are in **(B)** and the net synaptic activation of the populations in **(C)**. Here, the original network is plotted in (blue) and the lumped model in (red).

E_2. that the neurons in E_2 are not subject to bursts, it remains unclear why the difference in amplitude between the network and corresponding reduction is still in place. A partial explanation is that the excitatory background activity in E_1, which is missed by the population reduction, lowers the threshold of neurons, such that network bursts are generated faster than without input. With a lower inter-burst time, the slow recovery variable u has not decayed as much as it would have if the bursts were further apart. As a result of this, the number of spikes within the burst is expected to be lower. Although this effect plays a role, it is expected that other, yet unidentified mechanisms, are also involved.

Discussion

The work presented here shows how the global activity of networks of intrinsically bursting Izhikevich neurons can be described with a lumped model. This lumped model relies, as the majority of reduced models, on the firing rate of individual neurons. Assisted by phase plane analysis of a single neuron an additional term is proposed, which abolishes the need for discontinuous resets. Effectively, the new term divides the phase space in two regions, corresponding with bursting and non-bursting. Within the bursting regime, the neuron's firing can, once a separation of timescales is assumed, be determined analytically, while in the non-bursting regime, the neuron's original subthreshold behavior is kept intact. We note that the resulting model is Lipschitz smooth and its phase portrait shares key features with the Fitzhugh-Nagumo [18,19] and Morris-Lecar [20] reductions. In our case, however, 'active periods' correspond to an entire burst of action potentials rather than a single pulse. Using simulations, the proposed single cell reduction is shown to match well with the original Izhikevich model. Although a small mismatch is present between the models' bursting frequency, both duration and number of action potentials within the burst are accurately represented.

With the single cell model into place, the scene is set for lumping networks of bursting neurons. The cells' planar dynamics are augmented with another state variable, which corresponds to the activation of synapses within the population. By employment of assumptions common in neural mass modeling, a reduced network model is formulated, which consists of three state variables representing the population averages of membrane potential, slow current, and synaptic activation. Similar as for the single cell model, simulations are performed to assess the quality of the newly proposed model. For one population of excitatory intrinsically bursting neurons, the reduced model matches closely with the network dynamics. Indeed, the network bursts, which last longer than the bursts of isolated cells, are mimicked by the lumped model in terms of amplitude, duration, and shape. While the model's burst frequency is lower than the corresponding spiking neuron network, both models' bursting frequencies in the network are notably lower than the isolated cells.

Here it is important to point out that a very similar result for a single population has been obtained in [15]. The fact that their formulation is expressed in only two state variables, suggests that the third state variable in our formulation is superfluous. We believe, however, that this third state variable, which corresponds to the mean membrane potential, is essential for keeping the neurons' subthreshold behavior intact. To demonstrate this particular trait of our reduction, a network is simulated whose rhythm depends critically on the subthreshold behavior (c.f. Figure 5). More precisely, neurons belonging to one of the three populations are capable of generating post-inhibitory rebound bursts, which affect the network's behavior at a global scope. Apart from a discrepancy in amplitude, simulations show that the reduced model is able to accurately describe the network activity. By raising the inhibitory reversal potential, such that the inhibition becomes shunting, we show that this particular rhythm is, in both the original spiking neuron network and the reduced model, indeed reliant on the presence of post-inhibitory rebound bursts (c.f. Figure 6).

The single cell reduction proposed in this article considers the bursting Izhikevich neuron merely as an example. We indeed assumed the reset value c to be higher than the local minimum of the V nullcline (i.e. $c > -62.5$) and $d > 0$. Although this particular formulation might not be applicable when the parameters at hand do not correspond with the bursting type, we believe that the procedure here can be generalized for all relevant choices of parameters. Going even further, it appears to us that, besides the Izhikevich and other integrate-and-fire models, also higher dimensional models can be reduced in a similar manner; that is, leaving the subthreshold dynamics intact by only adjusting the dynamics in the suprathreshold regime. In this regime, one should seek to eliminate the (very) fast dynamics and determine their contribution on the slow variables. Yet, these fast variables can play a prominent role in the subthreshold dynamics (e.g. to generate post-inhibitory rebound potentials). Elimination of *all* fast variables could therefore have an unanticipated (and hence undesirable) effect on the subthreshold behavior. So caution is required when constructing firing rate reductions.

Endnote

[a]Software freely available via ModelDB entry 154739: http://senselab.med.yale.edu/modeldb/showmodel.asp?model=154739.

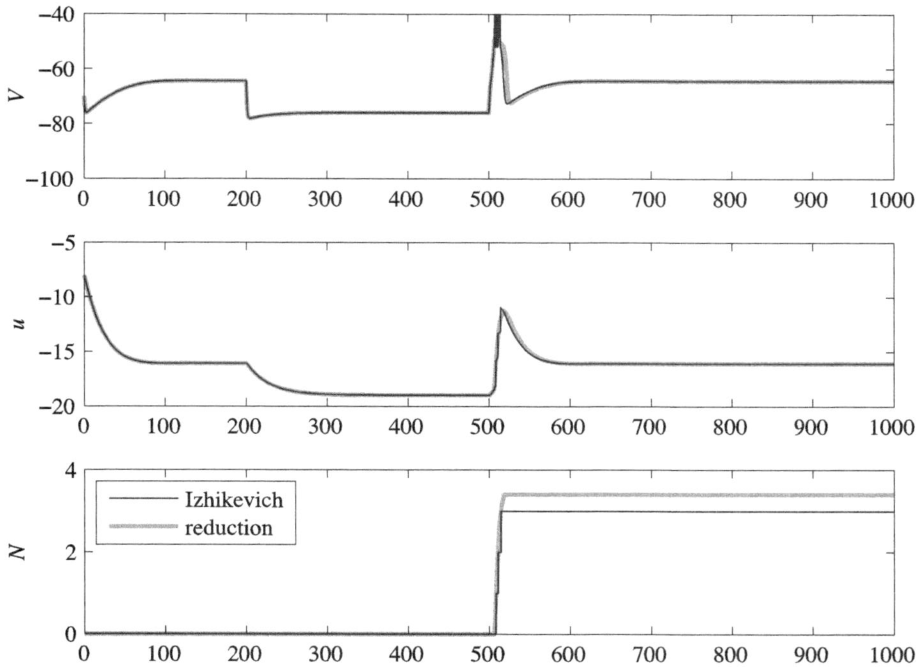

Figure 7 Negative current induces post-inhibitory rebound potentials in both the spiking neuron and the reduction. Top, middle, and bottom diagrams correspond with V, u and N (c.f. (7)) respectively. Parameters as follows: $a = 0.03, b = 0.25, c = -52, d = 2, \kappa = 0.8$, and $f_{\text{min}} = 0.1$. Furthermore, $I_{\text{ext}} = -10$ between 200–500 ms, after which both models fire (the equivalent of) three action potentials.

Appendix A: Single cell comparison with post-inhibitory rebound potentials

Figure 7 shows a concrete example of post-inhibitory rebound potentials in both the spiking model (black) and rate model (gray). An negative current of -10 is injected between 200–500 ms, which lowers the membrane potential. After the inhibiting current is taken away, the spiking neuron fires three action potentials. Not only the count of action potentials is accurately reproduced by the rate model, also the time course of all other state variables is mimicked precisely.

Appendix B: Details of three population network

This appendix provides the details of the three population network described in Section 'Network with rebound bursts'. The spiking neuron is characterized first, followed by the reduced model.

Table 1 Population parameters

Param.	I	E_1	E_2	Description
N	100	300	300	Number of cells in population
a	$N(0.02, 0.001)$	$N(0.03, 0.001)$	$N(0.01, 0.001)$	Decay rate of recovery variable
b	$N(0.2, 0.01)$	$N(0.3, 0.01)$	$N(0.2, 0.01)$	Slope parameter of u nullcline
c	$N(-55, 1)$	$N(-55, 1)$	$N(-55, 1)$	Reset value after spike
d	$N(2, 0.1)$	$N(1, 0.1)$	$N(2, 0.1)$	Increment of u after spike
I_{ext}	$N(0, 0.2)$	-3	$N(6, 0.2)$	Constant applied background current

Table 2 Connectivity parameters

	Source		
Dest.	***I***	***E*₁**	***E*₂**
I			$p = 0.1$ $\bar{g} = 0.02$
E_1	$p = 0.3$ $\bar{g} = 0.02$		
E_2		$p = 0.1$ $\bar{g} = 0.02$	$p = 0.1$ $\bar{g} = 0.005$

B.1 Spiking neuron network

Three population of neurons are generated, one inhibitory I and two excitatory E_1 and E_2. Parameters for neurons within each population are i.i.d. according to the normal distributions given in Table 1.

Connections between neurons are determined randomly, depending on which population they belong to. The population-specific probability p of forming a connection from a source neuron to a destination neuron is denoted in Table 2; as is the related maximal conductance $\bar{g}$. All connections formed have synaptic weight $w_0 = 1$, hence weight distribution W is given by

$$W = \begin{cases} 1 & \text{w.p.}p \\ 0 & \text{w.p.}1 - p \end{cases}$$

Synaptic dynamics of all cells, both excitatory and inhibitory, are described by the first order process

$$\frac{ds}{dt} + \alpha s = \delta(t - T)$$

with T the time of a presynaptic spike. Values of α, as well as the corresponding reversal potential of the resulting postsynaptic current, are in Table 3. The total synaptic current arriving at a particular cell is now determined as in (8).

B.2 Reduction

Following the reduction procedure outlined in Section 'A population of Izhikevich neurons', each population of neurons is reduced to a system of three differential equations. For $k \in \{i, 1, 2\}$ representing the populations I, E_1, E_2 respectively, the state variables are denoted by V_k, u_k, and S_k. Parameters for the right hand side functions f_k, g_k, and h_k, c.f. (11), are for each population set to their expected values given in Table 1. Interactions between populations are exclusively via the synaptic current I_{syn}, which we

Table 3 Synaptic parameters

	Exc.	**Inh.**	**Descriptions**
α	5^{-1}	10^{-1}	Synaptic decay rate
E	0	−80	Reversal potential

state in full detail for clarity. Let $I_{\text{syn},k}$ represent the postsynaptic current arriving at cells of population $k = \{i, 1, 2\}$, then one finds:

$$I_{\text{syn},i} = \bar{g}_{i2} p_{i2} S_2 (V_i - E_{\text{exc}})$$
$$I_{\text{syn},1} = \bar{g}_{1i} p_{1i} S_i (V_1 - E_{\text{inh}})$$
$$I_{\text{syn},2} = (\bar{g}_{21} p_{21} S_1 + \bar{g}_{22} p_{22} S_2)(V_2 - E_{\text{exc}})$$

where p_{kl} and g_{kl} for $k, l \in \{i, 1, 2\}$ are as in Table 2 and we used (12).

Competing interests
The authors declare that they have no competing interests.

Authors' contributions
SV conceived the study, performed the analysis and drafted the manuscript. SvG offered valuable counsel and refined the manuscript. Both authors read and approved the final manuscript.

Acknowledgements
The authors like to thank both reviewers for their constructive feedback and suggestions.

Author details
[1]School of Mathematical Sciences, University of Nottingham, Nottingham, UK. [2]Department of Applied Mathematics, University of Twente, Enschede, The Netherlands. [3]MIRA — Institute for Biomedical Technology and Technical Medicine, University of Twente, Enschede, The Netherlands.

References
1. Izhikevich EM: **Simple model of spiking neurons.** *IEEE Trans Neural Netw* 2003, **14:**1569–1572.
2. Izhikevich EM: *Dynamical Systems in Neuroscience*. Cambridge: The MIT press; 2007.
3. Touboul J: **Bifurcation analysis of a general class of nonlinear integrate-and-fire neurons.** *M J Appl Math* 2008, **68**(4):1045–1079.
4. Touboul J, Brette R: **Spiking dynamics of bidimensional integrate-and-fire neurons.** *SIAM J Appl Dyn Syst* 2009, **8**(4):1462–1506.
5. Shlizerman E, Holmes P: **Neural dynamics, bifurcations, and firing rates in a quadratic integrate-and-fire model with a recovery variable. i: Deterministic behavior.** *Neural Comput* 2012, **24**(8):2078–2118.
6. Wilson HR, Cowan JD: **Excitatory and inhibitory interactions in localized populations of model neurons.** *Biophys J* 1972, **12**(1):1–24.
7. Wilson HR, Cowan JD: **A mathematical theory of the functional dynamics of cortical and thalamic nervous tissue.** *Biol Cybern* 1973, **13**(2):55–80.
8. Lopes da Silva FH, Hoeks A, Smits H, Zetterberg LH: **Model of brain rhythmic activity.** *Biol Cybern* 1974, **15:**27–37. doi:10.1007/BF00270757.
9. Freeman WJ: *Mass Action in the Nervous System*. New York: Academic Press; 1975.
10. Amari S: **Dynamics of pattern formation in lateral-inhibition type neural fields.** *Biol Cybern* 1977, **27**(2):77–87.
11. Omurtag A, Knight BW, Sirovich L: **On the simulation of large populations of neurons.** *J Comput Neurosci* 2000, **8**(1):51–63.
12. Nykamp DQ, Tranchina D: **A population density approach that facilitates large-scale modeling of neural networks: analysis and an application to orientation tuning.** *J Comput Neurosci* 2000, **8**(1):19–50.
13. Abbott L, van Vreeswijk C: **Asynchronous states in networks of pulse-coupled oscillators.** *Phys Rev E* 1993, **48**(2):1483.
14. Ly C, Tranchina D: **Critical analysis of dimension reduction by a moment closure method in a population density approach to neural network modeling.** *Neural Comput* 2007, **19**(8):2032–2092.
15. Nicola W, Campbell SA: **Bifurcations of large networks of two-dimensional integrate and fire neurons.** *J Comput Neurosci* 2013, **35**(1):87–108.
16. Dhooge A, Govaerts W, Kuznetsov YA: **MATCONT: a MATLAB package for numerical bifurcation analysis of ODEs.** *ACM Trans Math Softw* 2003, **29**(2):141–164.
17. Bressloff PC: **Spatiotemporal dynamics of continuum neural fields.** *J Phys A Math Theor* 2011, **45**(3):033001.
18. FitzHugh R: **Impulses and physiological states in theoretical models of nerve membrane.** *Biophys J* 1961, **1**(6):445–466.
19. Nagumo J, Arimoto S, Yoshizawa S: **An active pulse transmission line simulating nerve axon.** *Proc IRE* 1962, **50**(10):2061–2070.
20. Morris C, Lecar H: **Voltage oscillations in the barnacle giant muscle fiber.** *Biophys J* 1981, **35**(1):193–213.

9

Nonlinear analysis of EEG in chess players

Pawel Stepien[1*], Wlodzimierz Klonowski[1] and Nikolay Suvorov[2]

*Correspondence:
pstepien@ibib.waw.pl
[1] Nalecz Institute of Biocybernetics and Biomedical Engineering PAS, Warsaw, Poland
Full list of author information is available at the end of the article

Abstract

Background: The chess game is a good example of cognitive task which needs a lot of training and experience. The aim of this work is to compare applicability of two nonlinear methods - Higuchi Fractal Dimension and Empirical Mode Decomposition - in analysis of EEG data recorded during chess match. We analyzed data of three master chess players registered during their matches with computer program.

Methods: We used two nonlinear methods: Higuchi Fractal Dimension that is a good and fast tool for analyzing signal complexity and modification of Empirical Mode Decomposition, called Sliding Window Empirical Mode Decomposition, that breaks down a signal into its monocomponents. Obtained results are compared with the resting state i.e. EEG during relax witch closed eyes.

Results: The analysis shows higher values of Higuchi Fractal Dimension during the thinking over chess moves than in the players' rest state. There are no statistically significant differences in contribution of EEG bands to total power of EEG calculated with Sliding Window Empirical Mode Decomposition.

Conclusions: Our results show beter applicability of Higuchi Fractal Dimension method for analysis of EEG signals related to chess tasks than that of Sliding Window Empirical Mode Decomposition.

Keywords: Chess; EEG; Higuchi fractal dimension; Fractal analysis; Empirical mode decomposition; Sliding window empirical mode decomposition

Background

The chess game, as expert task which needs a lot of training and experience, interested neuroscientists for a long time. For example Volke *et al.* (1998) [1] investigated 4 simple chess problems like: "pattern recognition" (if the King was or not on the board); if the King was in check; if the King was checkmated; checkmate in one move. They noticed that evoked coherences of EEG signals were sensitive both to sensory and mental activity (respectively theta and beta coherences), and beta coherences depended on the type of task. Later, Volke *et al.* (2002) [2] compared differences in evoked coherences between experts and novices - the experts showed higher values during tasks compared to the resting condition, whereas the novices - lower values.

The same chess problems were studied by Wright *et al.* (2013) [3]. They found differences in amplitudes of N2 and P3 components of event-related potential signals between professionals and beginners.

The game similar to chess, shogi, was studied by Wan *et al.* (2011) [4] with a functional magnetic resonance imaging (fMRI). They found two activations specific to experts: one in the precuneus of the parietal lobe during perception of board patterns, and the other in the caudate nucleus of the basal ganglia during quick generation of the best next move.

Hänggi *et al.* (2014) [5] showed (with fMRI) that chess experts recruit different psychological functions and activate different brain areas during chess-related tasks: "Grey matter volume and cortical thickness were reduced in chess players compared with those of control men in the OTJ and precunei. The volumes of both caudate nuclei were not different between groups, but correlated inversely with the years of chess playing experience".

Data

Data of three chess master players: one of them is the Grandmaster (player A - 22 years old male) and others are the FIDE Masters (players B - 23 years old male and C - 22 years old male) were analyzed. The data were recorded by the group of Prof. Nikolay Suvorov (Institute of Experimental Medicine NWB RAMS, St.Petersburg, Russia) [6]. Players took the chess matches with computer program Fritz-10 (player A - two matches, B - two matches and C - one match). During (and before) each match the EEG signal from each player was recorded in 10-20 system (the following electrodes were used: Fp1, Fpz, Fp2, F7, F3, Fz, F4, F8, T3, C3, Cz, C4, T4, T5, P3, Pz, P4, T6, O1, Oz, O2 and A1, A2 as reference electrodes; sampling frequency was 500 Hz). To avoid the artifacts players had closed eyes during the all match (and in control EEG recording before the play). The players had to store the actual arrangement on a chessboard in memory and they were informed about opponent's moves by the physician's voice. Every match was won by the computer program, and had about 40 moves and took 40 ± 5 minutes except first player A's first match which took 66 minutes.

EEG signal was filtered with band pass 0.53 - 150 Hz and with notch 45 - 55 Hz filter.

Methods

Higuchi fractal dimension

Higuchi Fractal Dimension (*DF*) [7] is calculated directly from the signal, in time domain, without reconstruction of strange attractors in multidimensional phase space. Value of *DF* is always between 1 and 2 (simple curve has dimension equal 1 and a plane has dimension equal 2). Fractional part of *DF* shows what fraction of the plane is "filled up" by the curve, so it is a measure of the signal's complexity. Further description of Higuchi Fractal Dimension is in Klonowski (2007) [8].

DF of the resting state (relaxed with closed eyes) EEG record (about 250 seconds long) was calculated in 0.5 second length time windows and then averaged for each channel. *DF* of EEG during the match was calculated at time intervals corresponding to the thinking on each chess movement by the player (it gave about 40 values of *DF* for specific match) and then averaged for each channel. The maps of mean *DF* for each channel were constructed. The standard deviation was used as a parameter of statistical significance.

Empirical mode decomposition

Empirical Mode Decomposition (*EMD*) is an entirely data-driven algorithm which breaks down nonstationary, multicomponent signal into its monocomponents. Such monocomponents are called Intrinsic Mode Function*s (IMFs)*. Each *IMF* must fulfill the following criteria:

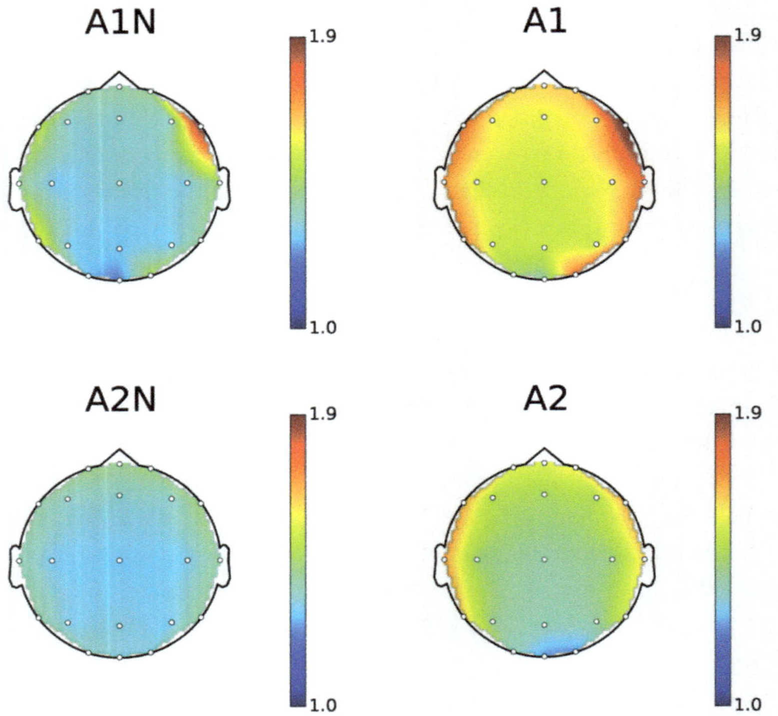

Figure 1 Mean *DF* for player A: before the first match (A1N); during the first match (A1); before the second match (A2N) and during the second match (A2).

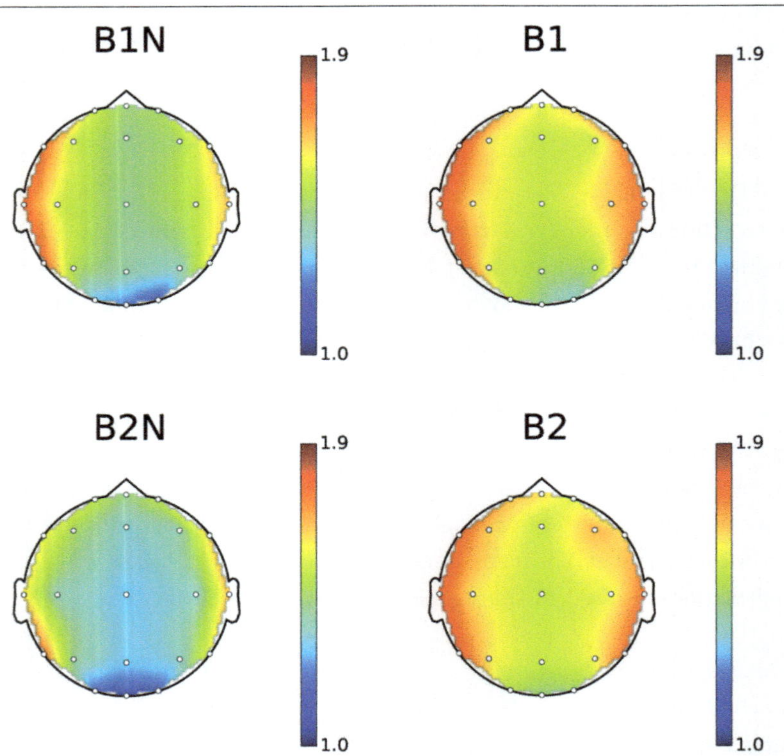

Figure 2 Mean *DF* for player B: before the first match (B1N); during the first match (B1); before the second match (B2N) and during the second match (B2).

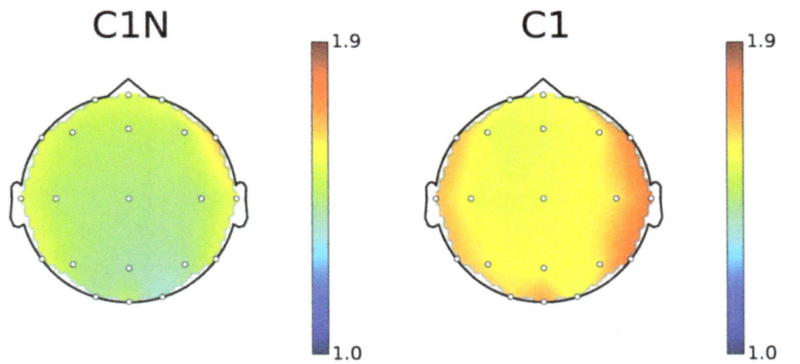

Figure 3 Mean *DF* for player C: before the first match (C1N); during the first match (C1); before the second match (C2N) and during the second match (C2).

- the number of extrema and zeros are equal or their difference is not greater than 1,
- the signal has "zero mean" - the mean value of the envelope determined by maxima and the envelope determined by minima is equal 0 at every point.

Further description of *EMD* is in Huang *et al.* (1998) [9]. We used modified algorithm called Sliding Window Empirical Mode Decomposition (*SWEMD*, see [10]) which speeds up the calculation about 10 times when compared with 'classical' *EMD* method.

Signal decomposed by *SWEMD* can be further analyzed by obtaining the marginal Hilbert-Huang Spectrum *hhs*(*f*), see equation (17) in [10].

The information about the contribution of a frequency range to the total power of signal can be obtained by integration of the marginal Hilbert-Huang Spectrum with frequency

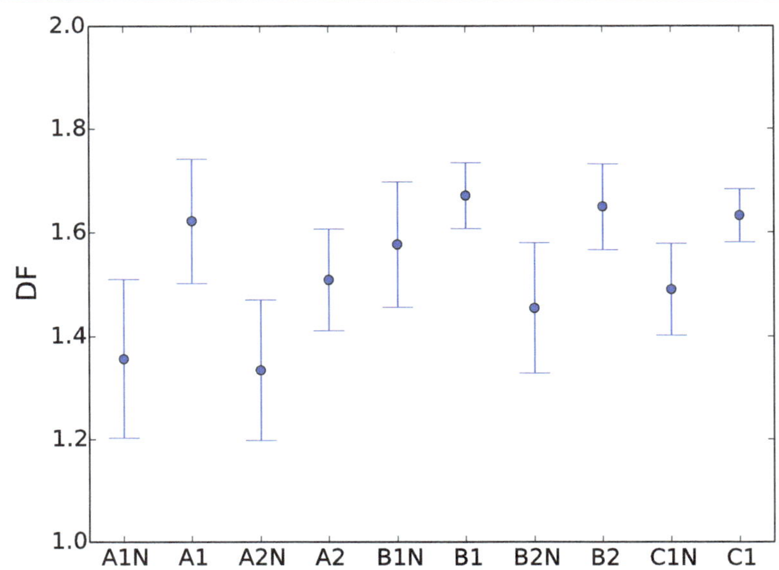

Figure 4 Mean *DF* and standard deviation of EEG on electrodes T3, C3, Cz, C4 and T4 before matches and during matches for all players. The difference of mean *DF* is visible between matches and resting state for A1, B2 and C1 cases.

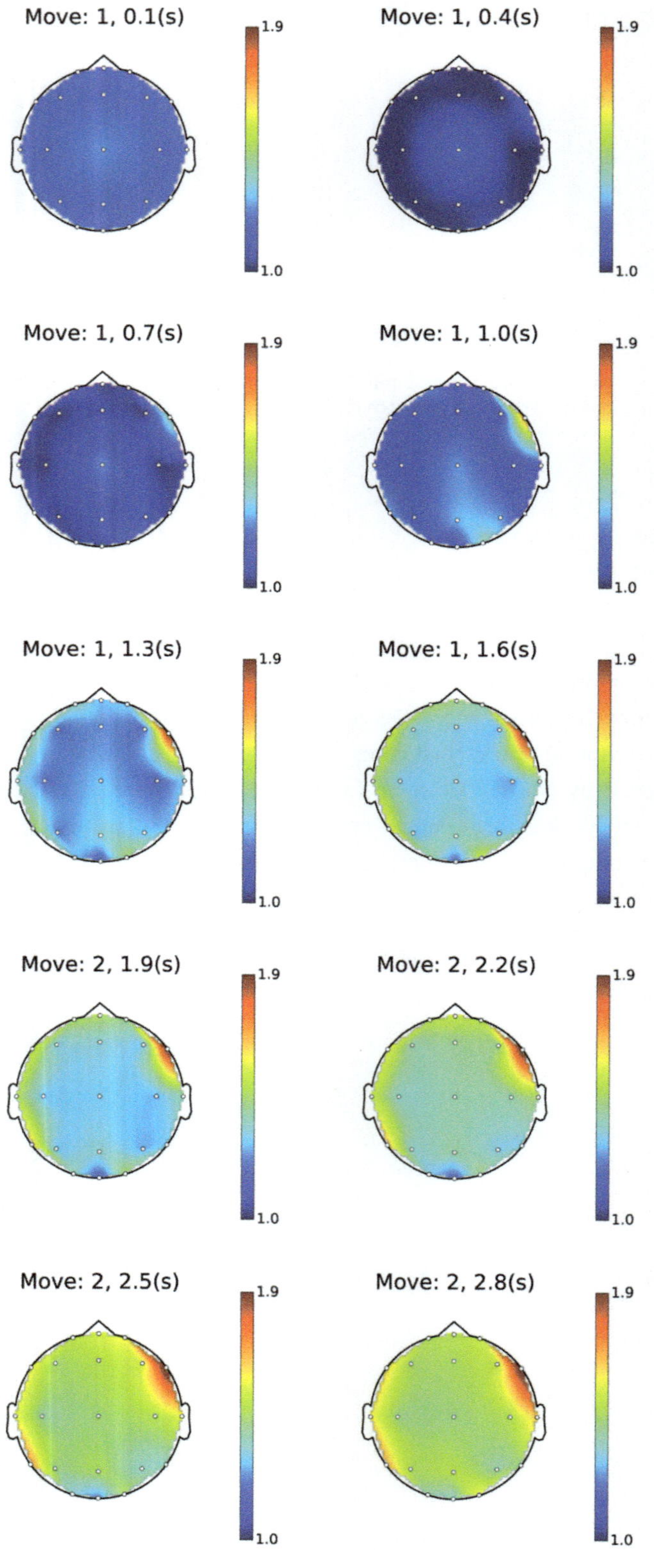

Figure 5 Dynamics of *DF* at the beginning of the first game (A1) of player A.

range limits. For example the contribution of the alpha band in EEG can be calculated with:

$$\alpha\ power = \frac{\int_8^{13} hss(f)df}{\int_0^{\infty} hss(f)df}$$

Contribution of each EEG band was calculated using *SWEMD* and the marginal Hilbert-Huang Spectrum in similar way as Higuchi Fractal Dimension. For the resting state contribution of each band was calculated in 0.5 second length time windows and then averaged for each channel and standard deviation was calculated. Bands' contributions during the match was calculated at time intervals corresponding to the thinking on each chess movement by the player and then averaged for each channel and standard deviation was obtained.

Results

Figures 1, 2 and 3 show (respectively for players A, B and C) differences in *DF* between the resting condition with closed eyes before the match and during the chess play. Figure 4 shows mean *DF* and standard deviation of EEG on electrodes T3, C3, Cz, C4 and T4 before matches and during matches for all players.

Attached movie (Additional file 1) shows dynamics of Higuchi Fractal Dimension during the first match (A1) of player A (first ten moves). Interesting is the first move, when increasing of *DF* is noticed. After about 2.5 seconds (Figure 5) the pattern of *DF* becomes similar to Figure 1 A1.

Figures 6, 7, 8, 9 and 10 show mean contribution and standard deviation of each EEG (respectively: theta, delta, alpha, beta and gamma), calculated using *SWEMD*, on electrodes T3, C3, Cz, C4 and T4 before matches and during matches for all players.

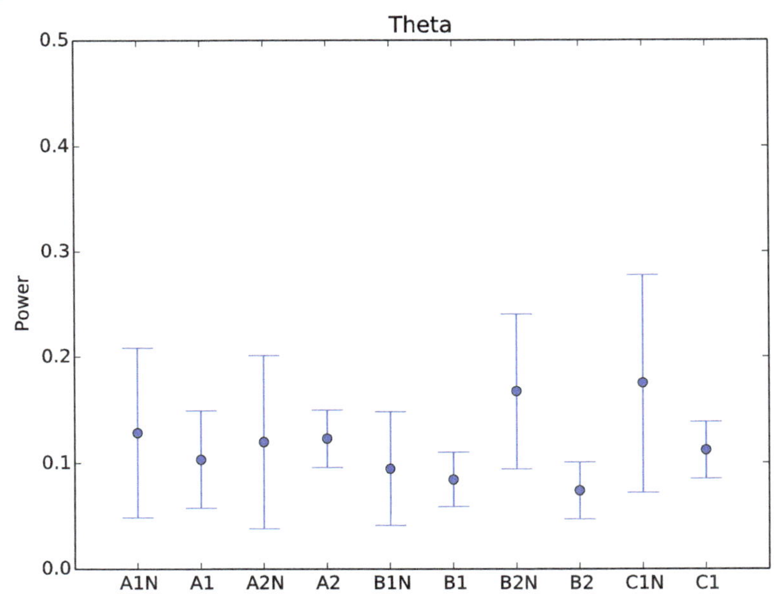

Figure 6 Mean contribution and standard deviation of theta EEG band on electrodes T3, C3, Cz, C4 and T4 before matches and during matches for all players.

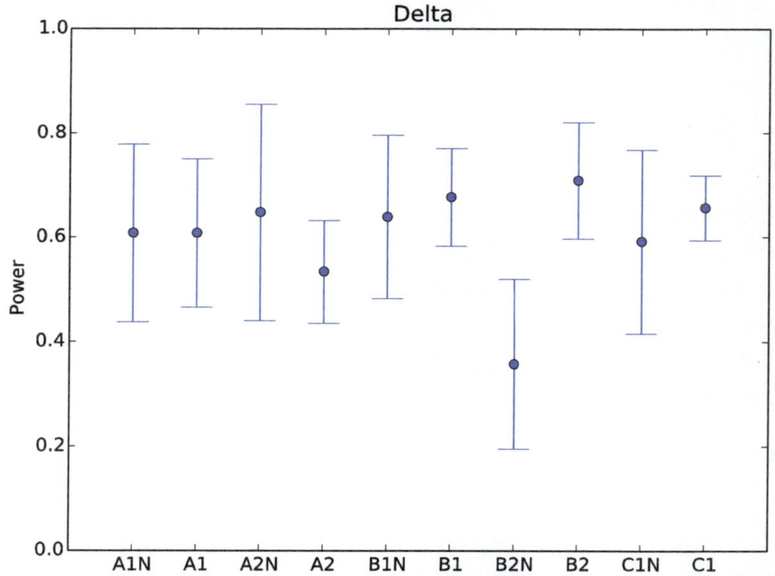

Figure 7 Mean contribution and standard deviation of delta EEG band on electrodes T3, C3, Cz, C4 and T4 before matches and during matches for all players.

Discussion

The results show that *DF* is higher during the match than in the resting state, mainly in lateral electrodes. Although this trend is clearly visible, it is not always statistically significant, in particular in games A2 and B1, as shown in Figure 4. Analysis of EEG bands' contributions, calculated with *SWEMD*, to total power of EEG gives no statistically significant differences between the resting state and chess match near in every case (Figures 6, 7, 8, 9 and 10).

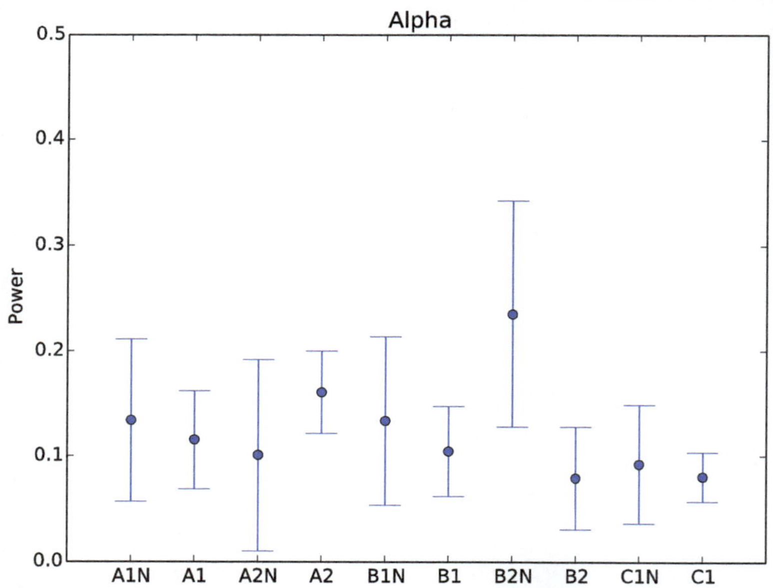

Figure 8 Mean contribution and standard deviation of alpha EEG band on electrodes T3, C3, Cz, C4 and T4 before matches and during matches for all players.

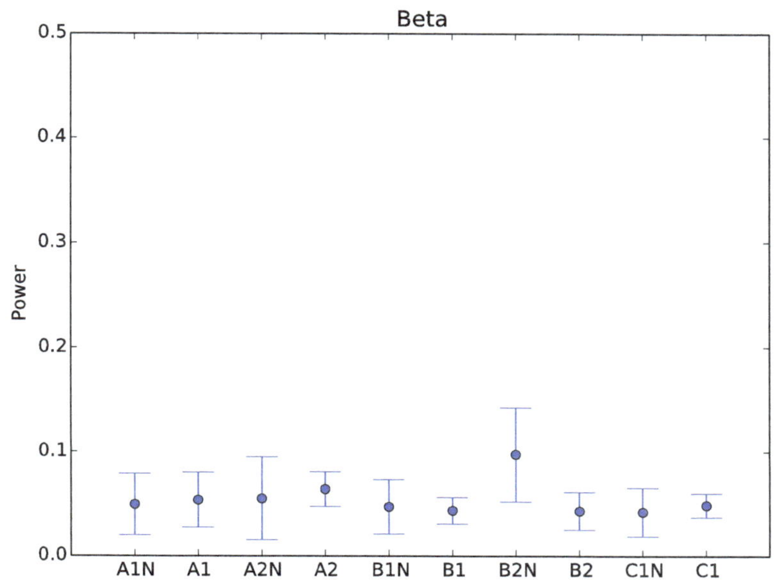

Figure 9 Mean contribution and standard deviation of beta EEG band on electrodes T3, C3, Cz, C4 and T4 before matches and during matches for all players.

Conclusions

Presented results show better applicability of Higuchi Fractal Dimension method for analysis of EEG signals related to chess tasks than that of Sliding Window Empirical Mode Decomposition. *DF* shows that the EEG signal during the game is more complex, non-linear, and non-stationary even when there are no significant differences between game and relaxed state in contribution of different EEG bands to total power of the signal. There

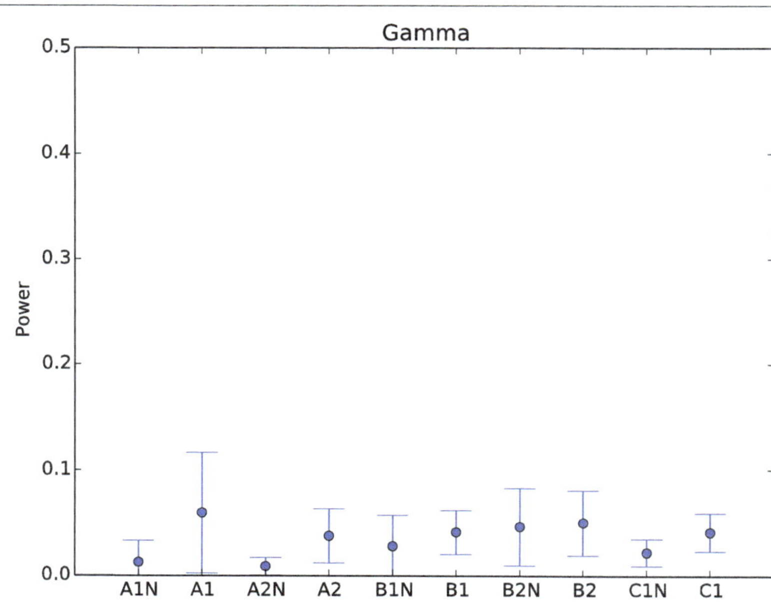

Figure 10 Mean contribution and standard deviation of gamma EEG band on electrodes T3, C3, Cz, C4 and T4 before matches and during matches for all players.

is the need of gathering more data from more chess experts and of comparing them with data from novice chess players.

Additional file

Additional file 1: Dynamics of Higuchi Fractal Dimension during the first match (A1) of player A (first ten moves of the match). The installed MPEG4 decoder is needed to open this movie.

Competing interests
The authors declare that they have no competing interests.

Authors' contributions
PS has performed the data analysis. PS and WK have written the paper. NS has provided data. All authors read and approved the final manuscript.

Acknowledgements
This work was supported by Nalecz Institute of Biocybernetics and Biomedical Engineering, Polish Academy of Sciences, Warsaw, through statutory project 4.4/st/14.

Author details
[1]Nalecz Institute of Biocybernetics and Biomedical Engineering PAS, Warsaw, Poland. [2]Institute of Experimental Medicine NWB RAMS, St. Petersburg, Russia.

References
1. Volke H-J, Dettmar P, Richter P, Rudolf M. Evoked coherences of eeg in chess playing. Int J Psychophysiol. 1998;30(1-2):225–6.
2. Volke H-J, Dettmar P, Richter P, Rudolf M, Buhss U. On-coupling and off-coupling of neocortical areas in chess experts and novices as revealed by evoked eeg coherence measures and factor-based topological analysis–a pilot study. J Psychophysiol. 2002;16(1):23.
3. Wright MJ, Gobet F, Chassy P, Ramchandani PN. Erp to chess stimuli reveal expert-novice differences in the amplitudes of n2 and p3 components. Psychophysiology. 2013;50(10):1023–33.
4. Wan X, Nakatani H, Ueno K, Asamizuya T, Cheng K, Tanaka K. The neural basis of intuitive best next-move generation in board game experts. Science. 2011;331(6015):341–6.
5. Hänggi J, Brütsch K, Siegel AM, Jäncke L. The architecture of the chess player's brain. Neuropsychologia. 2014;62(0): 152–62.
6. Suvorov NB, Bozhokin SV, Polonski YuZ. Electrophysiological correlates of human intellectual activity during a chess-game with the computer program blindly. Wavelet analysis. Inf Control Syst. 2012;3:71–76.
7. Higuchi T. Approach to an irregular time series on the basis of the fractal theory. Physica D: Nonlinear Phenom. 1988;31(2):277–83.
8. Klonowski W. From conformons to human brains: an informal overview of nonlinear dynamics and its applications in biomedicine. Nonlinear Biomed Phys. 2007;1(1):5.
9. Huang NE, Shen Z, Long SR, Wu MC, Shih HH, Zheng Q, et al. The empirical mode decomposition and the Hilbert spectrum for nonlinear and non-stationary time series analysis. Proc R Soc London. Ser A: Math Phys Eng Sci. 1998;454(1971):903–95.
10. Stepien P. Sliding window empirical mode decomposition - its performance and quality. EPJ Nonlinear Biomed Phys. 2014;2(1):14.

10

Sliding Window Empirical Mode Decomposition -its performance and quality

Pawel Stepien

Correspondence:
pstepien@ibib.waw.pl
Nalecz Institute of Biocybernetics and Biomedical Engineering PAS, Warsaw, Poland

Abstract

Background: In analysis of nonstationary nonlinear signals the classical notion of frequency is meaningless. Instead one may use *Instantaneous Frequency (IF)* that can be interpreted as the frequency of a sine wave which locally fits the signal. *IF* is meaningful for monocomponent nonstationary signals and may be calculated by *Hilbert transform (HT)*.

Methods: A multicomponent signal may be decomposed into its monocomponents. *Empirical Mode Decomposition (EMD)*, developed by Norden E. Huang, is a new method of such breaking down of a signal into its monocomponents. *EMD* combined with *HT* (called *Hilbert-Huang Transform*) is a good tool for analyzing nonstationary signals, but unfortunately the traditional *EMD* algorithm consumes a lot of time and computer resources. I propose a modified *EMD* algorithm - *Sliding Window EMD, SWEMD*.

Results: Proposed algorithm speeds up (about 10 times) the computation with acceptable quality of decomposition.

Conclusions: *Sliding Window EMD* algorithm is suitable for decomposition of long signals with high sampling frequency.

Keywords: Empirical Mode Decomposition; EMD; Fast EMD; Sliding Window Empirical Mode Decomposition; SWEMD; Signal analysis

Background

Empirical Mode Decomposition (*EMD*) is a new method of analyzing nonlinear nonstationary signals. It was proposed by Norden E. Huang in 1996 [1]. *EMD* is an entirely data-driven algorithm and it does not depend on any predefined basis functions. *EMD* breaks down nonstationary, multicomponent signal into its monocomponents. Such monocomponents are called *Intrinsic Mode Function*s *(IMFs)* (see subsection 'Intrinsic mode function'). Analyzing long signals with *EMD* is time consuming or even impossible in reasonable time due to the fact, that spline interpolation of large number of points takes a lot of computer resources. For such signals 'classical algorithm' needs some modifications. There are several modifications of *EMD* improving its performance. C. D. Blakely developed *FastEMD* method [2]; instead of cubic spline interpolation, matrix-free moving least-squares approximation is used. Q. Chen et al. proposed representing of *IMFs* as B-splines [3]. This method was tested on earthquake records. It decomposes signal into more modes than 'classical' *EMD* so it gives better frequency resolution. Unfortunately authors did not give information about time performance of this method. S. Meignen and V. Perrier [4] eliminate sifting process (subsection 'EMD algorithm') by calculating *IMFs* through the resolution of a quadratic programming problem with equality and inequality constraints. This method should greatly accelerate decomposition of short signals (no sifting iterations), but for long signals it can take longer than 'classical' *EMD* (quadratic optimization problem is solved in polynomial time, spline interpolation is solved in linear time). *A.* Roy and J. F. Doherty [5] use the raised cosine filter based interpolation instead splines, *RCEMD*. They tested it on signal being a superposition of two harmonics with different frequencies. Depending on harmonic separation complexity (the ratio of the frequencies), calculations were performed from 0 to 10 times faster than when using the 'classical' method. Their method also needs lower sampling rates of signal than the 'classical' *EMD* [6]. Further investigation of this method [7] shows that *RCEMD* algorithm is generally faster than *EMD* for short signal, so the authors developed windowed version of *RCEMD* which speeds up calculation about 10 times.

For long data sets, I suggest to calculate *EMD* in a small window and to slide this window along the time axis (*Sliding Window EMD, SWEMD*) [8]. In this paper I describe proposed algorithm (subsection 'Sliding Window Empirical Mode Decomposition'), test its performance (subsection 'Performance of SWEMD') and quality (subsection 'Quality of SWEMD'). Then I compare time-frequency characteristics of specimen signals obtained by my algorithm, with those obtained by 'classical' *EMD* and those obtained by typical calculation of *EMD* in time windows (subsection 'Quality of SWEMD').

Methods

Empirical Mode Decomposition

Intrinsic mode function

EMD decomposes signal into, so called, *intrinsic mode functions* (*IMFs*). Each *IMF* is a signal that must meet the following criteria [1]:

- the number of extrema and zeros are equal or their difference is not greater than 1,
- the signal has "zero mean" - the mean value of the envelope determined by maxima and the envelope determined by minima is equal 0 at every point.

These conditions show the essence of the *EMD* decomposition: non-stationary signal is decomposed into symmetrical oscillations around zero amplitude, which are easy for further analysis.

EMD algorithm

The first step of the *EMD* algorithm [1,9] is extraction of extrema from original signal $x(t)$ end creation of the upper envelope e_{max} by cubic spline interpolation of maxima and of the lower envelope e_{min} by interpolation of minima (Figure 1a). Then mean value of envelopes is calculated:

$$m(t) = \frac{e_{max} + e_{min}}{2}. \tag{1}$$

The mean value from Eq. 1 is subtracted from original data:

$$imf_1(t) = x(t) - m(t). \tag{2}$$

Above procedure is called *sifting process.*

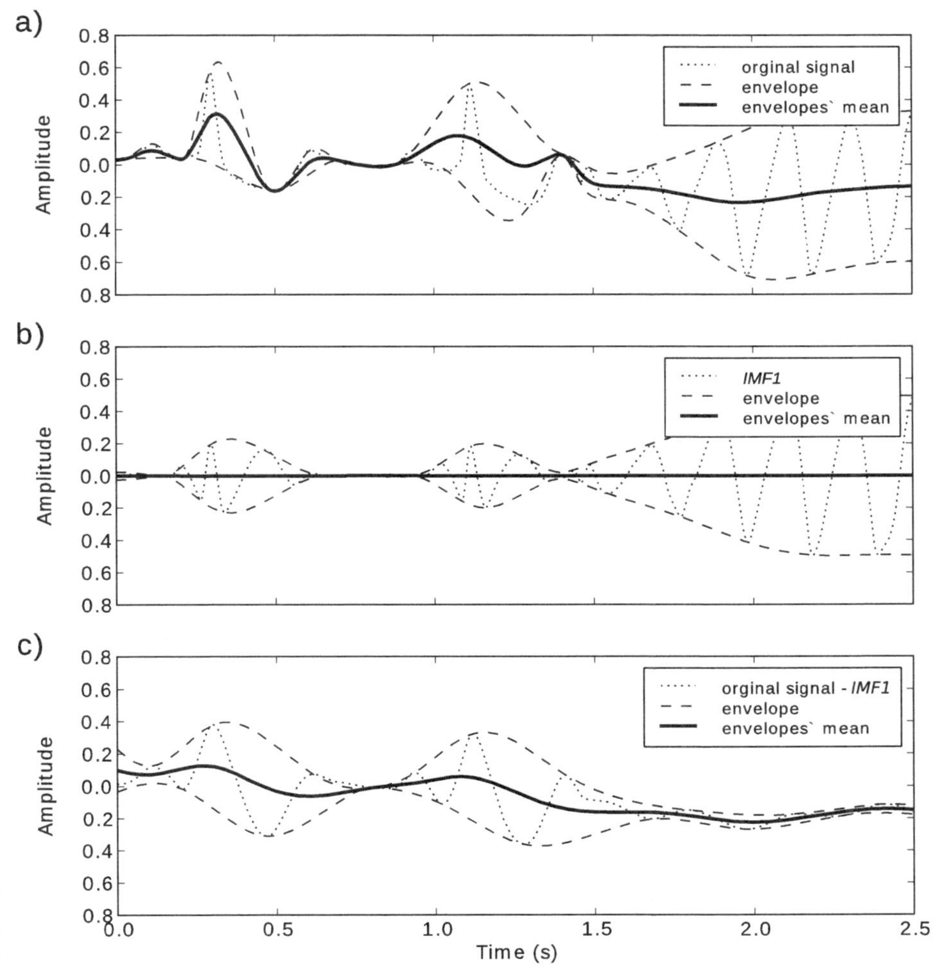

Figure 1 Main steps of *EMD*: a) start of decomposition, b) end of decomposition of first *IMF*, c) start of second *IMF* decomposition [10].

Then $imf_1(t)$ is treated as input data for next *sifting process*. Mean value $m(t)$ of envelopes of $imf_1(t)$ is calculated and this value is subtracted from $imf_1(t)$:

$$imf_1(t) := imf_1(t) - m(t), \tag{3}$$

where $' :=' $ means *it becomes*.

Sifting process is iterated till $imf_1(t)$ fulfills conditions of *IMF* signal ((Figure 1b), see subsection 'Intrinsic mode function'). When the first *sifting process* is finished (see subsection 'Sifting process stopping criteria') then the original signal is reduced by the first mode (Figure 1c):

$$IMF_1(t) := imf_1(t) \tag{4}$$

$$r_1(t) = x(t) - IMF(t). \tag{5}$$

The residue $r_1(t)$ of subtraction (5) is treated as input data for extraction of second *IMF* (next sifting loop). Procedure is looped to extract all *IMFs:*

$$r_i(t) = r_{i-1}(t) - imf_i(t), \tag{6}$$

where i is index of current mode. Decomposition is finished when residue $r_i(t)$ has less than three extrema (there is no possibility to construct envelopes) or all its points are nearly equal zero.

The sum of all *IMF* components and the residue gives the original signal:

$$r_n + \sum_{i=1}^{n} imf_i(t) = x(t), \tag{7}$$

where n is number of all modes.

Algorithm of *EMD* is shown, as a block diagram, in Figure 2.

Sifting process stopping criteria

According to the second criteria of *IMF*, the mean of its envelope is equal to 0 at every point of *IMF*. In fact, it is very difficult to get numerically this value equal to zero at every point. What is needed is an approximation. In my work I use the stop conditions of the sifting procedure proposed by Rilling et al. [9]. In each iteration ratio of the mean value of the envelope of iterated mode and the amplitude of this envelope is checked:

$$\sigma(t) = \left| \frac{m(t)}{a(t)} \right|, \tag{8}$$

where:

$$m(t) = \frac{e_{max}(t) + e_{min}(t)}{2}; \; a(t) = \frac{e_{max}(t) - e_{min}(t)}{2} \tag{9}$$

and e_{max} is the upper envelope, e_{min} - the lower envelope.

There are two thresholds:

- θ_1 - guaranteeing globally small fluctuations of the envelopes' mean around zero,
- θ_2 - locally allowing higher fluctuations.

Sifting process is stopped if $\sigma(t) < \theta_1$ is true for $(1 - \alpha)$ part of the number of signal's points, and if $\sigma(t) < \theta_2$ is true for the rest of points. Typical values for these parameters are as follows:

$$\alpha = 0,05; \;\; \theta_1 = 0,05; \;\; \theta_2 = 10 \cdot \theta_1. \tag{10}$$

These values of parameters allow a compromise between quality and speed of decomposition process.

Hilbert-Huang spectrum

Signal decomposed into *IMFs* can be further analyzed with time-frequency characteristic by obtaining *Hilbert-Huang spectrum*. For this purpose, the analytic signal is created from each *EMD* mode:

$$imf_{a_k}(t) = imf_k(t) + jH[\,imf_k(t)] \tag{11}$$

where $H(imf_k(t))$ is *Hilbert transform* of k-th *IMF* and j means imaginary part. *Hilbert transform*is defined as:

$$H[\,x(t)] = \frac{1}{\pi} P \int_{-\infty}^{\infty} \frac{x(t')}{t - t'} dt', \tag{12}$$

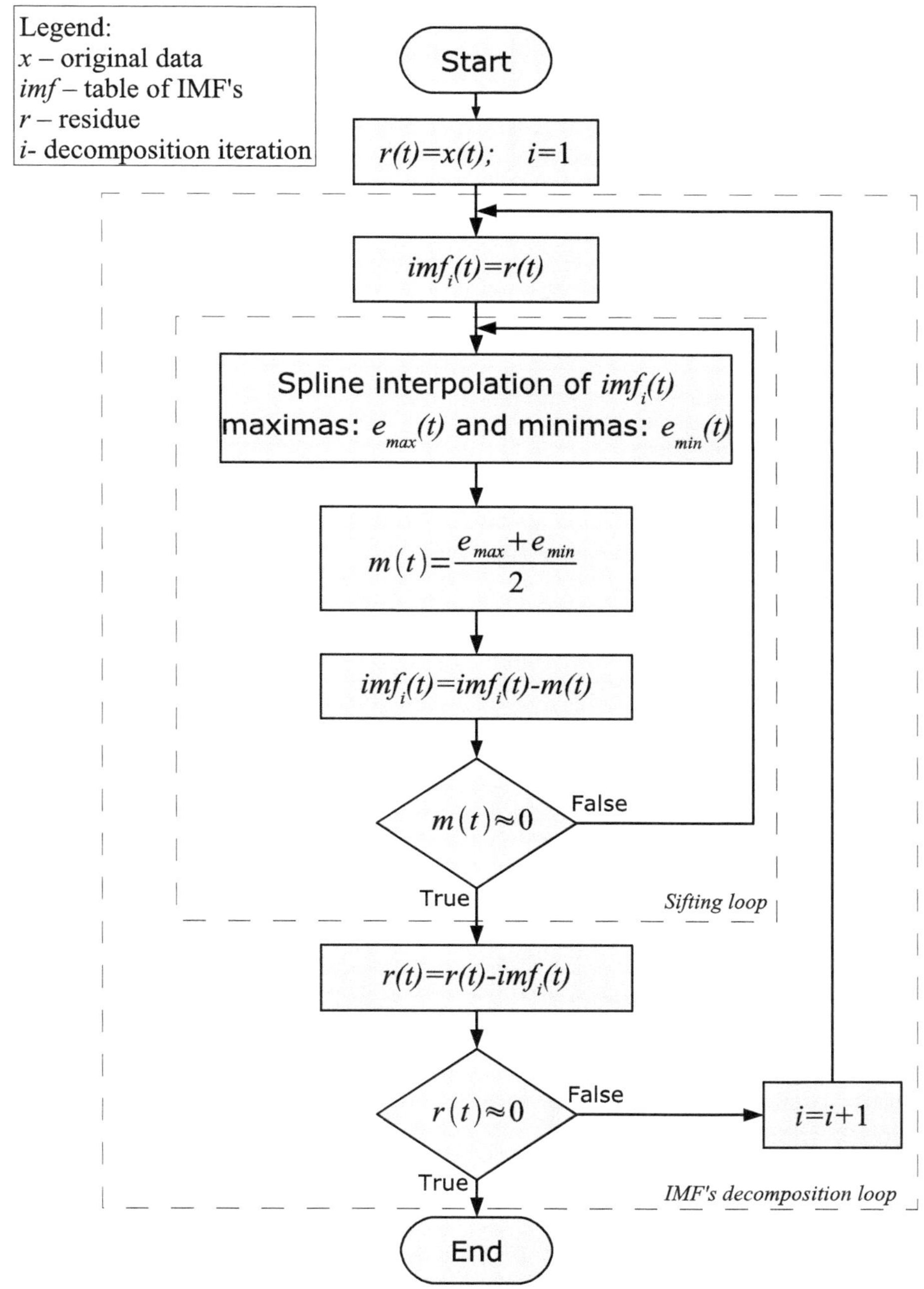

Figure 2 Flowchart of *EMD* algorithm [10].

where P denotes Cauchy principal value:

$$P \int_{-\infty}^{\infty} f(x)dx = \lim_{\rho \to \infty} \int_{-\rho}^{\rho} f(x)dx. \tag{13}$$

From analytic signal the instantaneous amplitude can be obtained as a module of this signal:

$$a_k(t) = \left| imf_{a_k}(t) \right| \tag{14}$$

and the instantaneous frequency as a differential of argument of this signal:

$$f_k(t) = \frac{1}{2\pi}\frac{d\,\arg(imf_{a_k}(t))}{dt}. \tag{15}$$

The sum of all modes' instantaneous frequencies and amplitudes gives the *Hilbert-Huang Spectrum* (*HSS*):

$$HHS(f_k, t) = Re\left[\sum_{k=1}^{N} a_k(t)^{j\int f_k(t)dt}\right]. \tag{16}$$

With *HHS*, the *marginal Hilbert-Huang Spectrum* can be calculated. It gives information about the contribution of a given frequency to the total power and is defined as follows:

$$hhs(f_k) = \int_{-\infty}^{\infty} HHS(f_k, t)dt. \tag{17}$$

Sliding Window Empirical Mode Decomposition

Sliding Window Empirical Mode Decomposition, SWEMD, is based on calculation of *EMD* in a relatively small window and sliding this window along the time axis [8,10]. Size of the window depends on the signal's frequency band - there must be at least 5 maxima and 5 minima in the window for correct spline interpolation. To prevent possible discontinuities in *IMFs* between windows, the total number of modes and also the number of sifting iterations must be set *a priori* [9]. The number of sifting steps should be tailored for each module. These parameter depends on the sampling frequency and on analyzed signal, its complexity and spectrum. The empirical determination of these values can be difficult. Thus, I propose a simple algorithm for determining the number of sifting iterations: decomposition of several signal's windows by classic algorithm and counting the average number of sifting steps for each mod.

The block diagram of *SWEMD* algorithm is displayed in Figure 3. To eliminate possible windows' boundary effects, surroundings points (*S* in Figure 3) are added to the beginning and end of the window. I suggest to use the number of surroundings points equal to the size of the window or its multiple. *EMD* sifting process is done on the window and surroundings part of the data. Calculation of all the modes in the window at time t is called *SWEMD* step.

During calculation performed on the subsequent modes in a *SWEMD step*, it may occur (especially for modes containing low frequencies) that the window size is too small for spline interpolation (not enough extrema). In such a case, the calculation for this mode (and the next modes) is canceled in this *SWEMD step*. In the next step these modes will be calculated in a longer window. Therefore, in the implementation of the algorithm, the table for storing indices of the windows' start point is used (*Ind* in Figure 3). If extraction of mode M in *SWEMD step* at time t is successful then the array element corresponding to the given mode (*Ind[M]*) is assigned the value of the last window point: $t + win$. In a case of too small number of extrema, *Ind[M]* remains unchanged, therefore, the size of the window in the next step will be greater by *win* points from the basic size: *win*.

The calculation of the mode M in the *SWEMD step* begins with extracting from the analyzed signal y points corresponding to the window in time t and its surroundings (in Figure 3 these data are assigned to the variable R). Then, in order to prepare input data for

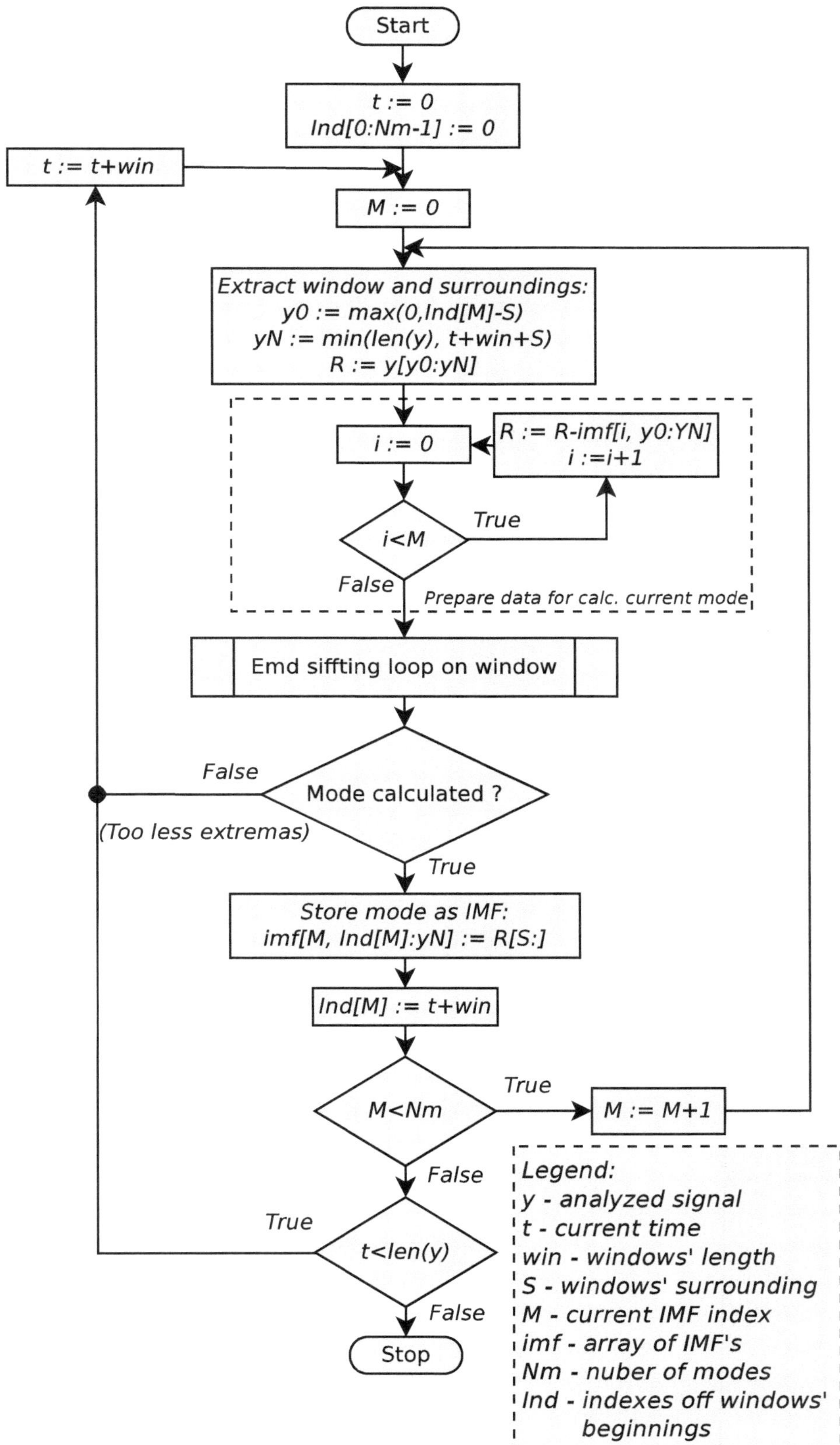

Figure 3 Flowchart of *SWEMD* algorithm [10].

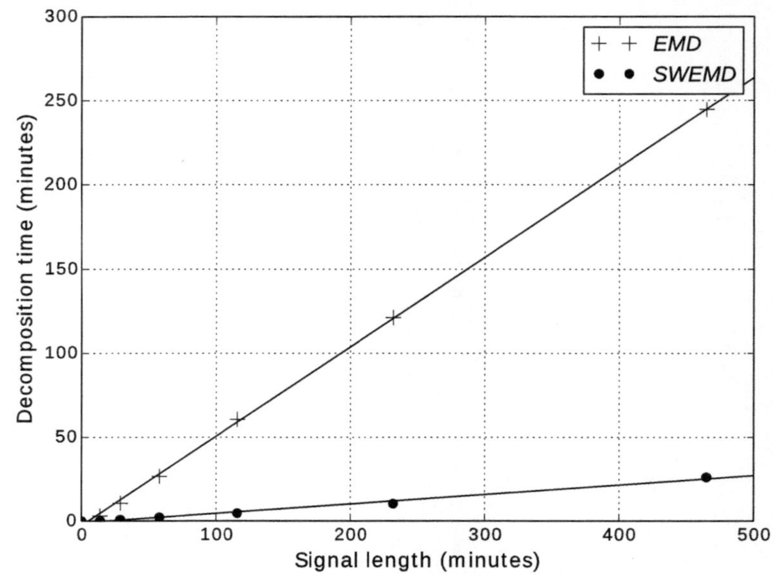

Figure 4 Performance of *SWEMD*.

calculation of the mode M, the signal R is reduced by the *IMFs* of orders lower than M (in Figure 3 block labeled "*Prepare data for calc. Current mode*"). Prepared data are subjected to the *sifting process*. If there is not enough extrema, calculations in this *SWEMD* step are interrupted and the counting of the first *IMF* in the next step ($t + win$) is started. If the *sifting process* is completed correctly, the resulting mode (after discarding surroundings) is placed in the array *imf[.,.]* (Figure 3) that stores *IMFs* at appropriate indices. Then, the value of the start point of the window for the mode M in the next *SWEMD* step is changed and a process of calculating the module $M + 1$ is started or current step is completed (if mode M was the last to achieve).

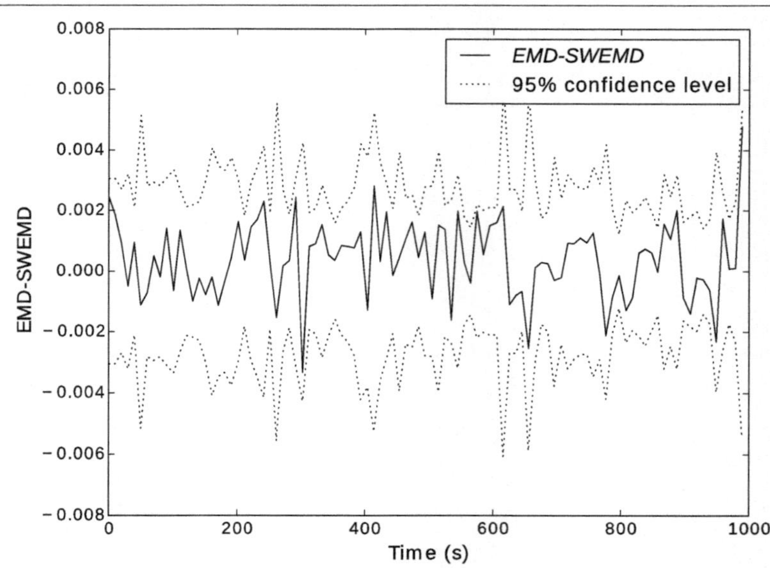

Figure 5 Quality of *SWEMD* algorithm: difference in mean amplitude *marginal spectra* of *SWEMD* and *EMD* and 95% confidence interval.

Results and discussion

Performance of SWEMD

Performance was tested on several *EEG* signals with different length. Figure 4 shows the comparison of the time of calculation *SWEMD* algorithm to 'classical' *EMD* depending on the length of the analyzed signal. Proposed algorithm performs calculations about 10 times faster.

Quality of SWEMD

In order to verify the quality of the proposed algorithm I compared the results *SWEMD* and 'classical' *EMD*. The test consisted of the following steps:

- decomposition of the test EEG signal using *EMD* and *SWEMD*;

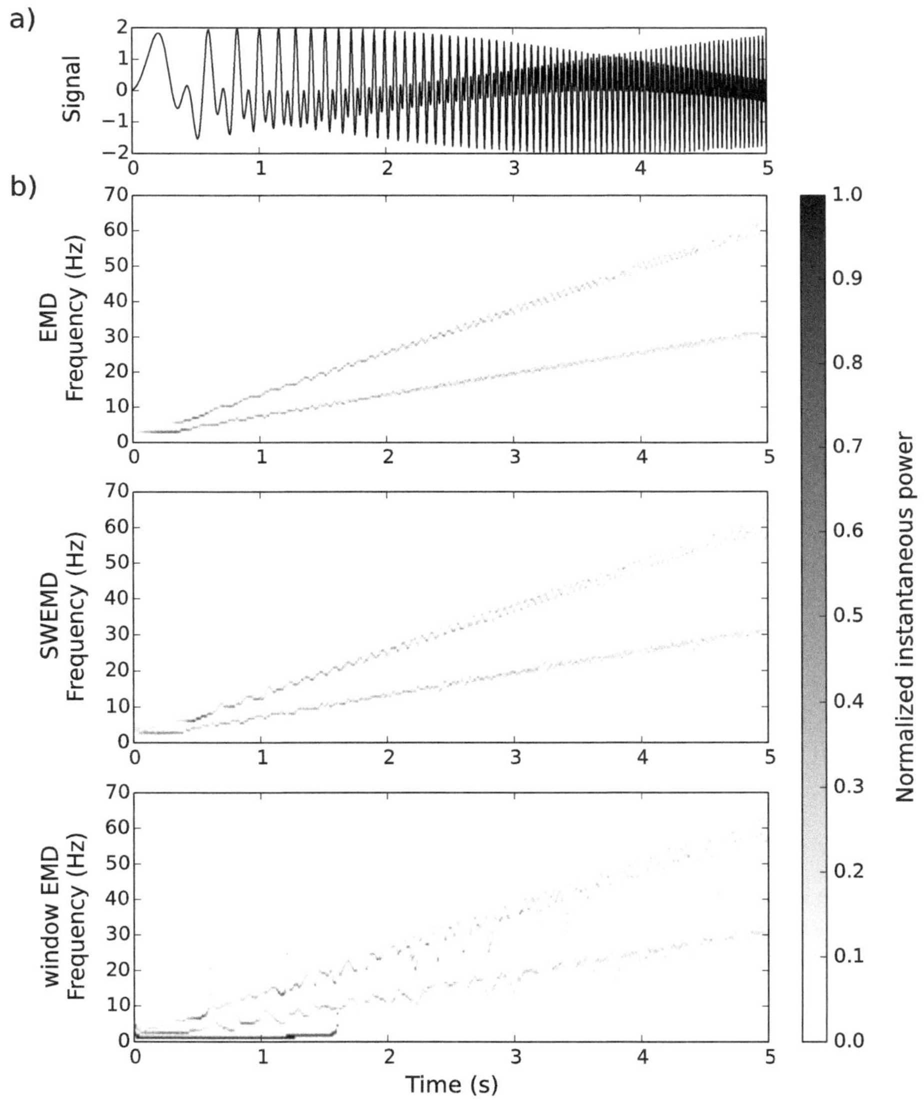

Figure 6 Comparison of *SWEMD* analysis with classical *EMD* and with *window EMD*. (a) Superposition of two chirps (signals with a linear frequency change), **(b)** its *HSS* obtained with three *EMD* algorithms.

- calculation of the *marginal Hilbert-Huang spectrum* for both *EMD* and *SWEMD* in time windows length of 10 seconds (the same window was used in the calculations of *SWEMD*);
- calculation of the difference of the *EMD* and *SWEMD* marginal spectra's mean amplitudes and the standard deviation of this difference;
- designate 95% confidence interval [11] for this difference

The test was performed on several *EEG* signals. The specimen results of the quality test of the algorithm *SWEMD* is shown in Figure 5. As it is seen in Figure 5 the difference in the amplitudes of marginal spectra of *EMD* and *SWEMD* at any point is approximately zero (solid line in Figure 5), and lies in the 95% confidence interval (dotted line in Figure 5). Therefore the results of both algorithms are statistically consistent.

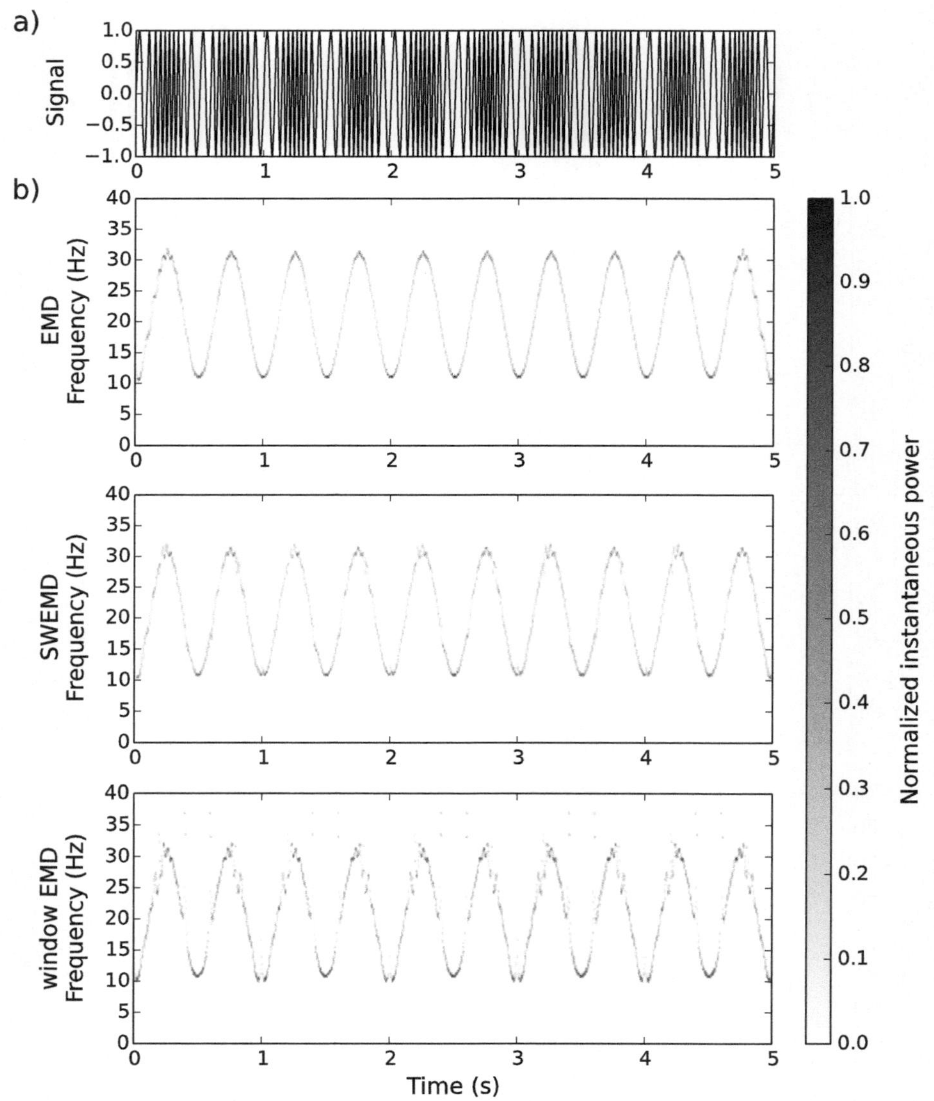

Figure 7 Comparison of *SWEMD* analysis with classical *EMD* and with *window EMD*. (a) Frequency modulated signal, **(b)** its *HSS* obtained with three *EMD* algorithms.

I also compared results of *SWEMD* analysis with classical *EMD* and with concatenated results of calculation of *EMD* in windows (*window EMD*) on simple signals.

Superposition of two chirp signals

Superposition of two chirps (signals with a linear frequency change) is shown in Figure 6a. *Hilbert-Huang spectra* calculated with three *EMD* algorithms: classical, *SWEMD*, and calculation of *EMD* in windows are compared in Figure 6b. Results of *SWEMD* are the same as *EMD*. Simple calculation in windows shows very strong boundary effects.

Frequency modulated signal

Frequency-modulated signal is shown in Figure 7a. The carrier frequency is 20 Hz, signal is modulated with 2 Hz sine. Figure 7b compares results of *Hilbert-Huang spectra*

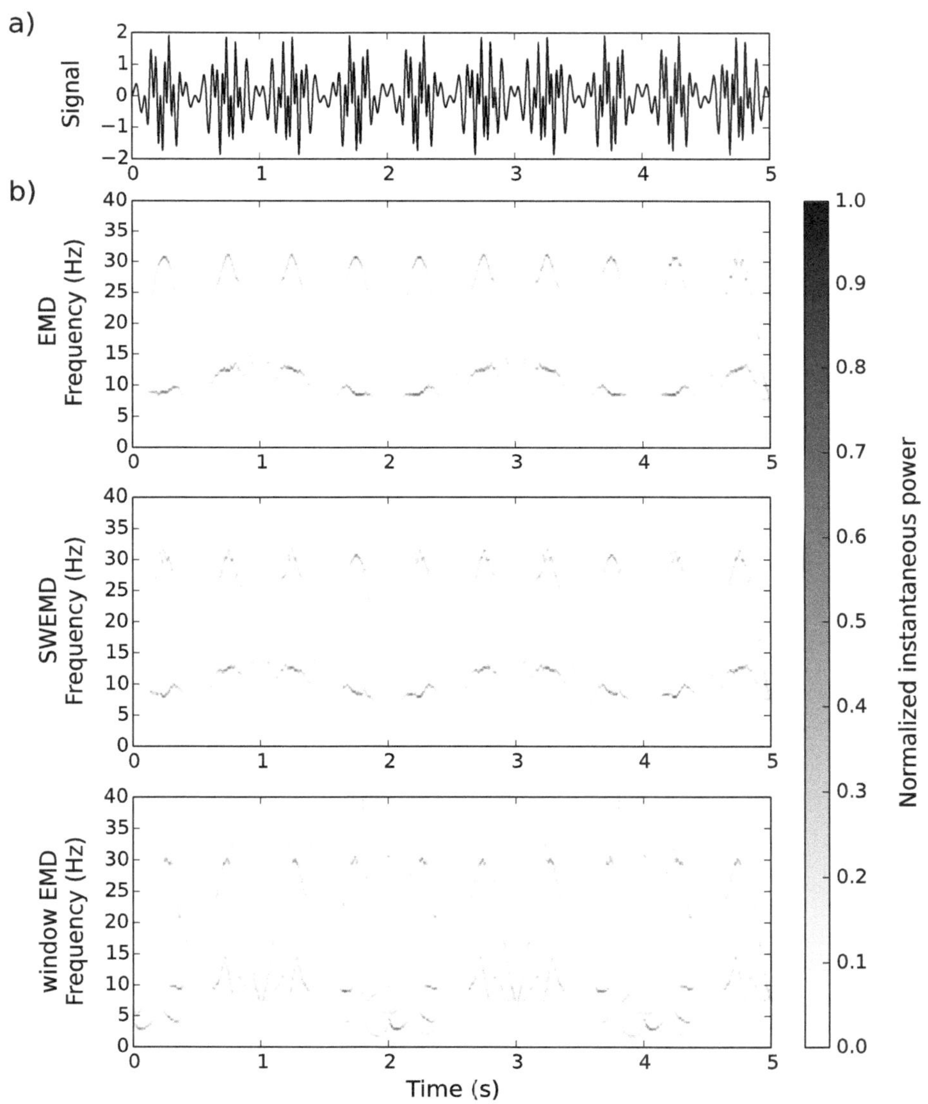

Figure 8 Comparison of *SWEMD* analysis with classical *EMD* and with *window EMD*. (a) Superposition of two frequency modulated signals, **(b)** its *HSS* obtained with three *EMD* algorithms.

obtained using three different *EMD* algorithms. *SWEMD* results show very small window boundary effects compared to typical windowing of *EMD.*

Superposition of two FM signals

Figure 8a shows superposition of signal from Figure 7a with another *FM* signal (the carrier frequency is 10 Hz modulated with 0.5 Hz). Results in Figure 8b are similar for *SWEMD* and *EMD* but *window EMD* gives greater errors, especially in decomposition of signal with 10 Hz carrier frequency.

Superposition of two harmonics

Superposition of two sines 101 Hzand 103 Hz is shown in Figure 9a. Such superposition gives 102 Hz signal amplitude modulated with 1 Hz. Also for this signal *SWEMD,* unlike typical calculation in windows, gives the same results as 'classical' *EMD* (Figure 9b).

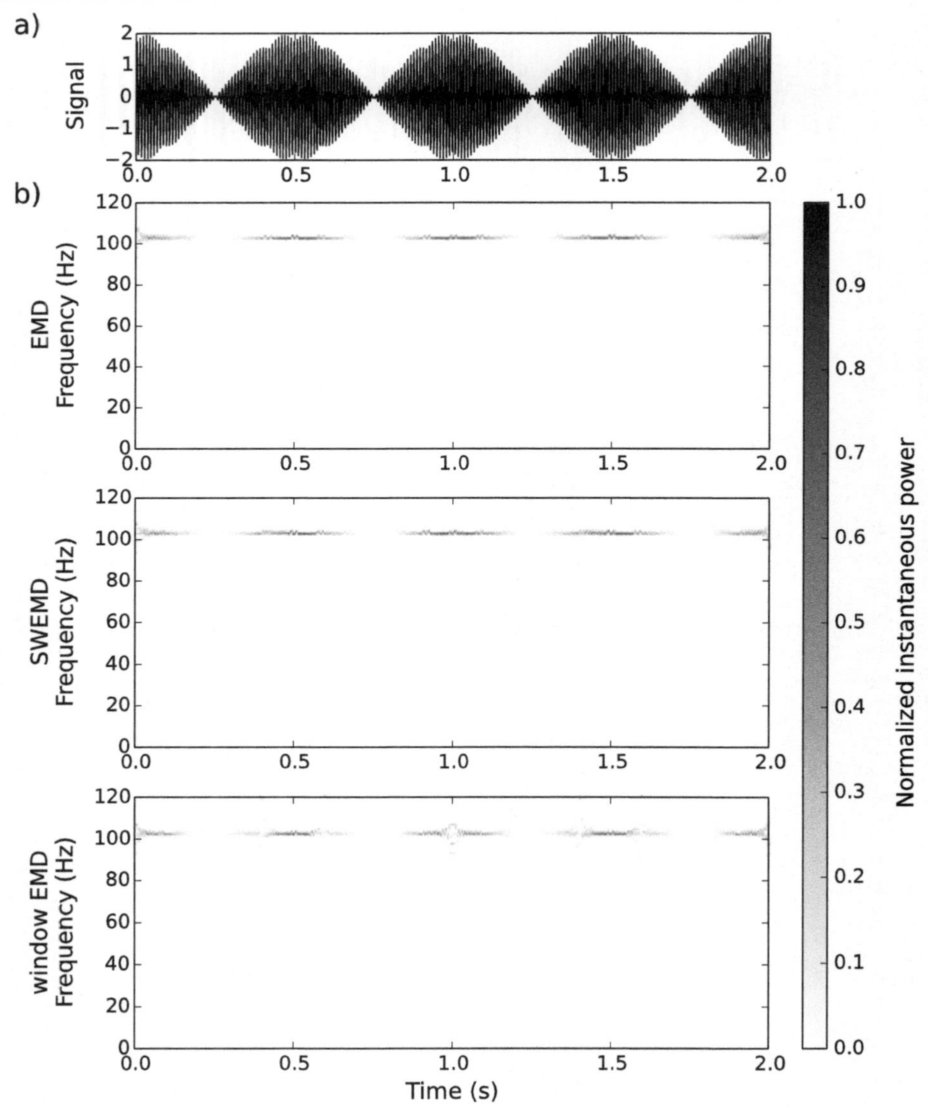

Figure 9 Comparison of *SWEMD* analysis with classical *EMD* and with *window EMD*. (a) Superposition of two harmonics, **(b)** its HSS obtained with three EMD algorithms.

Conclusions

Before using the classical *EMD* one needs to consider whether multiple spline interpolation does not take too much time. In particular this applies to analysis of long signals or acquisition systems with high sampling frequency or a large number of channels. Proposed algorithm, *Sliding Window EMD,* can speed up computation about 10 times. Examples of using *SWEMD* presented in this article show that it has similar quality as classical *EMD* - boundary effects in typical windowing are nearly completely reduced in *SWEMD.* Proposed algorithm is suitable also in real time analysis of data.

To compare the performance of 'clasicall' *EMD* method and its modifications [2-5] including *SWEMD* it would be necessary to apply all these methods to the same set of signals.

Competing interests

The author declares that he has no competing interests.

Acknowledgements

This work was supported by Nalecz Institute of Biocybernetics and Biomedical Engineering, Polish Academy of Sciences, Warsaw, through statutory project.

References

1. Huang NE, Shen Z, Long SR, Wu MC, Shih HH, Zheng Q, Yen N-C, Tung CC, Liu HH: **The empirical mode decomposition and the Hilbert spectrum for nonlinear and non-stationary time series analysis.** *Proc Roy Soc Lond Math Phys Sci* 1971, **454:**903–995.
2. Blakely CD: **A fast Empirical Mode Decomposition technique for nonstationary nonliear time series.** In *Center for Scientific Computation and Mathematical Modeling*. College Park, USA: University of Maryland. Preprint CSCAMM-05-10; 2005. Online: [http://www.cscamm.umd.edu/publications/fastEMDcBlakelyFinal_CS-05-10.pdf]
3. Chen Q, Huang N, Riemenschneider S, Xu Y: **A B-spline approach for empirical mode decompositions.** *Adv Comput Math* 2006, **24:**171–195.
4. Meignen S, Perrier V: **A new formulation for empirical mode decomposition based on constrained optimization.** *IEEE Signal Process Lett* 2007, **14**(12):932–935.
5. Roy A, Doherty JF: **Raised cosine interpolation for empirical mode decomposition.** In *Information Sciences and Systems, 2009. CISS 2009. 43rd Annual Conference on*. Baltimore: IEEE; 2009.
6. Roy A, Doherty JF: **Improved signal analysis performance at low sampling rates using raised cosine empirical mode decomposition.** *Electron Lett* 2010, **46**(2):176–177.
7. Roy A, Doherty JF: **Raised cosine filter-based empirical mode decomposition.** *IET Signal Process* 2011, **5**(2):121–129.
8. Stepien P: **Application of empirical mode decomposition method in biomedical signal analysis (in polish).** *PhD thesis*. Warsaw University of Technology; 2011. Supervisor: W. Klonowski.
9. Rilling G, Flandrin P, Gonçalvès P: **On empirical mode decomposition and its algorithms.** In *IEEE-EURASIP Workshop on Nonlinear Signal and Image Processing NSIP-03*. Grado, Italy; 2003. Online: [http://perso.ens-lyon.fr/paulo.goncalves/pub/NSIP03_GRPFPG.pdf]
10. Pierzchalski MP, Stepien RA, Stepien P: **New nonlinear methods of heart rate variability analysis in diagnostics of atrial fibrillation.** *Int J Biol Biomed Eng* 2011, **5**(4):201–208.
11. Huang NE, Wu MLC, Long SR, Shen SSP, Qu W, Gloersen P, Fan KL: **A confidence limit for the empirical mode decomposition and Hilbert spectral analysis.** *Proc R Soc Lond Math Phys Sci* 2003, **459**(2037):2317.

11

Attractor and saddle node dynamics in heterogeneous neural fields

Peter beim Graben[1,2] and Axel Hutt[3*]

*Correspondence: Axel.Hutt@inria.fr
[3]Bernstein Center for Computational Neuroscience Berlin, Berlin, Germany
Full list of author information is available at the end of the article

Abstract

Background: We present analytical and numerical studies on the linear stability of spatially non-constant stationary states in heterogeneous neural fields for specific synaptic interaction kernels.

Methods: The work shows the linear stabiliy analysis of stationary states and the implementation of a nonlinear heteroclinic orbit.

Results: We find that the stationary state obeys the Hammerstein equation and that the neural field dynamics may obey a saddle-node bifurcation. Moreover our work takes up this finding and shows how to construct heteroclinic orbits built on a sequence of saddle nodes on multiple hierarchical levels on the basis of a Lotka-Volterra population dynamics.

Conclusions: The work represents the basis for future implementation of meta-stable attractor dynamics observed experimentally in neural population activity, such as Local Field Potentials and EEG.

Keywords: Chaotic itinerancy; Linear stability; Heteroclinic orbits; Lotka-Volterra model

Background

Neural field models, such as [1,2] are continuum limits of large-scale neural networks. Typically, their dynamic variables describe either mean voltage [2] or mean firing rate [1,3] of a population element of neural tissue (see [4,5] for recent reviews).

The present article considers the paradigmatic Amari equation [2] describing the spatiotemporal dynamics of mean potential $V(x,t)$ over a cortical d-dimensional manifold $\Omega \subset \mathbb{R}^d$:

$$\frac{\partial V(x,t)}{\partial t} = -V(x,t) + \int_\Omega K(x,y)S[V(y,t)]\,\mathrm{d}y\,, \tag{1}$$

where $K(x,y)$ is the spatial synaptic connectivity between site $y \in \Omega$ and site $x \in \Omega$, and S is a nonlinear, typically sigmoidal, transfer function. This model neglects external inputs for simplicity but without constraining the generality of the subsequent results. Possible synaptic time scales are supposed to be included in the kernel function K and can be introduced by a simple scaling of time.

In general, the connectivity kernel $K(x,y)$ fully depends on both sites x and y, which case is referred to as *spatial heterogeneity*. If the connectivity solely depends on the difference between x and y, i.e. $K(x,y) = K(x-y)$, the kernel is called *spatially homogeneous*

[2]. Furthermore, if the connectivity depends on the distance between x and y only, i.e. $K(x,y) = K(||x-y||)$, with $||x||$ as some norm in Ω, the kernel is *spatially homogeneous and isotropic* [6].

Spatially homogeneous (respectively isotropic) kernels have been intensively studied in the literature due to their nice analytical properties. In this case, the evolution equations have exact solutions such as bumps [2,7], breathers [8-10] or traveling waves [7,11]. Moreover, such kernels allow the application of the technique of Green's functions for deriving partial neural wave equations [7,12,13].

The present work focuses on spatially heterogeneous neural fields which have been discussed to a much lesser extent in previous studies than homogeneous neural fields [14-22]. This study resumes these attempts by investigating stationary states of the Amari equation (1) with heterogeneous kernels and their stability. Such a theory would be mandatory for modeling transient neurodynamics as is characteristic, e.g., for human cognitive phenomena [23], human early evoked potentials [24] or, among many other phenomena, for bird songs [25].

The article is structured in the following way. In the "Results" section we present new analytical results on stationary solutions of the Amari equation (1) and their stability in the presence of heterogeneous connectivity kernels. Moreover, we present numerical simulation results for the kernel construction and its stability analysis. The "Methods" section is devoted to construct such kernels through dyadic products of desired stationary states (cf. the previous work of Veltz and Faugeras [26]). A subsequent linear stability analysis reveals that these stationary solutions could be either attractors or saddles, depending on the chosen parametrization. Finally, we present a way to connect such saddle state solutions via heteroclinic sequences [27,28] in order to construct transient processes.

Results

In this section we present the main results of our study on heterogeneous neural fields.

Stationary states and their stability

Analytical study

The Amari equation (1) has a trivial solution $V_0(x) = 0$ and non-trivial solutions $V_0(x) \neq 0$ that obey the Hammerstein integral equation [29]

$$V_0(x) = \int_\Omega K(x,y)S[V_0(y)]\,\mathrm{d}y\,. \tag{2}$$

Inspired by Hebbian learning rules for the synaptic connectivity kernel $K(x,y)$ which found successful applications, e.g., in bi-directional associative memory [30], we consider symmetric spatially heterogeneous kernels $K(x,y) = K(y,x)$ that can be constructed from dyadic products of the system's non-trivial stationary states

$$K(x,y) = (V_0 \otimes V_0)(x,y) = V_0(x)V_0(y)\,. \tag{3}$$

Together with Eq. (2), this choice yields the additional condition for non-trivial stationary states

$$1 = \int_\Omega V_0(y)S[V_0(y)]\,\mathrm{d}y\,, \tag{4}$$

which is a nonlinear integral equation of Fredholm type. Since $0 < S(x) < 1$ for a logistic transfer function, a necessary condition for non-trivial stationary states is

$$1 < \int_{\Omega} V_0(y)\,\mathrm{d}y \tag{5}$$

which indicates immediately a method to find a non-trivial solution numerically as shown below.

Small deviations $u(x,t) = V(x,t) - V_0(x)$ from a non-trivial stationary state $V_0(x)$ obey the linear integro-differential equation

$$\frac{\partial u(x,t)}{\partial t} = -u(x,t) + \int_{\Omega} L(x,y)u(y,t)\,\mathrm{d}y, \tag{6}$$

where $L(x,y) = K(x,y)S'(y)$ and $S'(y) = \mathrm{d}S[V_0(y)]/\mathrm{d}V_0(y)$. A linear stability analysis of (6) carried out in the next section shows that a non-trivial stationary state $V_0(x)$ is either a fixed point attractor, or (neglecting a singular case) a saddle with one-dimensional unstable manifold. Such saddles can be connected to form stable heteroclinic sequences.

Numerical study

To gain deeper insight into possible stationary solutions of the Amari equation (1) and their stability, the subsequent section presents the numerical integration of equation (1) in one spatial dimension for a specific spatial synaptic connectivity kernel.

Since previous experimental studies [31] have revealed Gaussian distributed probability densities of neuron interactions in the visual cortex of rats, it is reasonable to look for spatially discretized stationary states in the family of Gaussian functions

$$V_{0,i} = W_0 \exp\left(-(i-i_0)^2\Delta^2 x/2\sigma^2\right)/\sqrt{2\pi}\sigma + \kappa\eta_i \tag{7}$$

parameterized by the amplitude W_0, the variance σ^2, the noise level κ and the spatial discretization interval Δx.

By virtue of this parametrization of the discrete Hammerstein equation, it is sufficient to fit the model parameters optimally in such a way that the Hammerstein equation holds. Figure 1(a) illustrates a noisy kernel K and (b) shows the corresponding stationary state $V_0(x)$ for certain parameters.

For each stationary state, one obtains a kernel $L(x,y)$ of the linear stability analysis whose spectrum characterizes the stability of the system in the vicinity of the stationary state. If the eigenvalue with the maximum real part $\epsilon_1 > 1$, then the stationary state $V_0(x)$ is exponentially unstable whereas $\epsilon_1 < 1$ guarantees exponential stability. Figure 1(b) shows the eigenmode $e_1(x)$ corresponding to the eigenvalue with maximum real part which has a similar shape as the stationary state.

Moreover Figure 2 presents parameters for which $V_0(x)$ fulfills the Hammerstein equation (2), i.e. for which $V_0(x)$ is the stationary solution. We observe that some parameter sets exhibit a change of stability, i.e. the eigenvalue with maximum real part may be $\epsilon_1 > 1$ or $\epsilon_1 < 1$ for certain parameter subsets.

Taking a closer look at the stability of $V_0(x)$, the computation of the eigenvalues ϵ_k, $k = 1, \ldots, n$ reveals a dramatic gap in the spectrum: the eigenvalue with maximum real part ϵ_1 is well isolated from the rest of the spectrum $\{\epsilon_{k>1}\}$ with $|\epsilon_{k>1}| < 10^{-14}$. This is in accordance to the discussion of Eq. (17) on the linear spectrum.

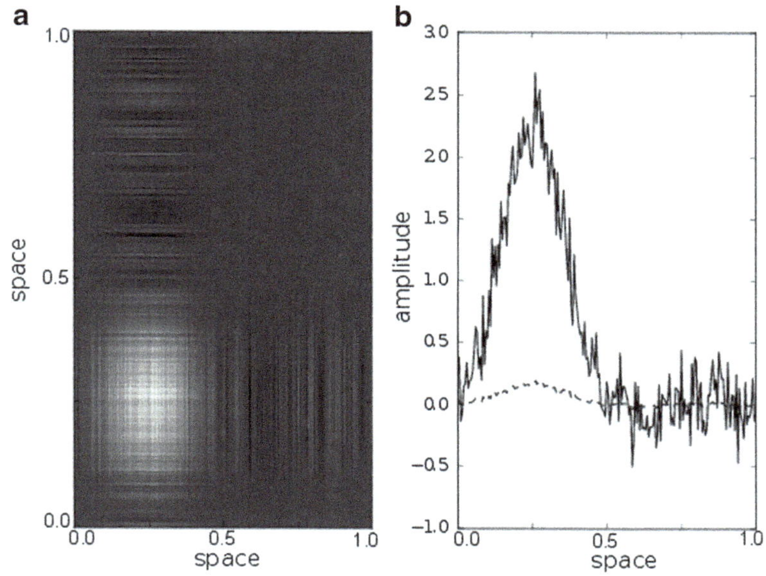

Figure 1 Illustration of (a) a noisy kernel $K(x, y)$ and (b) the corresponding solution of the Hammerstein equation (solid line) together with the first eigenmode $e_1(x)$ corresponding to the eigenvalue with the maximum real part ϵ_1. Parameters are $\alpha = 0.86$, $W_0 = 1.76$, $\sigma = 0.15$, $\theta = 3.0$, $\kappa = 0.2$ and $n = 200$, $L = 1$.

Figure 3 presents the spatiotemporal evolution of the heterogeneous neural field starting close to the stable stationary state $V_0(x)$, see point (1) in Figure 2. As expected, the field activity remains in the vicinity of the stable state.

In contrast, for the system starting close to an unstable stationary state, cf. point (2) in Figure 2, the field activity moves away from $V_0(x)$ and approaches a new stationary

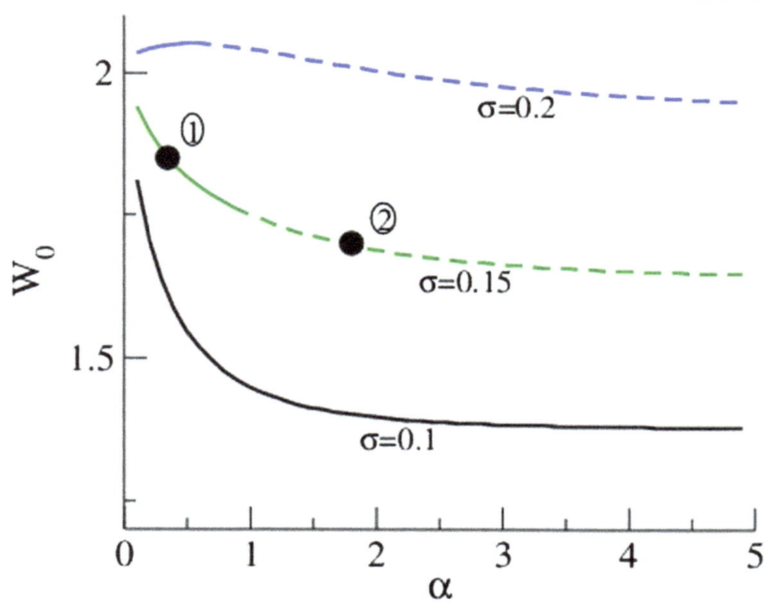

Figure 2 Parameters which guarantee the solution of the Hammerstein equation (2) which are the stationary solutions of Eq. (1) and the line styles encode their exponential stability (stable: solid line, unstable: dashed line). Further parameters are $\kappa = 0.0, \theta = 3.0$.

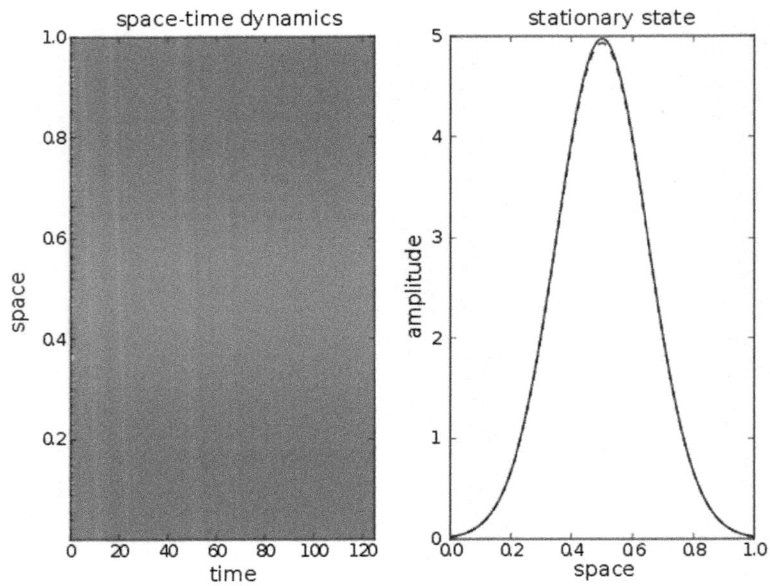

Figure 3 Numerically simulated spatio-temporal dynamics of the heterogeneous neural field (left) and stationary states (right). Here the stationary state $V_0(x)$ is exponentially stable with the parameters of point (1) in Figure 2. The simulation starts close to the stationary state $V(x,0) = V_0(x) + \xi(x)$ with the random numbers $\xi(x)$ taken from a zero-mean Gaussian distribution with variance 0.3. The plot on the left hand side shows the deviation from the stationary state $V_0(x)$. The plot on the right hand side shows the stationary states $V_0(x)$ (solid line) and the final field activity at large times $V(x, t = 125)$ (dashed line) which is almost identical to $V_0(x)$. The spatial domain has length $L = 1$ with $n = 300$.

state close to but different from $V_0(x)$, cf. Figure 4. This new stationary state obeys the Hammerstein equation (2).

Recalling the presence of the trivial stable solution $V = 0$, the activity shown in Figure 4 indicates the bistability of the system for the given parameter set.

Figure 5 supports this bistability for the same parameter set but different initial conditions, which presents the jump from the unstable stationary state $V_0(x)$ to the trivial stable stationary state $V = 0$. The choice whether the system approaches the upper or lower stable stationary state depends on the initial condition of the simulation and is random for random initial conditions as implemented in Figures 3, 4 and 5. Hence, this example reveals the existence of a saddle-node bifurcation in heterogeneous neural fields.

Finally, we would like to stress that the analysis presented above does not depend on the smoothness of the kernel and stationary state. For a strong noise level in the synaptic interaction kernel K, the analytical discussion above still describes the stationary state and the linear stability quite well as shown in Figure 6 for a stable stationary state $V_0(x)$ close to the stability threshold.

Heteroclinic orbits

The previous section has shown that heterogeneous neural fields may exhibit various stationary states with different stability properties. In particular we found that stationary states could be saddles with one-dimensional unstable manifolds that could be connected to stable heteroclinic sequences (SHS: [27,28]) which is supported by experimental

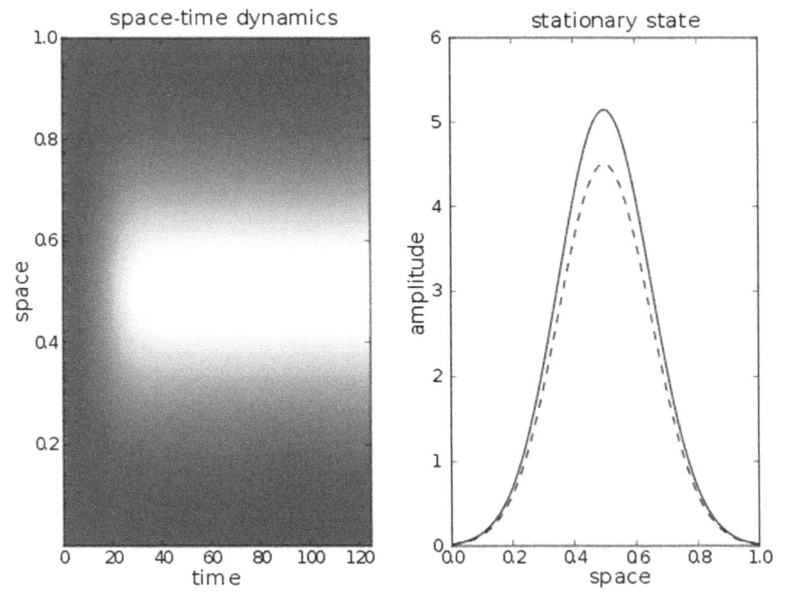

Figure 4 Numerically simulated spatio-temporal dynamics of the heterogeneous neural field (left) and stationary states (right). Here the stationary state $V_0(x)$ is exponentially unstable with the parameters of point (2) in Figure 2. The plot on the left hand side shows the deviation from the stationary state $V_0(x)$, the right hand side panel presents the stationary states $V_0(x)$ (solid line) and the final field activity at large times $V(x, t = 125)$ (dashed line) which is close to $V_0(x)$. Other parameters are taken from Figure 3.

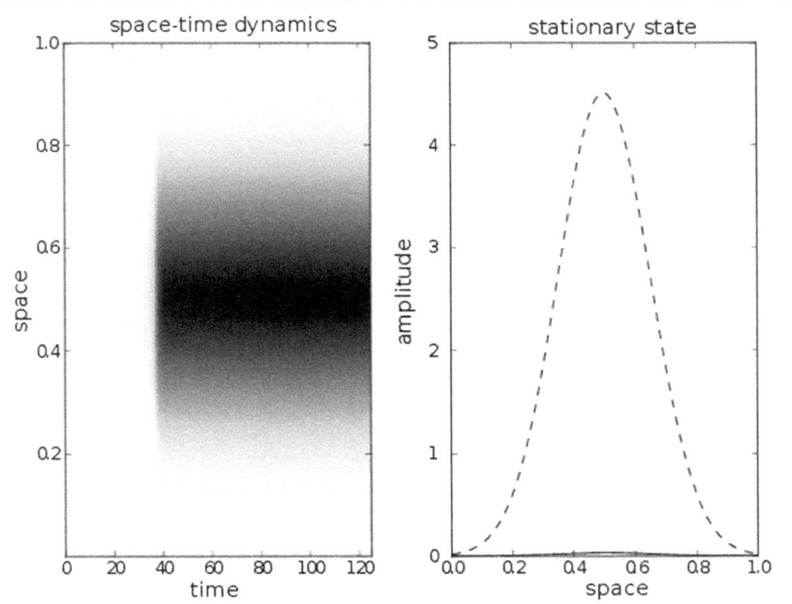

Figure 5 Numerically simulated spatio-temporal dynamics of the heterogeneous neural field (left) and stationary states (right). Here the stationary state $V_0(x)$ is exponentially unstable with the parameters of point (2) in Figure 2. All parameters are identical to Figure 3, however the final field activity at large times $V(x, t = 125)$ (dashed line) is close to the trivial solution $V = 0$. This jump of activity recalls the presence of the trivial stationary solution $V = 0$ and hence reflects the multi-stability of the heterogeneous neural field.

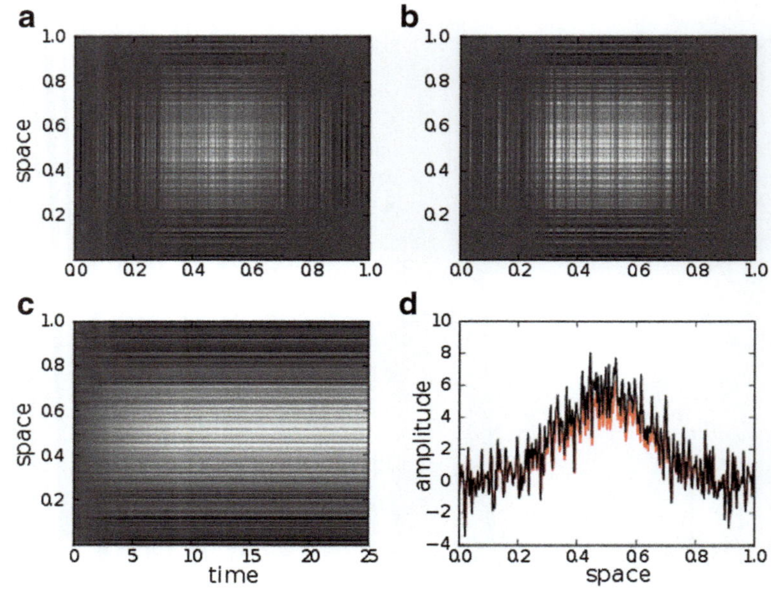

Figure 6 Strongly random connectivity kernels yielding noisy exponentially stable stationary states. Panel **(a)** shows $K(x,y)$, **(b)** $L(x,y) = K(x,y)S'(y)$, **(c)** spatio-temporal simulation starting close to the stationary state V_0, **(d)** stationary states $V_0(x)$ (red) and $V(x,t=25)$ (black). Parameters are $\kappa = 0.5, \alpha = 0.88, W_0 = 1.75, \sigma = 0.15, \theta = 3.0$ yielding a maximum eigenvalue $\epsilon_1 = 0.95$, i.e. close to the stability threshold.

evidence [24,32,33]. We present in the following paragraphs our main findings on heterogeneous neural fields exhibiting heteroclinic sequences and also hierarchies of such sequences.

One level heteroclinic sequence

It is possible to expand the integral in the Amari equation (1) into a power series yielding

$$\begin{aligned}&\frac{\partial V(x,t)}{\partial t} + V(x,t) \qquad (8)\\&= K_0 + \int_\Omega K_1(x,y)V(y,t)\,\mathrm{d}y + \frac{1}{2}\int_\Omega\int_\Omega K_2(x,y,z)V(y,t)V(z,t)\,\mathrm{d}y\,\mathrm{d}z\,.\end{aligned}$$

with kernels

$$\begin{aligned}K_0 &= 0 \qquad (9)\\K_1(x,y) &= \sum_k(\sigma_k+1)V_k^+(y)V_k(x)\\K_2(x,y,z) &= 2\sum_{kj}\rho_{kj}\sigma_j V_k^+(y)V_j^+(z)V_k(x)\,.\end{aligned}$$

The solution of Eq. (8) represents a heteroclinic sequence that connects saddle points $\{V_k(x)\}$ along their respective stable and unstable manifolds. Its transient evolution is described as winnerless competition in a Lotka-Volterra population dynamics governed by interaction weights ρ_{ik} between neural populations k and i and their respective growth rates σ_i.

In Eq. (9) the $\{V_k^+(x)\}$ comprise a bi-orthogonal system of the saddles $\{V_k(x)\}$. Therefore, the kernel $K_1(x,y)$ describes a Hebbian synapse between sites y and x that has been trained with pattern sequence V_k. This finding confirms the previous result of [18]. Moreover the three-point kernel $K_2(x, y, z)$ further generalizes Hebbian learning to interactions between three sites $x, y, z \in \Omega$. Note that the kernels K_i are linear combinations of dyadic product kernels, similar to those introduced in (3). Thus, our construction of heteroclinic orbits straightforwardly results in Pincherle-Goursat kernels used in [26].

Multi-level hierarchy of heteroclinic sequences

Now we assume that the neural field supports a hierarchy of stable heteroclinic sequences in the sense of [25,34]. For the general case one has to construct integral kernels for a much wider class of neural field equations which can be written as

$$V(x,t) = \int_{-\infty}^{t} \mathrm{d}\tau \, G(t,\tau) \int_{\Omega} \mathrm{d}y \, K(x,y) S[V(y,\tau)] \tag{10}$$

where the new temporal kernel G describing synaptic-dendritic filtering is usually the Green's function of a linear differential operator Q, such that $G(t,\tau) = G(t-\tau)$ and the temporal integration in Eq. (10) is temporal convolution. Equation (10) can be simplified by condensing space x and time t into spacetime $s = (x,t)$. Then, (10) becomes

$$V(s) = \int_{M} H(s,s') S[V(s')] \, \mathrm{d}s' \tag{11}$$

with a tensor product kernel $H(s,s') = (K \otimes G)(s,s')$ and integration domain $M = \Omega \times]-\infty, t]$.

For only two levels in such a hierarchy, we obtain

$$\begin{aligned}
V(x,t) = & \int_{\Omega} \mathrm{d}x_1 \int_{-\infty}^{t} \mathrm{d}t_1 K_1(x,x_1) V(x_1,t_1) \\
& - \int_{\Omega} \mathrm{d}x_1 \int_{\Omega} \mathrm{d}x_2 \int_{-\infty}^{t} \mathrm{d}t_2 K_2(x,x_1,x_2) V(x_1,t_1) V(x_2,t_1) \\
& - \int_{\Omega} \mathrm{d}x_1 \int_{\Omega} \mathrm{d}x_2 \int_{\Omega} \mathrm{d}x_3 \int_{-\infty}^{t} \mathrm{d}t_1 \int_{-\infty}^{t} \mathrm{d}t_2 \\
& \times K_3(x,x_1,x_2,x_3) V(x_1,t_1) \, V(x_2,t_1) \, V(x_3,t_2) \\
& + \int_{\Omega} \mathrm{d}x_1 \int_{\Omega} \mathrm{d}x_2 \int_{\Omega} \mathrm{d}x_3 \int_{\Omega} \mathrm{d}x_4 \int_{-\infty}^{t} \mathrm{d}t_1 \int_{-\infty}^{t} \mathrm{d}t_2 \\
& \times K_4(x,x_1,x_2,x_3,x_4) \, V(x_1,t_1) \, V(x_2,t_1) \, V(x_3,t_2) \, V(x_4,t_2)
\end{aligned} \tag{12}$$

with kernels

$$\begin{aligned}
K_1(x,x_1) &= \sum_{k_1=1}^{n_1} \frac{\sigma_{k_1}^{(1)}}{\tau^{(1)}} V_{k_1}^{(1)^+}(x_1) V_{k_1}^{(1)}(x) + \sum_{k_2=1}^{n_2} \frac{\sigma_{k_2}^{(2)}}{\tau^{(2)}} V_{k_2}^{(2)^+}(x_1) \, V_{k_2}^{(2)}(x) \\
K_2(x,x_1,x_2) &= \sum_{k_2=1}^{n_2} \sum_{j_2=1}^{n_2} \frac{\rho_{k_2 j_2}^{(2)}}{\tau^{(2)}} V_{k_2}^{(2)^+}(x_1) V_{j_2}^{(2)^+}(x_2) V_{k_2}^{(2)}(x) \\
K_3(x,x_1,x_2,x_3) &= \sum_{k_1=1}^{n_1} \sum_{j_1=1}^{n_1} \sum_{l_2=1}^{m_2} \frac{\sigma_{l_2}^{(2)} r_{k_1 j_1 l_2}^{(1)}}{\tau^{(1)} \tau^{(2)}} V_{k_1}^{(1)^+}(x_1) V_{j_1}^{(1)^+}(x_2) V_{l_2}^{(2)^+}(x_3) V_{k_1}^{(1)}(x) \\
K_4(x,x_1,x_2,x_3,x_4) &= \sum_{k_1=1}^{n_1} \sum_{j_1=1}^{n_1} \sum_{l_2=1}^{m_2} \sum_{j_2=1}^{n_2} \frac{r_{k_1 j_1 l_2}^{(1)} \rho_{l_2 j_2}^{(2)}}{\tau^{(1)} \tau^{(2)}} V_{k_1}^{(1)^+}(x_1) V_{j_1}^{(1)^+}(x_2) V_{l_2}^{(2)^+}(x_3) \\
& \times V_{j_2}^{(2)^+}(x_4) V_{k_1}^{(1)}(x) \, .
\end{aligned}$$

Here, the $V^{(\nu)}_{k_\nu}(x)$ denote the k_ν-th saddle in the ν-th level of the hierarchy (containing n_ν stationary states). Saddles are chosen again in such a way that they form a system of bi-orthogonal modes $V^{(\nu)^+}_{k_\nu}(x)$ whose behavior is determined by Lotka-Volterra dynamics with growth rates $\sigma^{(\nu)}_{k_\nu} > 0$ and time-dependent interaction weights $\rho^{(\nu)}_{k_\nu j_\nu}(t) > 0$, $\rho^{(\nu)}_{k_\nu k_\nu}(t) = 1$ that are given by linear superpositions of templates $r^{(\nu)}_{k_\nu j_\nu l_{\nu+1}}$. Additionally, $\tau^{(\nu)}$ are the characteristic time scales for level ν. Levels are temporally well separated through $\tau^{(\nu)} \gg \tau^{(\nu+1)}$ [35].

Interestingly these kernels are time-independent. Since neural field equations can be written in the same form as Eq. (12), this result shows that hierarchies of Lotka-Volterra systems are included in the neural field description. Again we point out that the resulting kernels are linear combinations of dyadic products as introduced in Eq. (3), see also the work of Veltz and Faugeras [26].

Discussion

This study considers spatially heterogeneous neural fields describing the mean potential in neural populations according to the Amari equation [2]. To our best knowledge this work is one of the first deriving the implicit conditions for stationary solutions and identifying the corresponding stability constraint as the Hammerstein integral equation. Moreover, as one of the first studies our work derives conditions for the linear stability of such stationary solutions and derives an analytical expression for stability subjected to the properties of heterogeneous synaptic interaction. The analytical results are complemented by numerical simulations illustrating the stability and instability of heterogeneous stationary states. We point out that the results obtained extend previous studies both on homogeneous neural fields and heterogeneous neural fields as other studies in this context have done before [21].

By virtue of the heterogeneity of the model, it is possible to consider more complex spatiotemporal dynamical behavior than the dynamics close to a single stationary state. We show how to construct hierarchical heteroclinic sequences built of multiple stationary states each exhibiting saddle node dynamics. The work demonstrates in detail how to construct such sequences given the stationary states and their saddle node dynamics involved. Motivated by a previous study on hierarchical heteroclinic sequences [25,34], we constructed such sequences in heterogeneous neural fields of the Amari type. Our results indicate that such a hierarchy may be present in a single heterogeneous neural field whereas previous studies [25,34] consider the presence of several neural populations to describe heteroclinic sequences. The kernels obtained from heteroclinic saddle node dynamics are linear superpositions of tensor product kernels, known as Pincherle-Goursat kernels in the literature [26].

Conclusion

The present work is strongly related to the literature hypothesizing the presence of chaotic itinerant neural activity, cf. previous work by [36,37]. This concept of chaotic itinerancy is attractive but yet lacking a realistic neural model. We admit that the present work represents just a first initial starting point for further model analysis of sequential neural activity in heterogeneous neural systems. It opens up an avenue of future research and promises to close the gap between the rather abstract concept of sequential, i.e. temporally transient, neural activity and corresponding mathematical neural models.

Methods

Stationary states and their stability

In order to learn more about the spatiotemporal dynamics of a neural field, in general it is reasonable to determine stationary states and to study their linear stability. This already gives some insight into the nonlinear dynamics of the system. Since the dynamics depends strongly on the interaction kernel and the corresponding stationary states, it is necessary to work out conditions for the existence and uniqueness of stationary states and their stability.

Analytical study

For stationary solutions, the left hand side of Eq. (1) vanishes and we obtain the Hammerstein equation (2) [29]. It has a trivial solution $V_0(x) = 0$ and further non-trivial solutions $V_0(x) \neq 0$ under certain conditions. The existence and number of solutions of Eq. (2) depends mainly on the operator $\mathcal{A} = \mathcal{I} + \mathcal{K}\mathcal{F}$ [38] and its monotonicity [39], with the unity operator $\mathcal{I}$, the linear integral operator

$$\mathcal{K}u = \int_\Omega K(x,y)u(y)\mathrm{d}y$$

and the Nemytskij operator $\mathcal{F}u(x) = S(u(x))$. For instance, a simple criterion for the existence of at least a single non-trivial solution is the symmetry and positive definiteness of the kernel $K(x,y)$ and the condition $S(u) \leq C_1 u + C_2$ [29] . Moreover previous studies have proposed analytical [40] and numerical [41] methods to find solutions of the Hammerstein equation (2).

To illustrate the non-uniqueness of solutions of the Hammerstein equation, let us expand the stationary solution into a set of bi-orthogonal spatial modes $\{\phi_n(x)\}$, $\{\psi_n(x)\}$

$$V_0(x) = \sum_{n=0}^{\infty} V_n \phi_n(x) = \sum_{n=0}^{\infty} U_n \psi_n(x)$$

with constant coefficients V_n, U_n and

$$(\psi_m, \phi_n) = \int_\Omega \psi_m(x)\phi_n(x)\,\mathrm{d}x = \delta_{n,m}\,.$$

with the Kronecker symbol $\delta_{n,m}$. Then the Hammerstein equation recasts to

$$V_m = S_0\kappa_m + S' \sum_{n=0}^{\infty} K_{m,n} V_n + f_m\left(\{V_n\}\right) \tag{13}$$

with

$$\kappa_m = (\psi_m, \mathcal{K}1)\ ,\ K_{m,n} = (\psi_m, \mathcal{K}\phi_n)\ ,\ f_m = \left(\psi_m, \mathcal{K}f\right)$$

and

$$f(x) = \sum_{p=2}^{\infty} \frac{1}{p!} \frac{\partial^p S(z)}{\partial z^p}\Big|_{z=0} \left(\sum_{n=0}^{\infty} V_n \phi_n(x)\right)^p .$$

Since f_m is a nonlinear function of V_n, Eq. (13) has multiple solutions $\{V_n\}$ for a given bi-orthogonal basis. Hence, $V_0(x)$ may not be unique.

Considering small deviations $u(x,t) = V(x,t) - V_0(x)$ from a non-trivial stationary state $V_0(x)$, these deviations obey the linear equation (6) with $L(x,y) = K(x,y)S'(y)$ and $S'(y) = \mathrm{d}S[V_0(y)]/\mathrm{d}V_0(y)$.

A solution of (6) is then $u(x,t) = \exp(\lambda t)e(x)$, $\lambda \in \mathbb{C}$ with mode $e(x)$. Inserting this solution into Eq. (6) yields the continuous spectrum of the corresponding linear operator $L(x,y)$ determined implicitly by the eigenvalue equation

$$(\lambda + 1)e(x) = \int_\Omega L(x,y)e(y)\,\mathrm{d}y \tag{14}$$

Under the above assumption of a dyadic product kernel (3), the eigenfunctions $\{e_k(x)\}$ of the kernel $L(x,y)$, defined through

$$\epsilon_k e_k(x) = \int_\Omega L(x,y)e_k(y)\,\mathrm{d}y\,, \tag{15}$$

with eigenvalues $\epsilon_k = \lambda_k + 1 \in \mathbb{C}$ form an orthonormal system with respect to the scalar product

$$\langle u, v\rangle_S = \int_\Omega u(x)v(x)S'(y)\,\mathrm{d}y \tag{16}$$

with weight $S'(y)$. This follows from the dyadic product kernel (3) through

$$\begin{aligned}
\int_\Omega L(x,y)e_k(y)\,\mathrm{d}y &= \int_\Omega K(x,y)S'(y)e_k(y)\,\mathrm{d}y \\
&= \int_\Omega V_0(x)V_0(y)S'(y)e_k(y)\,\mathrm{d}y \\
&= V_0(x)\int_\Omega V_0(y)e_k(y)S'(y)\,\mathrm{d}y \\
\epsilon_k e_k(x) &= V_0(x)\langle V_0, e_k\rangle_S\,.
\end{aligned} \tag{17}$$

From (17) we deduce two important results:

- a certain eigenmode $e_{k_0}(x)$ is proportional to the stationary state $V_0(x)$ with scaling factor $\langle V_0, e_{k_0}\rangle_S/\epsilon_{k_0}$, and
- other eigenmodes in the eigenbasis $e_{k\neq k_0}(x)$ are orthogonal to $V_0(x)$ yielding $\langle V_0, e_k\rangle_S = \epsilon_k = 0$.

Hence for dyadic product kernels (3) the spectrum of L includes one eigenmode with eigenvalue $\epsilon_1 \neq 0$, i.e. $\lambda_1 \neq -1$, while all other eigenmodes are stable with $\epsilon_{k\neq 1} = 0$ and thus $\lambda_{k\neq 1} = -1$. Therefore, a non-trivial stationary state could become either an asymptotically stable fixed point, i.e. an attractor, for $\epsilon_n < 1$, for all n, or a saddle with a one-dimensional unstable manifold, for $\epsilon_1 > 1$ and $\epsilon_{n\neq 1} < 1$ (neglecting the singular case $\epsilon_1 = 1$).

Finally we have to justify the necessary condition (5) which follows from $0 < S(x) < 1$ and

$$1 = \int_\Omega V_0(y)S[V_0(y)]\,\mathrm{d}y < \int_\Omega V_0(y)\,\mathrm{d}y\,,$$

inserted into the Hammerstein equation (2) for a dyadic product kernel (3).

Numerical study

In order to investigate stationary states of the Amari equation numerically, we choose a finite one-dimensional spatial domain of length L and discretize it into a regular grid of n

intervals with grid interval length $\Delta x = L/n$. Then the kernel function is $K(x = n\Delta x, y = m\Delta x) = K_{nm}/\Delta x$ and Eq. (1) reads

$$\frac{\partial V_i(t)}{\partial t} = -V_i(t) + \sum_{j=1}^{n} K_{ij} S[V_j(t)] \tag{18}$$

where $V_n(t) = V(n\Delta x, t)$. The corresponding Hammerstein equation (2) is given by

$$V_{0,i} = \sum_{j=1}^{n} K_{ij} S[V_{0,j}] \ . \tag{19}$$

Taking into account the insight from the discussion of Eq. (3) and its consequences for the eigenmodes, we have chosen the spatial kernel to

$$K_{ij} = V_{0,i} V_{0,j} \ . \tag{20}$$

We employ an Euler-forward integration scheme for the temporal evolution with discrete time step $\Delta t = 0.05$.

We render the stationary state random by adding the noise term η_i which are random numbers taken from a normal distribution with zero mean and unit variance. We point out that we choose η_i such like $|\sum_{i=1}^{n} \eta_i| < 0.05$ in order to not permit an amplitude increase in the dynamics by noise, but just as a modulation. The sigmoid function is chosen as $S(V) = 1/(1 + \exp(-\alpha(V - \theta)))$ parameterized by the slope parameter α and the mean threshold θ.

Heteroclinic orbits

The general neural field equation (10) supplies the Amari equation (1) as a special case for $G(t) = \mathrm{e}^{-t}\Theta(t)$ ($\Theta(t)$ as Heaviside's step function). Then $Q = \partial_t + 1$ is the Amari operator for the Green's function $G(t)$. For second-order synaptic dynamics [42] and for the filter properties of complex dendritc trees [43], more complicated kernels or differential operators, respectively, have to be taken into account.

As a first step for constructing heteroclinic sequences for (1) or (10), we expand the nonlinear transfer function S in Eq. (11) into a power series about a certain state $\bar{V}(s)$ [15],

$$\begin{aligned}
S[\,V(s)] &= \sum_{n=0}^{\infty} \frac{S^{(n)}(s)}{n!} (V(s) - \bar{V}(s))^n \\
&= \sum_{n=0}^{\infty} \frac{S^{(n)}(s)}{n!} (-1)^n (\bar{V}(s) - V(s))^n \\
&= \sum_{n=0}^{\infty} \frac{S^{(n)}(s)}{n!} (-1)^n \sum_{k=0}^{n} (-1)^k \binom{n}{k} \bar{V}(s)^{n-k} V(s)^k \\
&= \sum_{n=0}^{\infty} \sum_{k=0}^{n} (-1)^{n+k} \frac{S^{(n)}(s)}{n!} \binom{n}{k} \bar{V}(s)^{n-k} V(s)^k
\end{aligned}$$

Inserting $V(s)$ from Eq. (11) yields

$$
\begin{aligned}
S[V(s)] &= \sum_{n=0}^{\infty}\sum_{k=0}^{n}(-1)^{n+k}\frac{S^{(n)}(s)}{n!}\binom{n}{k}\bar{V}(s)^{n-k}\left(\int_M H(s,s')S[V(s')]\,\mathrm{d}s'\right)^k \\
&= \sum_{n=0}^{\infty}\sum_{k=0}^{n}(-1)^{n+k}\frac{S^{(n)}(s)}{n!}\binom{n}{k}\bar{V}(s)^{n-k} \\
&\quad \times \int_M \cdots \int_M H(s,s_1)H(s,s_2)\ldots H(s,s_k) \\
&\quad \times S[V(s_1)]\,S[V(s_2)]\ldots S[V(s_k)]\,\mathrm{d}s_1\cdots \mathrm{d}s_k \\
&= \sum_{n=0}^{\infty}\sum_{k=0}^{n}(-1)^{n+k}\frac{S^{(n)}(s)}{n!}\binom{n}{k}\bar{V}(s)^{n-k} \\
&\quad \times \int_M \cdots \int_M \prod_{j=1}^{k} H(s,s_j)S[V(s_j)]\,\mathrm{d}s_j\,.
\end{aligned}
$$

By eliminating the nonlinearities $S[V(s_j)]$ using the power series above again, we get

$$
\begin{aligned}
S[V(s)] &= \sum_{n=0}^{\infty}\sum_{k=0}^{n}(-1)^{n+k}\frac{S^{(n)}(s)}{n!}\binom{n}{k}\bar{V}(s)^{n-k} \\
&\quad \times \int_M \cdots \int_M \prod_{j=1}^{k} H(s,s_j)\left(\sum_{n_j=0}^{\infty}\frac{S^{(n_j)}(s_j)}{n_j!}(V(s_j)-\bar{V}(s_j))^{n_j}\right)\mathrm{d}s_j\,.
\end{aligned}
$$

Inserting this expression into Eq. (11) leads to a generalized Volterra series

$$
\begin{aligned}
&V(s) = H_0(s) \\
&+\sum_{m=1}^{\infty}\frac{1}{m!}\int_M \cdots \int_M H_m(s,s_1,\ldots,s_m)V(s_1)\ldots V(s_m)\,\mathrm{d}s_1\cdots \mathrm{d}s_m
\end{aligned} \tag{21}
$$

with a sequence of integral kernels H_m that can be read off after some further tedious rearrangements [44].

One-level hierarchy

In a previous work, we have derived a one-level hierarchy of stable heterogenic sequences [44], which we briefly recapitulate here. We assume a family of n stationary states $V_k(x)$, $1 \leq k \leq n$ that are to be connected along a heteroclinic sequence. Each state is assumed to be realized by a population in the neural field governed by (10), that is characterized by a population activity $\alpha_k(t) \in [0,1]$. Then the overall field quantity is obtained through an order parameter expansion [15,45]

$$
V(x,t) = \sum_{k=1}^{n}\alpha_k(t)V_k(x)\,. \tag{22}
$$

These population amplitudes result from a winnerless competition in a generalized Lotka-Volterra system approximating a Wilson-Cowan model [46,47]

$$
\begin{aligned}
\frac{\mathrm{d}\xi_k(t)}{\mathrm{d}t} &= \xi_k(t)\left(\sigma_k - \sum_{j=1}^{n}\rho_{kj}\xi_j(t)\right) \\
\alpha_k(t) &= \frac{\xi_k(t)}{\sigma_k}
\end{aligned} \tag{23}
$$

with growth rates $\sigma_k > 0$, and interaction weights $\rho_{kj} > 0$, $\rho_{kk} = 1$, that are trained by the algorithm of [27] and [28] for the desired sequence of transitions.

For the following construction, we also assume that the system of stationary states $\{V_k(x)\}$ is linearly independent such that there is a bi-orthogonal system $\{V_k^+(x)\}$, obeying

$$\int_\Omega V_j^+(x) V_k(x) \, \mathrm{d}x = \delta_{jk} \, . \tag{24}$$

Then, we obtain from (22)

$$\begin{aligned}\int_\Omega V_j^+(x) V(x,t) \, \mathrm{d}x &= \sum_{k=1}^{n} \alpha_k(t) \int_\Omega V_j^+(x) V_k(x) \, \mathrm{d}x \\ &= \sum_{k=1}^{n} \alpha_k(t) \delta_{jk}\end{aligned}$$

and hence

$$\begin{aligned}\alpha_j(t) &= \int_\Omega V_j^+(x) V(x,t) \, \mathrm{d}x \\ \xi_j(t) &= \sigma_j \int_\Omega V_j^+(x) V(x,t) \, \mathrm{d}x \, .\end{aligned} \tag{25}$$

Next, we take the derivation of (22) with respect to time t, by exploiting (23)

$$\begin{aligned}\frac{\partial V(x,t)}{\partial t} &= \sum_k \frac{1}{\sigma_k} \frac{\mathrm{d}\xi_k}{\mathrm{d}t} V_k(x) \\ &= \sum_k \frac{1}{\sigma_k} \left[\xi_k \left(\sigma_k - \sum_j \rho_{kj} \xi_j \right) \right] V_k(x) \\ &= \sum_k \xi_k V_k(x) - \sum_{kj} \frac{\rho_{kj}}{\sigma_k} \xi_k \xi_j V_k(x) \, .\end{aligned} \tag{26}$$

Adding (22), we obtain

$$\begin{aligned}\frac{\partial V(x,t)}{\partial t} + V(x,t) &= \sum_k \xi_k V_k(x) - \sum_{kj} \frac{\rho_{kj}}{\sigma_k} \xi_k \xi_j V_k(x) + \sum_k \frac{1}{\sigma_k} \xi_k V_k(x) \\ &= \sum_k \left(1 + \frac{1}{\sigma_k} \right) \xi_k V_k(x) - \sum_{kj} \frac{\rho_{kj}}{\sigma_k} \xi_k \xi_j V_k(x)\end{aligned} \tag{27}$$

from which we eliminate all occurrences of ξ by means of (25). This yields

$$\begin{aligned}\frac{\partial V(x,t)}{\partial t} + V(x,t) &= \sum_k \left(1 + \frac{1}{\sigma_k} \right) \sigma_k \int_\Omega V_k^+(y) V(y,t) V_k(x) \, \mathrm{d}y \\ &\quad - \sum_{kj} \frac{\rho_{kj}}{\sigma_k} \sigma_k \sigma_j \int_\Omega \int_\Omega V_k^+(y) V(y,t) V_j^+(z) V(z,t) V_k(x) \, \mathrm{d}y \, \mathrm{d}z \\ &= \int_\Omega \sum_k (\sigma_k + 1) \, V_k^+(y) V_k(x) V(y,t) \, \mathrm{d}y \\ &\quad - \int_\Omega \int_\Omega \sum_{kj} \rho_{kj} \sigma_j V_k^+(y) V_j^+(z) V_k(x) V(y,t) V(z,t) \, \mathrm{d}y \, \mathrm{d}z \, .\end{aligned} \tag{28}$$

Moreover, the nonlinear spatial integral transformation in the Amari equation (1) may be written as a generalized Volterra series (21)

$$\mathcal{J}[V(\cdot,t)](x) = K_0(x) + \sum_{m=1}^{\infty} \frac{1}{m!} \int_\Omega \cdots \int_\Omega K_m(x, y_1, y_2, \ldots, y_m) V(y_1, t) V(y_2, t) \ldots V(y_m, t) \, \mathrm{d}y_1 \, \mathrm{d}y_2 \cdots \mathrm{d}y_m \,. \tag{29}$$

Comparison of the first three terms with (28) yields the result (9).

Multi-level hierarchy of heteroclinic sequences

Here we assume that the SHS are given by (possibly infinitely many) generalized Lotka-Volterra systems

$$\tau^{(\nu)} \frac{\mathrm{d}\alpha^{(\nu)}_{k_\nu}(t)}{\mathrm{d}t} = \alpha^{(\nu)}_{k_\nu}(t) \left(\sigma^{(\nu)}_{k_\nu} - \sum_{j_\nu=1}^{n_\nu} \rho^{(\nu)}_{k_\nu j_\nu}(t) \alpha^{(\nu)}_{k_\nu}(t) \right) \tag{30}$$

with growth rates $\sigma^{(\nu)}_{k_\nu} > 0$, and time-dependent interaction weights $\rho^{(\nu)}_{k_\nu j_\nu}(t) > 0$, $\rho^{(\nu)}_{k_\nu k_\nu}(t) = 1$. Again, $\nu \in \mathbb{N}$ indicates the level within the hierarchy, n_ν is the number of stationary states and $\tau^{(\nu)}$ represents the characteristic time scale of that level. Levels are temporally well separated through $\tau^{(\nu)} \gg \tau^{(\nu+1)}$ [35].

Following [34], the population amplitudes $\alpha^{(\nu)}_{k_\nu}(t)$ of level ν prescribe the control parameters of the higher level $\nu - 1$ by virtue of

$$\rho^{(\nu)}_{k_\nu j_\nu}(t) = \sum_{l_{\nu+1}=1}^{m_{\nu+1}} \alpha^{(\nu+1)}_{l_{\nu+1}}(t) r^{(\nu)}_{k_\nu j_\nu l_{\nu+1}} \tag{31}$$

where the constants $r^{(\nu)}_{k_\nu j_\nu l_{\nu+1}}$ serve as parameter templates.

Finally, the amplitudes $\alpha^{(\nu)}_{k_\nu}(t)$ recruit a hierarchy of modes $V^{(\nu)}_{k_\nu}(x)$ in the neural field such that

$$V(s) = V(x,t) = \sum_{\nu=1}^{\infty} \sum_{k_\nu=1}^{n_\nu} \alpha^{(\nu)}_{k_\nu}(t) V^{(\nu)}_{k_\nu}(x) \,. \tag{32}$$

Assuming a system of bi-orthogonal modes $V^{(\nu)^+}_{k_\nu}(x)$ with

$$\int_\Omega V^{(\nu)^+}_{k_\nu}(x) \, V^{(\nu)}_{j_\nu}(x) \, \mathrm{d}x = \delta_{k_\nu j_\nu} \tag{33}$$

we obtain from (32)

$$\alpha^{(\nu)}_{k_\nu}(t) = \int_\Omega V^{(\nu)^+}_{k_\nu}(x) \, V(x,t) \, \mathrm{d}x \,. \tag{34}$$

Next we convert the Lotka-Volterra differential equations (30) into their corresponding integral equations by formally integrating

$$\alpha^{(\nu)}_{k_\nu}(t) = \frac{1}{\tau^{(\nu)}} \int_{-\infty}^{t} \alpha^{(\nu)}_{k_\nu}(t_\nu) \left(\sigma^{(\nu)}_{k_\nu} - \sum_{j_\nu=1}^{n_\nu} \rho^{(\nu)}_{k_\nu j_\nu}(t_\nu) \alpha^{(\nu)}_{j_\nu}(t_\nu) \right) \mathrm{d}t_\nu \tag{35}$$

and obtain a recursion equation after inserting (31)

$$\alpha^{(\nu)}_{k_\nu}(t) = \frac{1}{\tau^{(\nu)}} \int_{-\infty}^{t} \alpha^{(\nu)}_{k_\nu}(t_\nu) \left(\sigma^{(\nu)}_{k_\nu} - \sum_{j_\nu=1}^{n_\nu} \sum_{l_{\nu+1}=1}^{m_{\nu+1}} r^{(\nu)}_{k_\nu j_\nu l_{\nu+1}} \alpha^{(\nu+1)}_{l_{\nu+1}}(t_\nu) \alpha^{(\nu)}_{j_\nu}(t_\nu) \right) \mathrm{d}t_\nu \,. \tag{36}$$

Eventually, we insert all results of the recursion of (36) into the field equation (32) and eliminate the amplitudes $\alpha_{k_\nu}^{(\nu)}(t)$ by means of (34) in order to get a series expansion of $V(s)$ in terms of spatiotemporal integrals. This series must equal the generalized Volterra series (21), such that the kernel $H(s, s')$ can be reconstructed to solve the neural field inverse problem [44]. For more mathematical details on the derivation of the two-level hierarchy, see Additional file 1.

Additional file

Additional file 1: Mathematical details on the two-level hierarchy.

Competing interests
The authors declare that they have no competing interests.

Authors' contributions
Both authors have developed the approach of the work. AH has worked out the analytical and numerical stability analysis part, while PbG has developed analytically the hierarchy of heteroclinic sequences. Both authors have written the manuscript. Both authors read and approved the final manuscript.

Acknowledgments
This research has been supported by the European Union's Seventh Framework Programme (FP7/2007-2013) ERC grant agreement No. 257253 awarded to AH, hosting PbG during fall 2013 in Nancy, and by a Heisenberg fellowship (GR 3711/1-2) of the German Research Foundation (DFG) awarded to PbG.

Author details
[1]Dept. of German Studies and Linguistics, Humboldt-Universität zu Berlin, Unter den Linden 6, 10099 Berlin, Germany. [2]Team Neurosys, INRIA Grand Est - Nancy, 615 rue du Jardin Botanique, 54602 Villers-les-Nancy, France. [3]Bernstein Center for Computational Neuroscience Berlin, Berlin, Germany.

References

1. Wilson H, Cowan J: **A mathematical theory of the functional dynamics of cortical and thalamic nervous tissue.** *Kybernetik* 1973, **13:**55–80.
2. Amari SI: **Dynamics of pattern formation in lateral-inhibition type neural fields.** *Biol Cybern* 1977, **27:**77–87.
3. Jancke D, Erlhagen W, Dinse HR, Akhavan AC, Giese M, Steinhage A, Schöner G: **Parametric population representation of retinal location: neuronal interaction dynamics in cat primary visual cortex.** *J Neurosci* 1999, **19**(20):9016–9028. [http://www.jneurosci.org/content/19/20/9016.abstract]
4. Bressloff PC: **Spatiotemporal dynamics of continuum neural fields.** *J Phys A* 2012, **45**(3):033001. [http://stacks.iop.org/1751-8121/45/i=3/a=033001]
5. Coombes S: **Large-scale neural dynamics: simple and complex.** *NeuroImage* 2010, **52**(3):731–739. [http://www.sciencedirect.com/science/article/B6WNP-4Y70C6H-3/2/334a01e2662e998a0fdd3e1bbe9087d7]
6. Coombes S, Venkov NA, Shiau L, Bojak I, Liley DTJ, Laing CR: **Modeling electrocortical activity through improved local approximations of integral neural field equations.** *Phys Rev E* 2007, **76**(5):051901.
7. Coombes S, Lord G, Owen M: **Waves and bumps in neuronal networks with axo-dendritic synaptic interactions.** *Physica D* 2003, **178:**219–241.
8. Folias S, Bressloff P: **Stimulus-locked waves and breathers in an excitatory neural network.** *SIAM J Appl Math* 2005, **65:**2067–2092.
9. Hutt A, Rougier N: **Activity spread and breathers induced by finite transmission speeds in two-dimensional neural fields.** *Phys Rev E* 2010, **82:**R055701.
10. Coombes S, Owen M: **Bumps, breathers, and waves in a neural network with spike frequency adaptation.** *Phys Rev Lett* 2005, **94:**148102.
11. Ermentrout GB, McLeod JB: **Existence and uniqueness of travelling waves for a neural network.** *Proc R Soc E* 1993, **123A:**461–478.
12. Jirsa VK, Haken H: **Field theory of electromagnetic brain activity.** *Phys Rev Lett* 1996, **77**(5):960–963.
13. Hutt A: **Generalization of the reaction-diffusion, Swift-Hohenberg, and Kuramoto-Sivashinsky equations and effects of finite propagation speeds.** *Phys Rev E* 2007, **75:**026214.
14. Bressloff PC: **Traveling fronts and wave propagation failure in an inhomogeneous neural network.** *Physica D* 2001, **155:**83–100.
15. Jirsa VK, Kelso JAS: **Spatiotemporal pattern formation in neural systems with heterogeneous connection toplogies.** *Phys Rev E* 2000, **62**(6):8462–8465.
16. Kilpatrick ZP, Folias SE, Bressloff PC: **Traveling pulses and wave propagation failure in inhomogeneous neural media.** *SIAM J Appl Dynanmical Syst* 2008, **7:**161–185.
17. Schmidt H, Hutt A, Schimansky-Geier L: **Wave fronts in inhomogeneous neural field models.** *Physica D* 2009, **238**(14):1101–1112.

18. Potthast R, beim Graben P: **Inverse problems in neural field theory.** *SIAM J Appl Dynamical Syst* 2009, **8**(4):1405–1433.
19. Potthast R, beim Graben P: **Existence and properties of solutions for neural field equations.** *Math Methods Appl Sci* 2010, **33**(8):935–949.
20. Coombes S, Laing C, Schmidt H, Svanstedt N, Wyller J: **Waves in random neural media.** *Discrete Contin Dyn Syst A* 2012, **32:**2951–2970.
21. Coombes S, Laing C: **Pulsating fronts in periodically modulated neural field models.** *Phys Rev E* 2011, **83:**011912.
22. Brackley C, Turner M: **Persistent fluctuations of activity in undriven continuum neural field models with power-law connections.** *Phys Rev E* 2009, **79:**011918.
23. beim Graben P, Potthast R: **Inverse problems in dynamic cognitive modeling.** *Chaos* 2009, **19:**015103.
24. Hutt A, Riedel H: **Analysis and modeling of quasi-stationary multivariate time series and their application to middle latency auditory evoked potentials.** *Physica D* 2003, **177**(1–4):203–232.
25. Yildiz I, Kiebel SJ: **A hierarchical neuronal model for generation and online recognition of birdsongs.** *PloS Comput Biol* 2011, **7**(12):e1002303.
26. Veltz R, Faugeras O: **Local/global analysis of the stationary solutions of some neural field equations.** *SIAM J Appl Dynamical Syst* 2010, **9:**954–998.
27. Afraimovich VS, Zhigulin VP, Rabinovich MI: **On the origin of reproducible sequential activity in neural circuits.** *Chaos* 2004, **14**(4):1123–1129.
28. Rabinovich MI, Huerta R, Varona P, Afraimovichs VS: **Transient cognitive dynamics, metastability, and decision making.** *PLoS Comput Biolog* 2008, **4**(5):e1000072.
29. Hammerstein A: **Nichtlineare Integralgleichungen nebst Anwendungen.** *Acta Math* 1930, **54:**117–176.
30. Kosko B: **Bidirectional associated memories.** *IEEE Trans Syst Man Cybernet* 1988, **18:**49–60.
31. Hellwig B: **A quantitative analysis of the local connectivity between pyramidal neurons in layers 2/3 of the rat visual cortex.** *Biol Cybernet* 2000, **82:**11–121.
32. Mazor O, Laurent G: **Transient dynamics versus fixed points in odor representations by locust antennal lobe projection neurons.** *Neuron* 2005, **48**(4):661–673.
33. Rabinovich MI, Huerta R, Laurent G: **Transient dynamics for neural processing.** *Science* 2008, **321**(5885):48–50.
34. Kiebel SJ, von Kriegstein K, Daunizeau J, Friston KJ: **Recognizing sequences of sequences.** *Plos Comp Biol* 2009, **5**(8):e1000464.
35. Desroches M, Guckenheimer J, Krauskopf B, Kuehn C, Osinga H, Wechselberger M: **Mixed-mode oscillations with multiple time scales.** *SIAM Rev* 2012, **54**(2):211–288. [http://epubs.siam.org/doi/abs/10.1137/100791233]
36. Tsuda I: **Toward an interpretation of dynamic neural activity in terms of chaotic dynamical systems.** *Behav Brain Sci* 2001, **24**(5):793–847.
37. Freeman W: **Evidence from human scalp EEG of global chaotic itinerancy.** *Chaos* 2003, **13**(3):1069.
38. Appell J, Chen CJ: **How to solve Hammerstein equations.** *J Integr Equat Appl* 2006, **18**(3):287–296.
39. Banas J: **Integrable solutions of Hammerstein and Urysohn integral equations.** *J Austral Math Soc (Series A)* 1989, **46:**61–68.
40. Lakestani M, Razzaghi M, Dehghan M: **Solution of nonlinear Fredholm-Hammerstein integral equations by using semiorthogonal spline wavelets.** *Math Problems Eng* 2005:113–121.
41. Djitteab N, Senea M: **An iterative algorithm for approximating solutions of Hammerstein integral equations.** *Numerical Funct Anal Optimization* 2013, **34**(12):1299–1316.
42. Hutt A, Longtin A: **Effects of the anesthetic agent propofol on neural populations.** *Cogn Neurodyn* 2010, **4:**37–59.
43. Bressloff PC, Coombes S: **Physics of the extended neuron.** *Int J Mod Phys* 1997, **B 11**(20):2343–2392.
44. beim Graben P, Potthast R: **A dynamic field account to language-related brain potentials.** In *Principles of Brain Dynamics: Global State Interactions.* Edited by Rabinovich M, Friston K, Varona P. Cambridge (MA): MIT Press; 2012:93–112.
45. Haken H: *Synergetics. An Introduction Volume 1 of Springer Series in Synergetics.* 3rd edition. Berlin: Springer; 1983. [1st edition 1977].
46. Fukai T, Tanaka S: **A simple neural network exhibiting selective activation of neuronal ensembles: from winner-take-all to winners-share-all.** *Neural Comp* 1997, **9:**77–97. [http://www.mitpressjournals.org/doi/abs/10.1162/neco.1997.9.1.77]
47. Wilson H, Cowan J: **Excitatory and inhibitory interactions in localized populations of model neurons.** *Biophys J* 1972, **12:**1–24.

12

Is it possible to predict electromagnetic resonances in proteins, DNA and RNA?

Irena Cosic[1,2*], Drasko Cosic[2] and Katarina Lazar[2]

* Correspondence:
irenacosic@me.com
[1]RMIT University, Melbourne 3000, VIC, Australia
[2]AMALNA Consulting, 46 Second St, Black Rock 3193, VIC, Australia

Abstract

Background: It has been shown that there are electromagnetic resonances in biological molecules (proteins, DNA and RNA) in the wide range of frequencies including THz, GHz, MHz and KHz. These resonances could be important for biological function of macromolecules, as well as could be used in development of devices like molecular computers. As experimental measurements of macromolecular resonances are timely and costly there is a need for computational methods that can reliably predict these resonances.

We have previously used the Resonant Recognition Model (RRM) to predict electromagnetic resonances in tubulin and microtubules. Consequently, these predictions were confirmed experimentally.

Methods: The RRM is developed by authors and is based on findings that protein, DNA and RNA electromagnetic resonances are related to the free electron energy distribution along the macromolecule.

Results: Here, we applied the Resonant Recognition Model (RRM) to predict possible electromagnetic resonances in telomerase as an example of protein, telomere as an example of DNA and TERT mRNA as an example of RNA macromolecules.

Conclusion: We propose that RRM is a powerful model that can computationally predict protein, DNA and RNA electromagnetic resonances.

Keywords: Molecular resonances; Proteins; DNA; RNA; Resonant recognition model

Background

Resonances are ever intriguing processes as they enable energy to be transferred selectively and with minimum loss. Resonances in biological macromolecules (proteins, DNA and RNA) are even more compelling, because they relate to macromolecular biological activity as well as have possible usage in technological developments (molecular computers) [1–11]. Recently, it has been shown that tubulin and microtubules have specific electromagnetic resonances that have a frequency range from THz to KHz [11]. Although there are similar examples where these resonances are measured experimentally [8–11], there is a need for theoretical models that can reliably predict these resonances in a variety of different macromolecules (proteins, DNA and RNA). Here we present the Resonant Recognition Model (RRM) [12–16] as a good candidate for such predictions. RRM is a revolutionary new approach proposing that macromolecular activity is based on electromagnetic resonances [12–16]. In our previous work we have used the RRM to predict resonances in tubulin and microtubules and shown that these resonances are in the frequency ranges from THz to KHz [17], which has been experimentally confirmed by a recently published paper [11]. Here, we are applying the RRM to predict possible electromagnetic resonances in other proteins (telomerase), DNA (telomere) and RNA (TERT mRNA). We propose that RRM is a powerful universal tool that can predict protein, RNA and DNA electromagnetic resonances, which is relevant for macromolecular biological function as well as applicable to new, innovative technological devices.

Methods

Resonant recognition model

The RRM is based on the findings that certain periodicities within the distribution of energy of delocalised electrons along a protein molecule are critical for protein DNA and RNA biological function and/or interaction with their targets. If possible charge transfer through these macromolecules is introduced, then charge moving through macromolecular backbone or through its 3D structure, like helical structure, can produce electromagnetic radiation, absorption and resonance with spectral characteristics corresponding to energy distribution along the protein. The RRM enables for these spectral characteristics to be determined.

All proteins, DNA and RNA can be considered as a linear sequence of their constitutive elements: amino acids or nucleotides. The RRM model interprets this linear information as a numerical series by assigning each amino acid a physical parameter representing the energy of delocalised electrons of each amino acid and then transforming this numerical series into the frequency domain using Fourier Transform. As the distance between amino acid in a polypeptide chain is 3.8 Å, it can be postulated that the points in the numerical sequence derived are equidistant. Therefore, for this numerical analysis the distance between points are set at an arbitrary value $d = 1$. The maximum frequency in the corresponding numerical spectrum is $F = 1/2d = 0.5$. The total number of points in the sequence influences the resolution of the spectrum. Thus, for N-point sequence the resolution in the spectrum is equal to $1/N$. The n-th point in the spectral function corresponds to the frequency of $f = n/N$.

The RRM uses cross-spectral function to extract common spectral characteristics for sequences with the same or similar biological function. The presence of a peak frequency in a multiple cross-spectral function implies that all of the analysed sequences within the group have this frequency component in common. The same approach can be used for DNA and RNA sequences. However, for the characteristic frequencies between proteins and DNA/RNA macromolecules to be compared, it is required to make adjustments for the differences in distances between nucleotides (3.4 Å) and amino acids (3.8 Å). These adjustments are made on nucleotide sequences spectrum so the final result could be compared with frequency calculations made for the proteins.

Our previous research has shown that all protein, DNA and RNA sequences with a common biological function have the common frequency component [12–16], that represents the characteristic feature for the observed function/interaction. This characteristic frequency is related to the protein biological function [12–16].

Furthermore, it was shown that the proteins and their targets, other proteins, DNA or RNA, have the same characteristic frequency in common. Thus, we propose that the RRM frequencies characterise not only a general function, but also a recognition and interaction between the particular protein and its target, which is based on resonant recognition [12–16].

Macromolecular interactions can be considered as resonant energy transfer between the interacting molecules. This energy can be transferred through oscillations of a physical field, possibly electromagnetic in nature. Since there is evidence that proteins, DNA and RNA have certain conducting or semi-conducting properties, a charge, moving through the macromolecular backbone and passing different energy stages caused by different amino acid or nucleotide side groups, can produce sufficient conditions for a specific electromagnetic radiation or absorption. The frequency range of this field depends on a charge velocity. The RRM proposes that charge is travelling through macromolecular backbone at the charge velocity estimated at 7.87×10^5m/s [12, 13]. For this velocity and the distance between amino acids in a protein molecule, which is 3.8 Å, the frequency range obtained for protein interactions was estimated to be in the range of 10^{13}Hz up to 10^{15}Hz. For nucleotide sequences, this frequency needs to be adjusted, as distance between nucleotides is 3.4 Å. The adjustment has been done within nucleotide sequences to match amino acid sequences. Therefore, the estimated range for both amino acid and nucleotide macromolecules includes infra-red, visible and ultra-violet light. These computational predictions were found to be related to biological function of the macromolecules by comparison with a number of experimental measurements [12–19].

However, if we take into account protein and DNA/RNA complex structures, and in particular alpha helices, the charge transfer is also possible through these structures in the form of solitons [20] (Davydov [21, 22], Hayman [23], Sinkala [24]), excitons (Davydov [21, 22], Sinkala [24], Pang [25],Yomosa [26]) and phonons (Pang [25], Yomosa [26], Ichinose [27], Pang [25]). These other forms of charge transfers are at different velocities ranging from 10^5 m/s for solitons and some exciton all the way down to the speed of sound and small fractions of the speed of sound for phonons. Thus, with the same periodicities within proteins sequences, as determined by the RRM, different modalities of charge transfer can produce different resonant frequencies, which are not necessarily related to their protein biological function, but could be related to the protein and DNA/RNA resonances, in general.

In our previous work, we have applied these charge moving modalities to tubulin and microtubule macromolecules and identified a number of possible electromagnetic resonance frequencies in their macromolecule structures. These results have been experimentally confirmed in research by Bandyopadhyay team, by Sahu et al. [11].

Results and discussion

Here we applied the RRM approach to three different groups of macromolecules: proteins (telomerases), coding nucleotide sequences (TERT mRNA) and DNA regulatory sequences (telomeres), with the aim to predict electromagnetic resonant frequencies for these three different groups of macromolecules.

Proteins

As an example of protein macromolecules, the following TERT telomerase proteins were analysed using the RRM approach: Q27ID4 - TERT_BOVIN, O14746 - TERT_HUMAN (1–230), O14746 - TERT_HUMAN (325–550), O70372 - TERT_MOUSE and Q673L6 - TERT_RAT. The characteristic common RRM frequency for analysed TERT telomerase proteins was found to be at frequency of f = 0.2930 ± 0.001, as presented in Fig. 1. This result is consistent with our previously published common frequency for human telomerase [28].

When RRM frequencies for telomerase were calculated and used in conjunction with different possible charge velocities through protein, the possible resonant frequencies were in the ranges of THz, GHz, MHz and KHz, as presented in Table 1a.

Coding mRNA

As an example of coding nucleotide sequences, we analysed the following telomerase TERT mRNA coding sequences, Homo sapiens telomerase reverse transcriptase isoforms: JF896280.1 - Delta2, JF896283.1 – Delta2-8, JF896282.1 – Delta4C, JF896285.1 – Delta4-13, JF896281.1 – Delta3p-12 and JF896286.1 – INTR1. The common numerical characteristic RRM frequency was found to be at frequency of f = 0.3340 ± 0.002, as presented in Fig. 2a. This is in accordance with our previous findings, that all coding nucleotide sequences have the same characteristic frequency of f = 0.3300 ± 0.002, describing the common function "coding". As three nucleotides code for each amino acid, it is reasonable to expect that characteristic frequency for "coding" function will reflect triplet codes and will represent periodicity of three nucleotides or frequency of 0.33 (1/3).

To be able to calculate the actual electromagnetic resonant frequencies in nucleotide sequence, it is necessary to adjust for the difference in distances between nucleotides and amino acids, as described above. Therefore, the main numerical frequency peak shifted to the frequency of f = 0.3730 ± 0.002, as presented in Fig. 2b. The analysis of the mRNA sequences have shown two distinct characteristic peaks at frequency of f = 0.3730 ± 0.002 and at frequency of f = 0.4014 ± 0.002. Keeping in mind the different modalities of possible charge transfer through the protein molecule these frequencies (periodicities) can produce specific resonances in the ranges of THz, GHz, MHz and KHz, as presented in Table 1b. It is interesting to note that TERT and TERT mRNA have similar the second most prominent peak frequency within the range from 0.40 to 0.41. This could be due to TERT telomerase activity which is also coded within TERT mRNA. Such similarity is causing overlaps between TERT and TERT mRNA across the whole range of electromagnetic resonances related to the second most prominent frequency.

Regulatory DNA

We have analysed the telomeres as an example of DNA sequences. The following homo sapiens telomeric repeat region isolates were analysed: HQ167745.1 - AE,

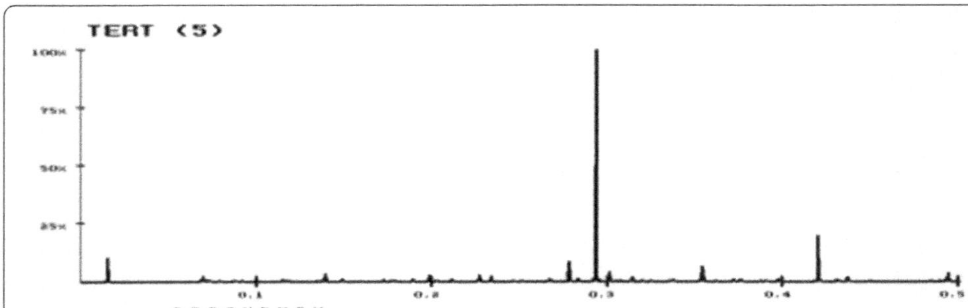

Fig. 1 Peak for five TERT telomerase proteins at frequency of f = 0.2930 ± 0.001

Table 1 Electromagnetic Resonance Frequencies taking into consideration as proposed by different mechanisms of charge velocities for: a. five TERT telomerase proteins; b. six TERT mRNA coding sequences; c. ten telomere sequences, after conversion

RRM Frequency	EMf THz per v = 787000 m/s (Cosic)	EMf THz per v = 180000m/s (Yomosa)	EMf GHz per v = 170 m/s (Davydov)	EMf GHz per v = 68 m/s (Pang)	EMf MHz per v = 3.2 m/s (Yomosa)	EMf MHz per v=0.34 m/s (Ichinose)	EMf KHz per v = 0.0005 m/s (Ichinose)
a. TERT (5)							
0.2930	299-308	46-47	65-67	26-27	1215-1252	129-133	190-196
0.4194	420-429	66-67	92-95	37-38	1747-1785	186-190	272-279
0.0161	12-21	2-2	2-5	1-2	49-86	5-9	8-14
0.2778	383-392	43-46	61-63	34-35	1151-1180	122-126	180-186
0.2525	261-271	55-57	70-80	21-22	1470-1507	156-160	220-225
b. TERT mRNA (6) conversion							
0.3730	881-891	58-60	82-85	33-34	1550-1592	165-169	242-249
0.4014	411-431	63-64	89-91	35-36	1669-1711	177-182	261-267
0.4902	503-513	77-78	109-111	43-44	2043-2085	217-222	319-326
c. TELOMERE (10)							
0.1875	181-207	28-32	39-45	16-18	737-842	78-89	115-132
0.1904	184-210	28-32	40-45	16-18	749-854	80-91	117-134
0.1768	170-196	26-30	37-42	15-17	692-797	78-85	108-125
0.1846	178-204	27-31	38-44	15-18	785-830	77-86	113-130

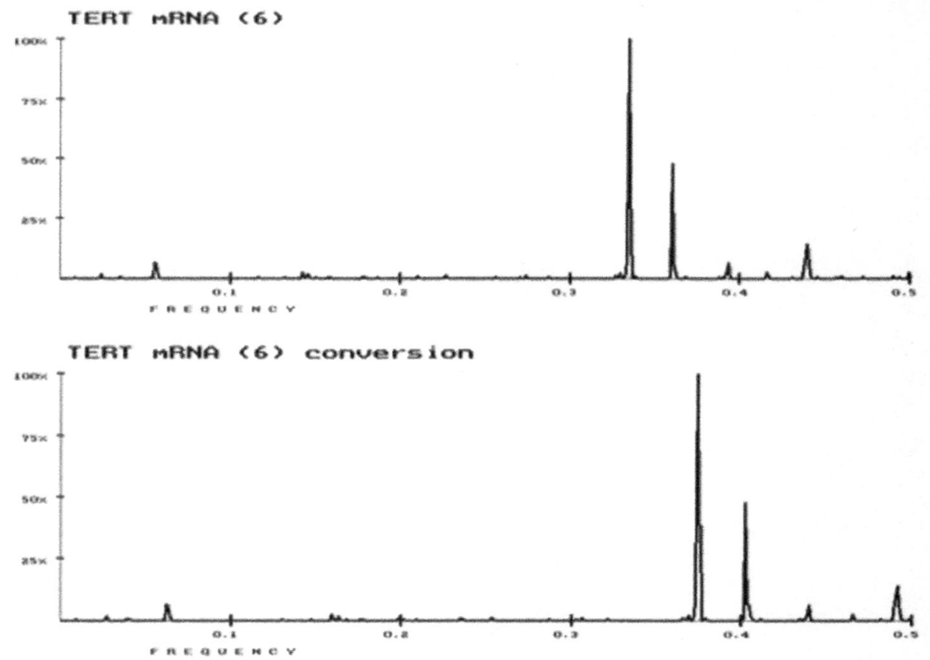

Fig. 2 a Peak for six TERT mRNA coding sequences at frequency of f = 0.3340 ± 0.002. **b**. Peak for six TERT mRNA coding sequences, after conversion, at frequency of f = 0.3730 ± 0.002

HQ167740.1 - DVW, HQ167744.1 - KA, HQ167746.1 - KM, HQ167741.1 - KP, HQ167742.1 - LH, HQ167748.1 - SA, HQ167747.1 - SF, HQ167743.1 - VDEC and AF020783.1 - chromosome 20. The main numerical characteristic RRM frequency was found to be at frequency of f = 0.1875 ± 0.002 [28], as presented in Fig. 3. This frequency has been adjusted for the differences in distances between the amino acids in protein and the nucleotides in DNA. When this adjusted numerical characteristic frequency was used in conjunction with different modalities of charge transfer, the possible electromagnetic resonances have been found to be in the ranges of THz, GHz, MHz and KHz, as presented in Table 1c. It is interesting to note that telomere have much lower characteristic frequencies than TERT telomerase and TERT mRNA.

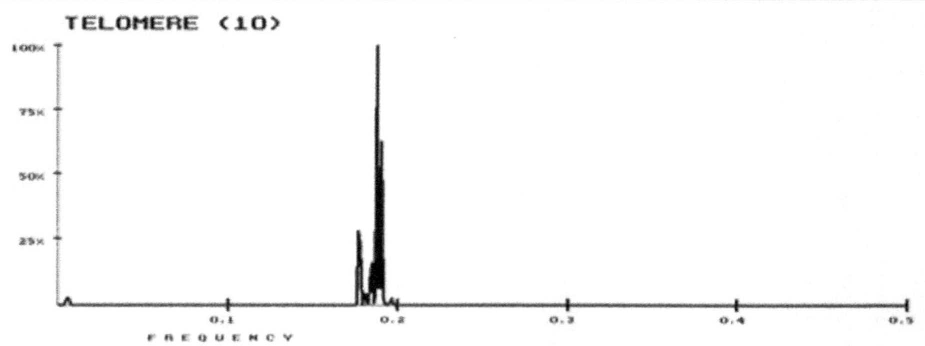

Fig. 3 Peak for ten telomere sequences, after conversion, at frequency of f = 0.1875 ± 0.002

Conclusion

We have shown here that the Resonant Recognition Model can be used as universal tool in predicting protein, RNA and DNA electromagnetic resonances in the wide frequency range from THz. GHz, MHz and KHz. Keeping in mind that our earlier predictions with tubulin molecules have been experimentally proved [11], we propose that the RRM could be used as a powerful universal method for predicting the electromagnetic resonances in biological macromolecules that could be used in experimental planning and in conjunction with experiments to minimise time and expenditure in exploring such complex macromolecular systems.

Abbreviations

RRM: Resonant recognition model; DNA: Deoxyribonucleic acid; RNA: Ribonucleic acid; TERT: Telomerase reverse transcriptase.

Competing interests

The authors declare that they have no competing interests.

Authors' contributions

IC developed the RRM model, designed the study, analysed data and wrote most of the paper, DC developed the RRM software, performed calculations and analysed data and KL was involved in writing the paper and reviewing the text. All authors discussed the results and commented on the manuscript. All authors read and approved the final manuscript.

Acknowledgments

The authors would like to thank Ms. Alison Kemp for proof reading this paper. The current research work is funded by AMALNA Consulting, registered consulting and research company, Australia.

References

1. Hameroff S. The brain is both neurocomputer and quantum computer. Cogn Sci. 2007;31:1035–45.
2. Tegmark M. The importance of quantum decoherence in brain processes. Physical Rev E. 2000;61:4194–206.
3. Hameroff S. Ultimate computing-biomolecular consciousness and nanotechnology. North Holland: Elsevier; 1987.
4. Saha AA, Craddock TJA, Tuszynski JA. An investigation of stochastic resonance in tubulin dimers. Biosystems. 2012;107(2):81–7.
5. Sahu S, Ghosh S, Ghosh B, Aswani K, Hirata K, Fujita D, et al. Atomic water channel controlling remarkable properties of a single brain microtubule: Correlating single protein to its supramolecular assembly. Biosens Bioelectron. 2013;47:141–8.
6. Sahu S, Ghosh S, Hirata K, Fujita D, Bandyopadhyay A. Multi-level memory-switching properties of a single brain microtubule. Appl Phys Lett. 2013;102(123701):1–4.
7. Hameroff S, Penrose R. Consciousness in the universe: a review of the 'Orch OR' theory. Physics Life Rev. 2014;11:39–78.
8. Craddock TJA, Tuszynski JA, Hameroff S. Cytoskeletal signaling: is memory encoded in microtubule lattices by CaMKII phosphorylation. PLoS Comput Biol. 2012;8(3):1–16. e1002421.
9. Ayoub AT, Craddock TJA, Tuszynski J. Analysis of the strength of interfacial hydrogen bonds between tubulin dimers quantum theory of atoms in molecules. Biophys J. 2014;2:740–50.
10. Dotta BT, Murugan NJ, Karbowski LM, Lafrenie RM, Persinger MA: Shifting wavelength of ultraweak photon emissions from dying melanoma cells: their chemical enhancement and blocking are predicted by Cosic's theory of resonant recognition model for macromolecules. Naturwissenschaften 2014; 101(2) doi:10.1007/s00114-013-1133-3.
11. Sahu S, Ghosh S, Fujita D, Bandyopadhyay A. Live visualizations of single isolated tubulin protein self-assembly via tunneling current: effect of electromagnetic pumping during spontaneous growth of microtubule. Scientific Reports 2014; 4: doi:10.1038/srep07303.
12. Cosic I. Macromolecular bioactivity: is it resonant interaction between macromolecules?-theory and applications. IEEE Trans Biomedical Engineer. 1994;41:1101–14.
13. Cosic I: The Resonant Recognition Model of Macromolecular Bioactivity: Theory and Applications. BirkhauserVerlag 1997.
14. Cosic I. Virtual spectroscopy for Fun and profit. Biotechnology. 1995;13:236–8.
15. Cosic I, Pirogova E:Bioactive Peptide Design using the Resonant Recognition Model. Nonlinear Biomedical Physics 2007; 1(7): doi:10.1186/1753-4631-1-7.
16. Cosic I, Vojisavljevic V, Pavlovic M. The relationship of the resonant recognition model to effects of Low-intensity light on cell growth. Int J Radiat Biol. 1989;56(2):179–91.
17. Cosic I, Lazar K, Cosic D. Prediction of Tubulin resonant frequencies using the Resonant Recognition Model (RRM). IEEE Trans. on NanoBioscience 2014; 12: doi:10.1109/TNB.2014.2365851.
18. Pirogova E, Vojisavljevic V, Istivan T, Coloe P, Cosic I. Review study: influence of electromagnetic radiation on enzyme activity and effects of synthetic peptides on cell trans- formation. MD-Medical Data. 2010;2(4):317–24.

19. Vojisavljevic V, Pirogova E, Cosic I. The effect of electromagnetic radiation (550 nm-850nm) on l-lactate dehydrogenase kinetics. Int J Radiat Biol. 2007;83(4):221–30.
20. Ciblis P, Cosic I. The possibility of soliton/exciton transfer in proteins. J Theor Biol. 1997;184:331–8.
21. Davydov AS. Excitons and solitons in molecular systems. Int Rev Cytol. 1987;106:183–225.
22. Davydov AS. Influence of electron–phonon interaction on the motion of an electron in a One-dimensional molecular system. Translated Teoreticheskaya i Matematicheskaya Fizika. 1979;40(3):408–21.
23. Hyman JM, McLaughlin DW, Scott AC: On Davydov's Alpha-Helix Solitons, Long-Time Prediction in Dynamics. John Wiley & sons 1983, 367–394.
24. Sinkala Z. Soliton/exciton transport in proteins. J Theor Biol. 2006;241:919–27.
25. Pang XF: Theory of Bio-Energy Transport in Protein Molecules and its Experimental Evidences as well as Applications. Higher Education Press and Springer-Verlag 2007.
26. Yomosa S. The exciton in protein. J Phys Soc Jpn. 1963;18(10):1494.
27. Ichinose S. Soliton excitations in alpha-helical protein structures. Chaos, Solitons Fractals. 1991;1(6):501–9.
28. Cosic I, Lazar K, Cosic D. Cellular ageing- telomere, telomerase and progerin analysed using resonant recognation model. MD-Medical Data. 2014;6(3):205–9.

13

Application of texture analysis to muscle MRI: 1-What kind of information should be expected from texture analysis?

Jacques D De Certaines[1,2*], Thibaut Larcher[3], Dorota Duda[4], Noura Azzabou[1,2], Pierre-Antoine Eliat[5], Luis M Escudero[6], Antonio MG Pinheiro[7], Guanyu Yang[8], Jean-Louis Coatrieux[9], Eduard Snezkho[10], Alexey Shukelovich[1], Manuela Pereira[7] and Richard A Lerski[11]

* Correspondence: jacques.de-certaines@orange.fr
[1]Institute of Myology, NMR laboratory, Paris, France
[2]CEA, I2BM, MIRCen, NMR laboratory, Paris, France
Full list of author information is available at the end of the article

Abstract

Several previous clinical or preclinical studies using computerized texture analysis of MR Images have demonstrated much more clinical discrimination than visual image analysis by the radiologist. In muscular dystrophy, a discriminating power has been already demonstrated with various methods of texture analysis of magnetic resonance images (MRI-TA). Unfortunately, a scale gap exists between the spatial resolutions of histological and MR images making a direct correlation impossible. Furthermore, the effect of the various histological modifications on the grey level of each pixel is complex and cannot be easily analyzed. Consequently, clinicians will not accept the use of MRI-TA in routine practice if TA remains a "black box" without clinical correspondence at a tissue level. A goal therefore of the multicenter European COST action MYO-MRI is to optimize MRI-TA methods in muscular dystrophy and to elucidate the histological meaning of MRI textures.

Review

Several methods of image post-processing have been developed for Texture Analysis of Magnetic Resonance images (MRI-TA) and have already demonstrated stimulating results in a large range of pre-clinical or clinical applications [1]; MRI-TA usually provides much more clinical discrimination than visual analysis of MR images by the radiologist [2]; visual analysis (about 100 grey level can be discriminated) is ten times less sensitive to fine and local grey level changes than computer analysis (about 1000 grey level by using computer detection). There is presently no well-established consensus concerning the limits of texture visual perception but it is clear that texture represented by higher orders statistics can only be discriminated by computerized image analysis.

In normal versus dystrophic muscles, a discriminating power at an early stage and during disease evolution has been demonstrated with various MRI-TA methods: some MRI texture parameters in healthy or diseased skeletal muscle have previously been related to fat infiltration [3,4] to the loss of orientation of muscle fibers [3], to the disturbance of perimysium [3], to the mean size of necrotic and regeneration foci [5], to the proportion of oxidative myofibers [5], to the endomysal fibrosis [5] and collagen content [4] or to the heterogeneity of myofiber size [5].

Unfortunately, a scale gap exists between the spatial resolutions of histological and MR images making a direct correlation impossible. Furthermore, the effect of the various histological modifications on the grey level of each pixel is complex and cannot be easily analyzed. Consequently, clinicians will not accept the use of MRI-TA in routine practice if TA remains a "black box" without clinical correspondence at a tissue level. The goal of this paper is then to try and present an early analysis of MRI-TA in relation to histological changes. Golden Retriever Muscular Dystrophy (GRMD) is the only animal model with relevance to human Duchenne muscular dystrophy (DMD): it has been selected for MRI-TA tests under the auspices of the European COST action MYO-MRI aiming to optimize TA methods and to try to elucidate the histological meaning of MRI textures.

Perception of visual textures

To Rosentholtz [6] a visual texture is the consequence of the reflection of a surface that leads to variations of the light that reaches our eyes. Typically these variations are considered a visual texture when they create a visual stimulus defined by repeated patterns. They can be regular (textiles, brick walls, honeycomb), random (forest floor) or in between (wood grain) [7]. Although a human observer has the ability to recognize textures with multiple properties, it is often difficult to define them. In fact the ability to recognize textures is a very important part of the visual system recognition mechanism. Moreover, the visual system easily performs the perceptual organization (definition of different regions, objects and structures of an image) using multiple visual cues. Texture similarity plays an important role on defining this segmentation process [8,9]. The earlier work on texture perception of Julesz [10] considers that human observers could only discriminate textures which differ on the first and second-order pixels statistics. However, Julesz himself later opposed that conjecture [11] and considered that some third [12] or forth [13] order pixel statistics could also be discriminated.

Sufficiently rich statistical models can capture texture structures up to a certain level. In fact, the most recent models of texture analysis are based on multiscale filtering models rather than higher-order pixels statistics [14], suggesting that the local processes used by human observers to compare textures might be more appropriate for texture detection [15]. As an example, Heeger and Bergen [16] defined a descriptor based on first order statistics of the outputs of multiscale filters, either with or without orientation. This work was extended by Portilla and Simoncelli [17] by adding second order statistics (joint statistics) of the multi scale responses. The successor of this model demonstrates that a multidimensional set of image statistics results in the capture of texture information that might be lost on the visual representation.

These theories all try to do texture segmentation with similar mechanisms to those that exist in the early vision. However, according to Rosenholtz [6] the visual system identifies texture boundaries in an intelligent way, not in the early vision system. Hence, texture identification is the result of statistical differences. The perception of different textures can be the result of different mean texture orientations, or because of the orientation variances. The perception of two different textures with the same mean and variance becomes typically difficult for human observers. This can happen for instance if one of the textures is unimodal and the other is bimodal [18]. This limitation results from the fact that texture segmentation involves local statistics instead of global statistics [19]. Tests using significance differences in mean orientation, mean contrast, orientation variance, and contrast variance reveal that the observers could define different regions. Rosenholtz even suggests that the human brain might have a mechanism comparable to a statistical test when defining a texture boundary. In fact the human brain might perform an earlier preprocessing of the information similar to the preprocessing of the texture image using a set of proper of multiscale filters. However, this processing step shall be complemented with sufficiently rich statistical models that provide the information for the detection of different textures and to define the image structure.

The "scale gap" between histology and MRI

Spatial resolution in MRI is limited by the NMR signal intensity depending on the number of detectable nuclear spins N(H) into the voxel, the magnetic field B_0 and the number of excitations NEX. Increasing the voxel size decreases the texture information into the image; fortunately, as muscle fibers are longitudinally oriented in skeletal muscles, 2D-MRI slice thickness can be increased without drastic loss of in-plane information when MRI is performed in the transverse plane with reference to the muscle fibers orientation. However, in 3D MRI-TA the voxel has to be isometric. The signal to noise ratio (SNR) increases with the static magnetic field with the relation $SNR = f(B_0^{3/2})$; however to increase B_0 also increases the number of artifacts such as the susceptibility artifact. SNR also depends on NEX with the relation $SNR = f(NEX^{1/2})$. Therefore the optimization of the SNR can only be a compromise between these different parameters; for instance, if we increase B_0, we increase SNR but the longitudinal relaxation time T1 is longer and then the repetition time TR has to be increased and NEX has to be decreased for the same total acquisition time. This subtle compromise between B_0 field, voxel size and NEX has presently induced an increase of the B_0 field of the marketed MRI devices up to 3T in clinical routine and 4 to 8 Tesla in research; the in-plane resolution is usually around $1mm^2$ in clinical routine and up to a few hundred μm^2 in research. For instance, previous muscle 2D MRI-TA studies have used rather similar voxel sizes: the pixel size was 0.8×0.8 mm^2 with a slice thickness of 3 mm [20] or 6 mm [21] at 1.5T, or 0.7×0.7 mm^2 with a slice thickness of 4 mm at 3T [22]. For the only one previous study on muscle 3D MRI-TA, the voxel size was 1mm × 1mm × 1mm [23]. Therefore in order to increase spatial resolution and contrast-to-noise ratio, MRI devices are often selected with a high magnetic field strength though field cycling experiments have demonstrated that the T1 contrast between different tissues usually decreases from 1 to 100 MHz. However, if the T1 contrast between muscle and fat mainly determines the MRI texture, then it has been demonstrated that there is a better result at high field as shown in Table 1.

There remains a large scale gap between MRI and histology, even when using a high field MRI device, because the resolution of MRI in routine practice cannot be sufficiently increased.

Table 1 Frequency dependence of T1 contrast between muscle and fat (calculated from the ex-vivo data collected by P.A.Bottomley et al. [24])

NMR frequency	20 MHz	40 MHz	60 MHZ	80 MHz	100 MHz
T1 contrast: (muscle-fat)/muscle	0.50	0.60	0.68	0.70	0.74

What MRI factors dictate the pixel grey level?

Several tissue NMR parameters determine the grey level in each voxel: the spin–lattice relaxation time T1, the spin-spin relaxation time T2, the spin density N(H) and the spin displacements (flow and diffusion). For instance, in the classical spin-echo (SE) pulse sequence, the MRI signal intensity I for solid tissues is given by:

$$I = f\left[k{v_o}^{1/3}N(H)\left(1 - e^{-TR/T1}\right)e^{-TE/T2}\right]$$

where:

k is a parameter related to the MRI device

v_o is the resonance frequency, varying from around 20 MHz (low field MRI devices) to around 90 MHz in current routine clinical usage and up to 350 MHz in pre-clinical and clinical research.

TR is the repetition time between two pulse sequences (i.e. the time for relaxing spin system).

TE is the echo time.

What are the values of T1 and T2 in human skeletal muscle and what is their frequency dependence? Mean T1 and T2 data collected from 13 MRI units clearly demonstrate a highly different field dependence between T1 and T2 measured in-vivo (Table 2). This in-vivo T1 field dependence is rather similar to the theoretical fits calculated from ex-vivo data by Fischer, Bottomley or Escanye (for review, see [25]).

Research conducted over the last forty years has shown that the T1 and T2 variations are related to the ratio between lipids and bound or free water in the tissues. Protons presenting very short relaxation times (for instance protons from proteins or from highly bound water) do not contribute to the MRI signal. Each one of these parameters is specifically more sensitive to some tissue modifications at a molecular level: for instance, fibrosis, corresponding to an increase of collagen fibers with an increase of associated bound water, decreases the T2; an oedema resulting from an increase of free water induces a T1 increase, etc. These specific effects of histological changes on T1, T2 and N(H) are not always in the same direction inducing a global result also depending on the amount of each histological variation; a previous study correlating quantitative histology with ex-vivo relaxometry in human lung tumours has clearly illustrated the difficulty in discussing NMR parameters with respect to histology [27].

Furthermore, the speed of exchange between bound and free water determines if there is a mean relaxation time or two or more separately detectable relaxation times: in healthy muscle, it has been established that two spin-spin relaxation times can be separately identified expressing a slow exchange between different water phases [28,29].

Table 2 T1 and T2 frequency dependence of human skeletal muscle in-vivo measured on 13 MRI units after correction resulting from a preliminary quality assessment using Eurospin TO5 test object [26]

NMR frequency	4.3 MHz	11.9 MHz	21 MHz	63 MHz
Number of in-vivo measurements	70	12	330	162
Mean T1 (ms)	223 +/− 31	557 +/− 26	557 +/− 41	1183 +/− 136
Mean T2 (ms)	34 +/− 3	42 +/− 2	34 +/− 4	33 +/− 5

All these considerations clearly demonstrate that the interpretation of voxel grey level is highly complex and must be prudently related to histological changes. The challenge is therefore to develop a pulse sequence potentially able to differentiate histological changes [30].

What texture parameters are mainly used in MR images analysis?

The first-order features are based only on the distribution of pixel grey levels and do not consider the relationships between neighboring pixels. They could contain some useful information in terms of tissue differentiation. They provide knowledge on the most frequent and the least frequently occurring grey levels, on the concentration of the grey levels around their average, or on the degree of asymmetry in their distribution.

When a MR image is suspected to contain a plurality of strips or beam-like structures, a method for its texture characterization seems to be the run-length matrix-based approach [31] (Figure 1 left). In such a method, the parameters are determined by the presence, the length, and the orientation of pixel runs (strings), composed of pixels with quite similar grey levels. With a certain adjustment of the method, it is possible to obtain the parameters that enhance (or not) a particular orientation of pixel runs (horizontal, vertical, or oblique, with $\theta = 45°$ or $\theta = 135°$). It is also possible to decide which differences in pixel grey levels within a single run are to be ignored. The run-length matrix-based features could emphasize the incidence of relatively short (or relatively long) runs, measure the non-uniformity of grey levels (or run lengths), or give the information about the fraction of image in runs.

Very important information could also be extracted with the co-occurrence matrix-based method [32] (Figure 1, right). This method analyses the co-occurrences of pairs of pixels, spaced apart a specified distance, *d*, and aligned with a given direction, θ. All the possible signed differences in grey levels of pixels are considered. A similar method, based on grey-level difference matrices [33], considers only the unsigned differences. With a properly chosen distance for pixel pairs (the parameter *d* of both methods) one

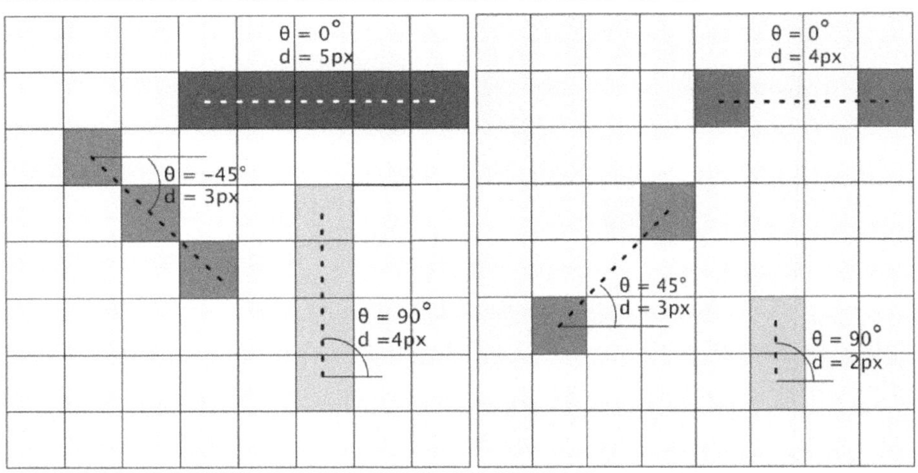

Figure 1 Illustration of the pixel relationship considered by the Run-length matrix approach (left) and the Cooccurrence matrix method (right).

can obtain the features sensitive to the presence or absence of different image elements (primitives). If certain texture structures appear in the image, or change their size, with the disease development, the features based on the pixel grey level difference matrices will also change (if the distance parameter, d, is constant). If two textures are characterized by different co-occurrence matrix-based parameters, it could be highly probable, that they correspond to different stages of the disease. As the parameter d depends on the image scale and the size of texture primitives, it is worthwhile to fix it experimentally.

Several texture parameters can be derived from the co-occurrence matrix with a given pair of parameters: (d,θ). Among them it is worth mentioning: energy or angular second moment (a measure of homogeneity of pixel grey levels, relatively high for homogeneous textures, and of small values for heterogeneous ones), contrast or inertia (with a quite large value for heterogeneous regions, characterized by a strong contrast, and quite small value for homogeneous regions), local homogeneity, and entropy (that quantifies the degree of randomness of pixel grey levels).

The ripples can also be extracted with a method based on image transformations using the Laws' filters. During the transformation, each image pixel is assigned a value that is a linear combination of initial grey levels of pixels belonging to the neighborhood of a transformed pixel. Typically neighborhoods of 3×3 or 5×5 pixels are considered. Weights of grey levels of neighboring pixels are determined by one of the Laws' masks. The masks are designed in order to detect different texture elements: ripples, edges, spots… The frequency rate of such elements determines the feature values calculated from the transformed image.

Another group of features, that could be used to indicate the presence of edges are those derived from the gradient matrices [34]. The grey-level gradient at a given image point depends on the differences between the grey levels of its neighboring pixels, located on lines (vertical and horizontal) intersecting at a point. The gradient matrix contains the values of the absolute gradient at each point of the image region considered. Based on the gradient matrix several first order statistics can be calculated. Apart of being edge-sensitive, such statistics allow us to draw conclusions about the uniformity (homogeneity) or the roughness of the texture. These latter texture properties are closely related to changes in the muscular tissue that occur during the development of the disease.

Moments, either continuous or discrete, rely on a very rich theory and encompass a wide family (Legendre, Zernike, Tchebichef, Racah, etc. [35]). They have been used in a large variety of problems including segmentation, pattern recognition, image reconstruction, watermarking. Although not often applied to texture analysis, they offer several interesting and unique properties among which their applicability to high dimensional space, their capability to capture low and high frequency components according to the selected order and the possibility to deal with noise and blurring. They not only allow the analysis of images but also allow the reconstruction of them using the computed moment values and thus permit the identification of the information they captured. Moment invariants to geometric transform (translation, rotation, scale) can be derived too [36,37] with a good stability to non-rigid and small local deformations.

Combining texture descriptors is a way to improve their discriminative power and can offer a tool to characterize a lesion.

Previous results in muscle MRI-TA

Sikio et al. [20] used co-occurrence matrix parameters on T2 weighted images such as the angular second moment, the inverse difference moment, the entropy and the difference entropy to describe the muscle tissue structure and the effect of different types of exercise. By mean of the Mann–Whitney statistical test, the study revealed exercise-load-associated differences in MRI textures of thigh soft tissues between various athlete groups and non-athletes. For each muscle of the thigh, each texture parameter was considered individually and the non-athletes group was compared to each of the five groups of athletes. The result showed that large variation in tests significance were observed depending on the muscle, the texture parameter used and the comparison performed. The authors did not show how the texture parameters evolve for athletes when compared to the non-athletes one, however they related the changes in texture descriptors to the change of the muscle architecture according to the type of exercise as well as the type of the muscle.

In the same way, Nketiah et al. [38] have detected texture differences in MRI of hip muscles associated with different level of long term exercise loading in female athletes (jumpers) and controls.

In pathological cases, texture analysis of MRI data can be a valuable tool to monitor the changes that occur to the muscle tissue. For instance, Skoch et al. [21] were interested in the classification of calf muscle of patients with different pathology. They scanned 93 volunteers (20 healthy and 73 with three types of pathology) and extracted texture parameters from T1 weighted images using MaZda software [39]. For dimensionality reduction, the author used Principal Component Analysis (PCA) to extract a group of features that best characterized each muscle. The projection on the two first principal components showed that no difference existed between control groups and healthy ones with hypertonic parents. In opposition, the healthy group and the patients with diabetes mellitus group could be linearly separated. Considering that different pathology had different impact on the muscle appearance, four groups scored by radiologists were created: 1) muscles without any visible signs of pathology, 2) muscles with mild pathological or nonspecific changes, 3) muscles with marked pathological or nonspecific changes such as apparent muscle atrophy or fat infiltration, 4) muscles with severe pathological changes or muscles almost complete replacement by fat. The projection of the different muscle on the two main axis of the PCA showed that the 4th muscle group is well separated from the others and PCA should be performed in subgroups 1, 2 and 3 to obtain better discrimination between muscles with moderate infiltration. The results obtained in the paper suggested that TA can be used as an objective tool for MR images evaluation and the results correlated well with the clinical status of patients.

Following almost the same scheme, a study carried out by Herlidou et al. [2] compared TA techniques to the assessment of several radiologists, on a set of data obtained from patients with muscular dystrophy and healthy volunteers. The texture descriptors such as histogram, co-occurrence matrix, gradient matrix and run length matrix, were extracted from T1 weighted images. In parallel, the radiologists were asked to characterize each ROI by rating the contrast, the texture coarseness, the texture complexity, the strength and the fat infiltration. The analysis of the texture parameters was achieved using Correspondence Factorial Analysis (CFA) resulting in an healthy and patient volunteer automatical classification with sensitivity of 70% and specificity of 86%. Visual inspection by radiologists that were blinded to clinical information provided less accurate classification (sensitivity of 56% and specificity of 71%). In addition to that, very significant differences were observed between the radiologist's assessments of texture features. Besides a large intra-observer variability was noted. All these studies confirm that TA can offer an objective and reliable tool for tissue characterization though the interpretation of the data still remains speculative. This is why the MYO-MRI European COST action recently has started a multicentre study based on a data set of MR images of Golden Retriever Muscular Dystrophy (GRMD) dogs.

Histological changes during the GRMD course

Duchenne Muscular Dystrophy (DMD) is a human muscular disease caused by mutations in dystrophin genes and characterized by a progressive muscle atrophy and weakness. GRMD dog is the commonly used animal model of human DMD with similar clinical and pathological disease evolution [40]. Necrosis, inflammation, fat infiltration and fibrosis progressively occur. The histological expression of the disease follows the clinical evolution up to death [41] (Figure 2).

From birth to 3 months of age, GRMD skeletal muscles displayed hypercontracted and degenerating isolated fibers [42]. Up to 3 months, these fibers are rather low in number, specially in extensor muscles, and are associated with many large clusters of regenerating fibers. After this first regenerative period, another period begins from about 3 to 6 months of age and is marked by a dramatic increase in the number of degenerating necrotic fibers, sometimes appearing in clusters [43,44]. This degenerative phase is associated with a low regenerative activity [45]. At this stage, skeletal muscle tissue displays early evidences of endomysial oedema and mild fibrosis with some minimal to mild infiltration by inflammatory cells. The last period begins from 6 months of age, and is a rather slow evolution period. Degeneration and regeneration still occurs at a lower level and intensity of fibrosis and fat tissue infiltration progressively increases [46] although some authors do not confirm this result. Scattered calcification foci may be present but their frequency strongly differs between GRMD dogs. When present, these mineral deposits elicit a strong local inflammatory response (Table 3). Clinically a dramatic decrease of muscular functions is observed from about 2 to 6 months of age corresponding to the tissue degenerative period and precedes a period of paradoxical stability during the next months until some major digestive or respiratory complications occurs.

How histological changes during GRMD disease modify NMR parameters?

As the histological variations (inflammation, fibrosis, necrosis, fat infiltration...) do not modify relaxation times in the same way, in the same proportion and at the same time, the final result (i.e. the pixel grey level) is rather difficult to be strictly interpreted in terms of histology, especially when two simultaneous tissue changes modify a relaxation time in opposite directions.

During GRMD disease evolution, T1 is first increased due to an increase of muscle free water concentration related to muscle degeneration-regeneration and inflammation. Later T1 decreases in relation to fibrosis and fat infiltration. T2 is longer in necrotic but also in fatty and connective tissue. Therefore the distinction between necrosis, fat infiltration or fibrosis cannot be easily performed. As the dystrophic process results

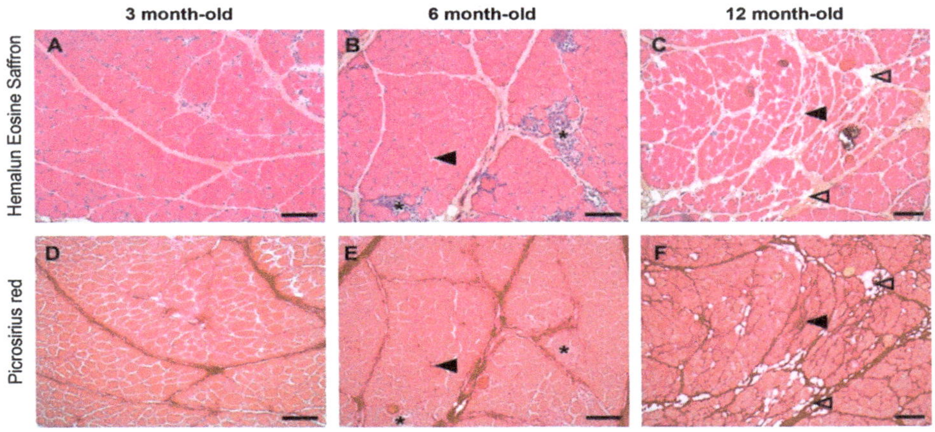

Figure 2 Typical histopathological presentation of skeletal muscles of GRMD dogs, 3, 6 and 12 month-old. Changes are mild at 3 months **(A)** with only some hypercontracted hyaline fibers, and a diffuse increase in cell number mainly due to a regenerative activity. At 6 months, degenerative and necrotic changes are more severe, focally associated with some inflammatory cell infiltration (*). At 12 months, fibrotic (black arrowhead) and fat tissues (open arrowhead) are proeminent. **(A-C)** Hemalun eosin saffron and (E-F) Picrosirius red stainings for fibrotic tissue. Bars = 100 μm **(A-B, D-E)** and 200 μm **(C and F)**.

in a muscle tissue disorganization, heterogeneous signal distribution within diseased muscle is expected, leading to the potential interest in MRI TA. This has been better observed in T2 weighted images.

Then, if the image grey level is rather difficult to analyze in terms of histology, it can be expected that the tissue texture can be more informative with MRI-TA. MRI texture results from the contrast between different parts of the tissue (for instance fibrosis, fat infiltration, inflammation, etc.), this contrast itself resulting from the effect of histological parameters on NMR parameters.

To illustrate the scale gap between histology and MRI, Figure 3 represent the same three histological images (Hemalun eosin saffron staining) at an MRI scale by addition of pixels fsrom the histological images.

MRI-TA results on GRMD dogs

The simplest texture parameter was the standard deviation or the heterogeneity parameter introduced by Thibaud et al. [22] in T2 weighted images. The authors compared the tissue heterogeneity between healthy and GRMD dogs at different age group (from 2

Table 3 Histological evolution during GRMD disease

Period	Regenerative	Degenerative	Slowly evolutive
Age (months)	2-3	5-6	9-12
Hyalin fibers	mild	marked to severe	marked
Necrosis	mild	marked to severe	marked
Regeneration	marked	mild	minimal
Centronucleated fibers	minimal	mild	marked
Inflammation	none	inconstant	mild
Fibrosis	none	mild to marked	marked to severe
Adiposis	none	minimal	mild

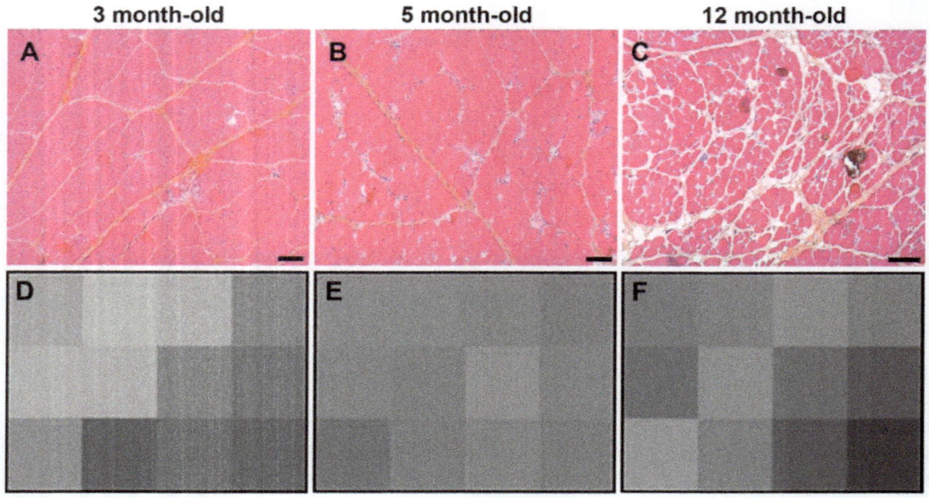

Figure 3 Typical histopathological presentation of skeletal muscles of GRMD dogs, 3, 5 and 12 month-old and corresponding images at an MRI scale. Progressive tissue changes characteristic of the disease evolution elicited some heterogeneity in the tissue aspect. This heterogeneity increases with age as the disease progress and the muscle tissue is replaced by fibrosis and some adipose tissue infiltration. **(A-C)** Hemalun eosin saffron and **(D-F)** Mean grey level of the histological images **A**, **B** and **C** were transformed to images **D**, **E** and **F** at an MRI resolution. Mean value was sampled using region of interest of 320 pixels **(A-B)** or 648 pixels **(C)** and consecutively converted to greyscale. Bars = 100 µm **(A-B)** and 200 µm **(C)**. Pixel size = 0.91 µm **(A-B)** and 0.78 µm **(C)** for histological images, 290 µm **(D-E)** and 500 µm **(F)** for MRI-simulated corresponding images.

months to 9 months). Comparison of groups was made using statistical tests to check if there were significant differences between them. The results showed that GRMD heterogeneity indices were significantly higher than in healthy dogs by at least 40% at all ages.

First order texture parameters such as mean grey levels, kurtosis or skewness clearly differ in histological Figure 3A, B and C and then are suspected to also differ in MRI-TA during the three steps of the GRMD disease as illustrated by the simulated MR images 3D, 3E and 3F.

Fan et al. [47] studied GRMD at 6 months of disease evolution: first order texture descriptor like entropy as well as higher order descriptors like the run length matrix were calculated using T2 weighted images to characterize the muscle tissue. The Mann–Whitney-Wilcoxon test showed that for all ages (3, 6 and 9–12 months), all texture features were significantly higher in GRMD when compared to healthy dogs. In addition to that, a linear discriminant analysis was performed where each NMR biomarker, including texture parameters, were used in addition with age as input variables and the group (GRMD or normal) as a response. The results showed that texture parameters extracted from the run length matrix had in average a better discriminant power than other NMR biomarkers such as the fat content and the relaxation time T2 that can be seen as an indicator of an inflammatory process. The best discriminant parameters were the ones related to muscle morphology. The entropy was not a good discriminant between healthy and GRMD dogs. For some muscles, a correlation between NMR biomarkers and histopathology indices (fibrosis, fiber number, fiber size etc.) was performed but texture parameters did not show a significant correlation with them; this lack of correlation has been explained by the scale difference because texture analysis was done at the macro level (minimal unit of voxel is $1 \times 1 \times 1$ mm^3), while histopathological data are represented at micro level (a scale of around 50 µm for fiber diameter).

In spite of the encouraging results obtained with a first order parameter, it is very important to consider other texture descriptors that provide information about the spatial organization of the tissue. Preliminary results on GRMD dogs from MYO-MRI COST action have already established that some texture methods discriminate early disease stages while others are more interesting for later phases of disease evolution. The MRI-TA study from Duda et al. [Differentiation based on MRI texture analysis between GRMD and healthy dogs at different phases of the disease evolution, in preparation] on GRMD dogs using textural features derived from different approaches (statistical, filtering-, and model-based) have demonstrated an early detection at 2–3 months with a percentage of correctly recognized cases reaching even nearly 99%. In this study, 8 methods for texture analysis were tested, including: grey-level histograms, autocorrelation coefficients, gradients, co-occurrence matrices, run length matrices, grey level difference matrices, factional Brownian motion model, and Laws' texture energy measures. In addition, a feature selection was performed, in order to assess the relative importance of each feature, in terms of the classification accuracies. The results differed between different types of muscles. Nevertheless, the application of some groups of features very often led to obtain satisfactory results. Among the top methods were: the co-occurrence matrix-based, grey-level differences matrix-based, gradient-based, histogram-based, Laws' filtering, and run-length matrix-based. The features from the co-occurrence matrices were top ranked for the early stage of GRMD development, while the run length-based method ensured the best tissue recognition for the final stage of canine life.

Conversely, another MRI-TA study of GRMD dogs from Yang et al. [Differentiation based on MRI texture analysis between GRMD and healthy dogs at different phases of the disease evolution, submitted], using orthogonal moments has not differentiated the early stage (2–3 months) but appeared efficient for disease follow-up during later stages.

Snezhko et al. [48], on the same set of GRMD image data, have extracted 313 texture features from MaZda software: a combination of several selected texture features have, for instance, have allowed an early (month 2) detection of the disease with an accuracy of 97% using cross-validation with leave-one-out strategy. For larger muscles, the efficiency of all the features was better than for smaller muscles. It was likely due to the size (number of voxels) of the available ROI.

The preliminary choice of TA methods and features should be done taking into account not only the particular development of GRMD-related pathological changes (detected on histological or MR images), but also other, not texture-related constraints. As regards the regions of interest, defined on GRMD muscles, they could be very small, in particular for the first stage of canine life. Moreover, they are usually thin and narrow. This excludes the use of many methods and parameters that have proven to be reliable in other texture-based classification tasks. For example, the run-length matrix-based method could be less useful for the small, tiny ROIs that have only few pixels of length or width. The larger is the ROI, the more useful will be the parameters based on pixel runs. This could explain why the run-length matrix-based features turned out to be reliable for tissue differentiation at the final stage of canine life. The same comments also apply to the co-occurrence matrix-based or grey level difference matrix-based methods with a parameter d of a large value. Here, a good solution would be to set a d parameter to 1 or, maximum, to 2 pixels. Finally, the first-order features, derived from the grey-level histograms, could be calculated for ROIs of any size.

As the ROI size could differ from one case to another, especially when it is determined by the age of the canine, an important task is the choice of features independent of ROI size. Admittedly, it can be predicted, that features obtained for a relatively large d will not be suitable to characterize small ROIs. It could even be problematic to calculate them. However, among the features considered potentially useful, some may be too unstable (dependent on the ROI size and/or position) to consider them as reliable texture descriptors [49]. Experimental analysis of the features stability, performed on MR images or phantoms, should therefore precede the classification experiments, in order to exclude *a priori* the features that could not really contribute to a muscular tissue recognition.

Conclusion

The MRI contrasts, depending on the incidence of histological components on relaxation parameters and spin density, determines the MRI texture. Then, two key problems have still to be more deeply discussed: i) How histological components modify MRI contrasts and then texture? and ii) How texture at the MRI scale can be related to histological texture at a very different scale? These two questions corresponds to the shared goals of our multicentre research group under the auspices of the European Union COST MYO-MRI action and is undoubtedly a prerequisite for a larger diffusion of clinical MRI-TA.

The way to bridge the scale gap should be based on the correlation of MRI-TA and the histological changes seen on the biopsies. However, a direct correlation between MRI and biopsy at the histological scale is probably not a reachable challenge. This complex comparison only would be possible under two premises. First, it will be necessary to use exactly the same samples for both analyses, requiring a stereotactic biopsy. Secondly, the histological analysis should be precise, objective and quantitative. Recently, a new computer based image analysis method has been developed to improve the evaluation of muscle biopsies [50,51]. This approach not only extracts important geometric parameters such as the size of the slow and fast fibers or the content of collagen, but also captures the organization of the tissue. The images are converted in a graph of fiber to fiber contacts. In this way it is possible to obtain certain global features that are relevant to objectively evaluate human muscular dystrophies. The quantification of the changes of these selected features will be a key to compare these images with their corresponding MRI-TA.

Competing interests

The authors declare that they have no competing interests.

Authors' contributions

JDC carried out introduction, conclusion and participated to "The scale gap between histology and MRI" and "What MRI factors dictate the pixel grey level" and drafted a large part of the manuscript. TL carried out "Histological changes during the GRMD course". TL and JDC carried out "How histological changes during GRMD disease modify NMR parameters". DD, NA, JLC and PAE carried out "What texture parameters are mainly used in MR images analysis". RAL carried out "Previous results in muscle MRI-TA". AMGP and MP carried out "Perception of visual texture". DD, LME, GY, ES and AS carried out "MRI-TA results on GRMD dogs". All authors read and approved the final manuscript.

Author details
[1]Institute of Myology, NMR laboratory, Paris, France. [2]CEA, I2BM, MIRCen, NMR laboratory, Paris, France. [3]INRA, UMR 703 PAnTher, Oniris, Nantes, France. [4]Faculty of Computer Science, Bialystok University of Technology, Bialystok, Poland. [5]PRISM-BioSCANs, University of Rennes, Rennes, France. [6]Instituto de Biomedicina, CSCIC Universidad de Sevilla, Sevilla, Spain. [7]University da Beira Interior, Covilha, Portugal. [8]LIST, Southeast University, Nanjing, China. [9]LTSI, INSERM, University of Rennes, Rennes, France. [10]Lab. Of Mathematical Cybernetics, UIIP, Belorussian Academy of Science, Minsk, Belarus. [11]Ninewells Hosp. and Medical School, Dundee, Scotland, United Kingdom.

References

1. For review, Hajek M, Dezortova M, Materka A, Lerski RA. Texture Analysis for Magnetic Resonance Imaging. Prague: Med4Publishing; 2006.
2. Herlidou S, Rolland Y, Bansard JY, Le Rumeur E, de Certaines JD. Comparison of automated and visual texture analysis in MRI: characterization of normal and diseased muscle. Magn Reson Imaging. 1999;17(9):1393–7.
3. Mahmoud-Ghoneim D, Cherel Y, Lesoeur J, Lemaire L, Rocher C, de Certaines JD, et al. Texture analysis of MR Images of rat muscles during atrophy and regeneration. Magn Reson Imaging. 2006;24(2):167–71.
4. Mamhoud-Ghoneim D, Bonny JM, Renou JP, de Certaines JD. Ex-vivo Magnetic Resonance Imaging Texture Analysis can identify genotypic origin in bovine meat. J Sci Food Agric. 2005;85:629–32.
5. Nguyen F, Eliat PA, Pinot M, Franconi F, Lemaire L, de Certaines JD, et al. Correlations between Magnetic Resonance Imaging and histopathology in mdx (X-linked Muscular Dystrophy) murine model of Duchenne Muscular Dystrophy, 24th congress of the European Society of Veterinary Pathology, Edinburgh. 2006.
6. Rosenholtz R. Texture perception. In: Wagemans J, editor. Oxford Handbook of Perceptual Organization (in press). Oxford, U.K: Oxford University Press; 2014.
7. Landy MS. Texture Perception. In: Adelman G, editor. Encyclopedia of Neuroscience. Amsterdam: Elsevier; 1996.
8. Giora E, Casco C. Region- and edge-based configurational effects in texture segmentation. Vis Res. 2007;47(7):879–86.
9. Machilsen B, Wagemans J. Integration of contour and surface information in shape detection. Vis Res. 2011;51:179–86.
10. Julesz B. Visual Pattern Discrimination. IRE Transactions on Information Theory. 1962;8(2):84–92.
11. Julesz B. Texture and Visual Perception. *Scientific American*, 212, 38–48. Julesz, B. (1975). Experiments in the visual perception of texture. Sci Am. 1965;232(4):34–43.
12. Julesz B, Gilbert EN, Victor JD. Visual discrimination of textures with identical third- order statistics. Biol Cybernet. 1978;31:137–40.
13. Klein SA, Tyler CW. Phase discrimination of compound gratings: generalized autocorrelation analysis. J Opt Soc Am A. 1986;3:868–79.
14. Malik J, Perona J. Preattentive texture discrimination with early vision mechanisms. J Opt Soc Am A. 1990;7:923–32.
15. Tyler CW. Theory of texture discrimination based on higher-order perturbations in individual texture samples. Vis Res. 2004;44:2179–86.
16. Heeger DJ, Bergen JR. Pyramid-based texture analysis/synthesis. In: Proceedings of the 22nd annual conference on Computer graphics and interactive techniques (SIGGRAPH'95). Los Angeles (USA): IEEE Comput. Soc. Press; 1995. p. 229–38.
17. Portilla J, Simoncelli EP. A Parametric Texture Model Based on Joint Statistics of Complex Wavelet Coefficients. Int J Comput Vis. 2000;40(1):49–71.
18. Rosenholtz R. Significantly different textures: A computational model of pre-attentive texture segmentation. In D. Vernon (Ed.), Proc. European Conf. on Computer Vision (ECCV'00). LNCS. 2000;1843:197–211.
19. Nothdurft HC. Texture segmentation and pop-out from orientation contrast. Vis Res. 1991;31(6):1073–8.
20. Sikio M, Harrison LCV, Nikander R, Ryymin P, Dastidar P, Eskola HI, et al. Influence of exercise loading on MRI texture of thigh soft tissue. Clin Physiol Funct Imaging. 2014;34:370–6.
21. Skoch A, Jirak D, Vyhnanovska P, Dezortova M, Fendrych P, Rolencova E, et al. Classification of calf muscle MR Images by texture analysis. Magma. 2004;16:259–67.
22. Thibaud JL, Monnet A, Bertoldi D, Barthélémy I, Blot S, Carlier PG. Characterization of dystrophic muscle in Golden Retriever Muscular Dystrophy dogs by nuclear magnetic resonance imaging. Neuromuscul Disord. 2007;17(7):575–84.
23. Wang J, Fan Z, Vandenborne K, Walter G, Shilhoh-Malmawsky Y, An H, et al. A computerized MRI biomarker quantification scheme for a canine model of Duchenne muscular dystrophy. Int J CARS. 2013;8:763–74.
24. Bottomley PA, Foster TH. Argensinger R.E?, Pfeiffer L.M., A review of normal tissue hydrogen NMR relaxation times and relaxation mechanisms from 1–100 MHz: Dependence on tissue type, NMR frequency, temperature, species, excision and age. Med Phys. 1984;11(4):425–48.
25. Henriksen O, de Certaines JD, Spisni A, Cortsen M, Muller RN, Ring PB. In-vivo field dependence of proton relaxation times in human brain, liver and skeletal muscle: a multicenter study: ii. Magn Reson Imaging. 1993;11:851–6.
26. de Certaines JD, Henriksen O, Spisni A, Cortsen M, Ring PB. In-vivo measurement of proton relaxation times in human brain, liver and skeletal muscle: a multicenter study: i. Magn Reson Imaging. 1993;11:841–50.
27. Bernard AM, De Certaines JD, Delaval P, Louvet M, Coetmeur D. Histological explanation of proton T1 and T2 variations in human lung tumors. In: Allen PS, Boisvert DPJ, Lentle BC, editors. Magnetic resonance in cancer. Toronto (Canada): Pergamon press; 1986. p. 49–50.
28. Le Rumeur E, de Certaines J, Toulouse P, Rochcongar P. Water phases in rat striated muscles as determined by T2 proton NMR relaxation times. Magn Reson Imaging. 1987;5(4):267–72.
29. Araujo ECA, Fromes Y, Carlier PG. New insights on human skeletal muscle tissue compartments revealed by in-vivo T2 NMR relaxometry. Biophys J. 2014;106:2267–74.
30. Lerski RA, de Certaines JD, Duda D, Klonowski W, Yang G, Coatrieux JL, et al. Application of texture analysis to muscle MRI: 2- Technical recommendations. EPJ Nonlinear Biomedical Physics 2015, 3:2.

31. Galloway MM. Texture analysis using grey level run lengths. Computer Graphics and Image Processing. 1975;4:172–9.
32. Haralick RM, Shanmugam K, Dinstein I. Textural features for image classification, IEEE Transactions on Systems. Man Cybernetics. 1973;3:610–21.
33. Weszka JS, Dyer CR, Rosenfeld A. A Comparative Study of Texture Measures for Terrain Classification, IEEE Trans. Systems, Man, Cybernetics. 1976;6:269–85.
34. Lerski RA, Straughan K, Shad L, Boyce D, Bluml S, Zuna I. MR Image Texture Analysis - An Approach to Tissue Characterization. Magn Reson Imaging. 1993;11:873–87.
35. Shu H, Luo L, Coatrieux J-L. Moment-based approaches in imaging. Part 1, basic features. IEEE EngMedBiol Magazine. 2007;5:70–4.
36. Shu H, Luo L, Coatrieux J-L. Derivation of moments invariants. In: Papakostas GA, editor. Moments and moments invariants. Xanthi (Greece): Science Gate Publishing; 2014.
37. Chen B, Shu H, Zhang H, Coatrieux G, Luo L, Coatrieux J-L. Combined invariants to similarity transformation and to blur using orthogonal Zernike moments. IEEE Trans Image Process. 2011;20(2):345–60.
38. Nketiah G, Savio S, Dastidar P, Nikander R, Eskola H, Sieväwen H. Detection of exercise load-associated differences in hip muscles by texture analysis. Scand J Med Sports. 2014;20:10.1111.
39. Szczypinski P, Strzelecki M, Materka A, Klepaczko A. MaZda-A software package for image texture analysis. Comput Methods Prog Biomed. 2009;94(1):66–76.
40. Kornegay JN, Bogan JR, Bogan DJ, Childers MK, Li J, Nghiem P, et al. Canine models of Duchenne muscular dystrophy and their use in therapeutic strategies. Mamm Genome. 2012;23(1–2):85–108.
41. Kornegay JN, Tuler SM, Miller DM, Levesque DC. Muscular dystyrophy in a litter of golden retriever dog. Muscle Nerve. 1988;11:1056–64.
42. Nguyen F. Muscle Lesions Associated with Dystrophin Deficiency in Neonatal Golden Retriever Puppies. J Comp Pathol. 2002;126(2–3):100–8.
43. Valentine BA, Cooper BJ, Cummings JF, de Lahunta A. Canine X-linked muscular dystrophy: morphologic lesions. J Neurol Sci. 1990;7(1):1–23.
44. Cooper BJ, Winand NJ, Stedman H, Valentine BA, Hoffman EP, Kunkel LM, et al. The homologue of the Duchenne locus is defective in X-linked muscular dystrophy of dogs. Nature. 1988;334(6178):154–6.
45. Cozzi F, Cerletti M, Luvoni GC, Lombardo R, Brambilla PG, Faverzani S, et al. Development of muscle pathology in canine X-linked muscular dystrophy. II. Quantitative characterization of histopathological progression during postnatal skeletal muscle development. Acta Neuropathol. 2001;101(5):469–78.
46. Nguyen F, Guigand L, Goubault-Leroux I, Wyers M, Cherel Y. Microvessel density in muscles of dogs with golden retriever muscular dystrophy. Neuromuscul Disord. 2005;15(2):154–63.
47. Fan Z, Wang J, Ahn M. Characteristics of magnetic resonance imaging biomarkers in a natural history study of golden retriever muscular dystrophy. Neuromuscul Disord. 2014;24:178–91.
48. Snezhko EV, Carlier P, Kovalev VA, Azzabou N, Dmitruk AA, Shukelovich AV. Application of Texture Analysis Techniques on NMR Images for Quantitative Assessment of Muscle Disorders. Informatics. 2014;3:5–13.
49. Duda D. Medical image classification based on texture analysis. PhD Thesis, University of Rennes 1, Rennes, France, 2009.
50. Sáez A, Acha B, Montero-Sánchez A, Rivas E, Escudero LM, Serrano C. A., Neuromuscular disease classification system. J Biomed Opt. 2013;18(6):66017.
51. Sáez A, Rivas E, Montero-Sánchez A, Paradas C, Acha B, Pascual A, et al. Quantifiable diagnosis of muscular dystrophies and neurogenic atrophies through network analysis. BMC Medicine. 2013;11:77.

14

Application of texture analysis to muscle MRI: 2 – technical recommendations

Richard A Lerski[1*], Jacques D de Certaines[2], Dorota Duda[3], Wlodzimierz Klonowski[4], Guanyu Yang[5,6], Jean Louis Coatrieux[7,8,9], Noura Azzabou[2] and Pierre-Antoine Eliat[10]

* Correspondence: r.a.lerski@dundee.ac.uk
[1]Medical Research Institute, Ninewells Hospital, Dundee, Scotland, UK
Full list of author information is available at the end of the article

Abstract

A goal of the multicenter European Cooperation in Science and Technology (COST) action MYO-MRI is to optimize Magnetic Resonance Imaging Texture Analysis (MRI-TA) methods for application in the study of muscle disease. This paper deals with recommendations on the optimal methodology to collect the MRI data, to analyse it via texture analysis and to make conclusions from the resultant texture parameter data. A full and detailed description is provided with respect to MR image quality control, sequence choice, image pre-processing, region of interest selection, texture analysis methods and data analysis. A series of conclusions are presented.

Keywords: Skeletal muscle; MRI; Texture analysis; Muscle dystrophy; GRMD

Review

Texture of an image region can be defined as describing the spatial relationship of pixel (or voxel) grey shades within that region. It was established in the 1970s [1] that textures could discriminate image regions and that high order properties not accessible to visual appreciation could be detected via computer analysis. Originally, the application was in aerial and geological image interpretation but, since then, there has been an expansion of application into many diverse fields.

Texture analysis (TA) of MR images has been demonstrated to be of clinical value in a wide range of situations [2]. This includes the brain, the liver, the kidney, the breast, the prostate, bone but also muscle [3-6]. There are also some other pre-clinical results in muscle MRI-TA [7-10]. All these results are presented in the First of these joint papers.

This article presents a series of recommendations for performing such MRI TA measurements in vivo for muscle and for analyzing and interpreting the results.

Fundamentally, texture analysis deals with regions of an image and often these are user defined as *regions of interest* (ROIs) for 2D TA, and *volumes of interest* (VOIs) for 3D TA; example applications and results from initial 3D MRI-TA have already been published [5]. The ROIs may be drawn to encompass single tissue types or placed on the image as simple squares/cubes or circles/spheres. Studies that use texture properties can distinguish changes as a disease, for example, progresses but must be trained to know what the significance of the change is against a gold standard eg. Pathology. A very wide range of TA procedures have been developed, providing many possibilities for investigation in different studies.

Preliminary MRI quality assessment

Quality assessment of MR imaging equipment has been widely developed over the last twenty years [11-15]. Techniques exist based on the use of test objects, which allow the measurement of the parameters of importance in ensuring a constant quality of imaging performance. Without a systematic programme of quality assessment it is impossible to ensure that diagnostic usefulness will be maintained. This is particularly the case when quantitative measurements (such as texture analysis) are attempted on a repeated basis over a period of time.

For the purposes of texture analysis the following quantities are the most important to assess: Firstly signal-to-noise, secondly spatial resolution, and thirdly image uniformity. Additionally it is useful to assess image contrast (dependence on relaxation time variation) and it is also possible to construct texture test objects. There are well established test object methods for investigating these parameters. Before commencement of a TA study a baseline of such measurements should be established. During the study regular checks of these parameters should be made and significant changes addressed.

The signal-to-noise (SNR) of an MR system is most easily assessed through the use of a flat field test object. The simplest technique is to place a square or circular ROI within the image of the test object (see Figure 1) and calculate the mean and standard deviation of the pixel values within it. The Ratio of Mean to Standard Deviation gives a measure of SNR.

However, it is common practice now that it is better to acquire two images under identical conditions (Figure 2) and subtract the two to determine the noise within the same ROI. This procedure removes the influence of image uniformity on the result.

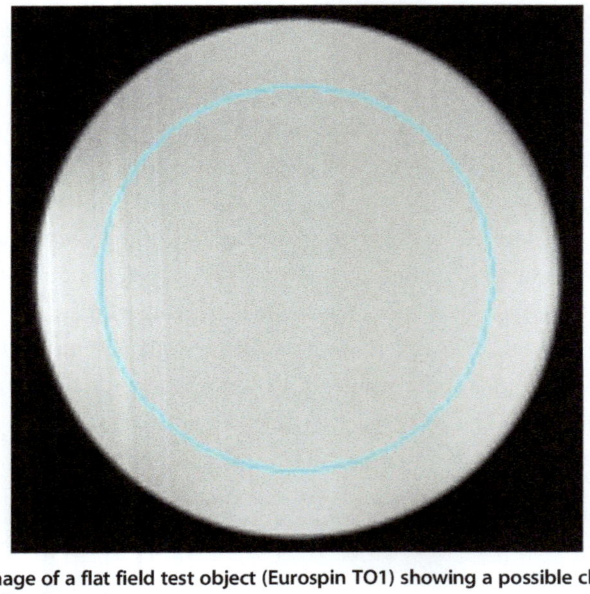

Figure 1 An MR image of a flat field test object (Eurospin TO1) showing a possible choice of circular ROI to calculate a signal-to-noise measure.

Figure 2 **Two MR images of a flat field test object (Eurospin TO1) which may be subtracted to calculate a signal-to-noise measure.**

The SNR is then calculated as $\sqrt{2}$ times the ratio of the Mean in the signal image divided by the standard deviation in the subtracted image. The problem with this can be the ability to calculate subtracted images or the standard deviation on the MRI scanner console.

The image uniformity (or homogeneity) can be assessed through use of the same flat field test object. Homogeneity may vary dependent on the precise MR imaging strategy used and this may affect texture results dependent on the normalisation strategy implemented.

It is common in modern scanners for image processing filters to be applied to improve the image uniformity so it is important to be consistent in the application of such filters during any study. Users may need assistance from equipment manufacturers to determine (or alter) the filters in use. Further, it is important to be sure that scanner software upgrades do not alter these filters and again the assistance of manufacturers is required. Quantitation of the uniformity may be done through the use of profiles drawn across these images. It is necessary to use horizontal and vertical profiles since there may be variation between the frequency encoding and phase encoding directions.

The spatial resolution can be investigated through the use of a test object that includes bar patterns (see Figure 3) with appropriate spacings. The Eurospin TO4 [12] is illustrated in Figure 3.

The resolution of the MR system should be determined by the field of view divided by number of pixels in the image matrix. For example a 250 mm field of view and pixel matrix of 256 x 256 should produce a resolution of just less than 1 mm.

The contrast of MR imaging systems can be investigated through the use of a test object such as the Eurospin TO5. This allows the use of 12 glass tubes with varied (but known) T1 and T2 relaxation times. The tubes contain agarose gels of varying 'thickness' (which alters T2) and varying paramagnetic doping (which changes T1). A typical image is shown in Figure 4. Dependent on the sequence used the contrast obtained will vary and can be measured and compared with expected values.

It is possible to construct texture test objects using foam materials with agarose gel filling material. Such a phantom does not completely mimic the structure of muscle (lack of directional properties) but is useful in testing MRI-TA methods.

In the examples shown in Figure 5, foam with specific porosities of 30, 45, 75 and 90 pores per inch (ppi) was sourced to use in a test phantom (Foam Engineers Ltd.;

Figure 3 An MR image of the Eurospin TO4 test object which allows the assessment of image spatial resolution.

Buckinghamshire, UK). The foam samples were submerged in tubes filled with a 2% agarose solution (Sigma Aldrich; St. Louis, MO) held at 70°C. The agarose was doped with 0.2% Magnevist (Bayer Healthcare; Germany) to shorten T1 relaxation times to values comparable with those measured in vivo.

It can be seen that there is no visual difference between the four foams in the MRI image given in Figure 6. Texture analysis could, however, discriminate perfectly between these foams. That is, a classifier using texture results could discriminate between the foams with no errors.

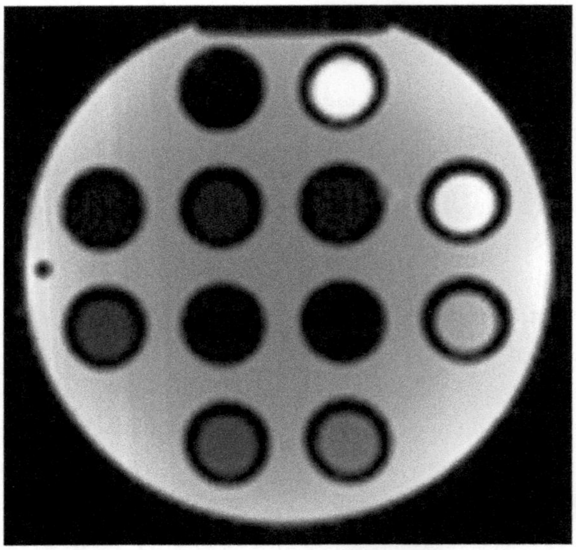

Figure 4 An MR image of the Eurospin TO5 test object which allows the assessment of image relaxation time contrast.

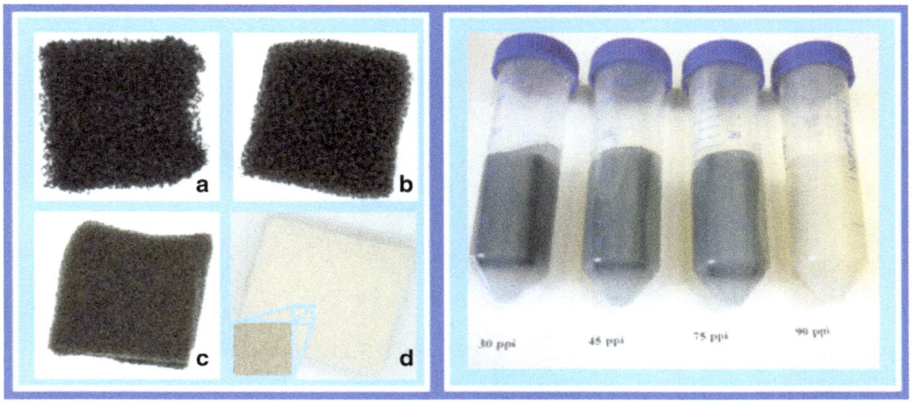

Figure 5 The use of reticulated foam samples to provide image texture variations in MR imaging. The differing foam samples are inserted in plastic tubes and filled with doped agarose gels. Panels on the left show close-ups of the actual foam samples and panels on the right show the foams inserted in the tubes.

Coils and reconstruction

Since the Signal to Noise Ratio (SNR) is a key feature when designing an MRI protocol, the choice of the most suited coils is really crucial. Depending on the body part that is explored several coils configurations can be used, including transmit/receive coils or separate coils for transmission and reception. This latter configuration is often used with surface coil to explore muscles around articulations (knee, elbow, wrist, etc.). These surface coils are an efficient solution to optimize SNR in a small region. Away from this specific area, the sensitivity of the coil decreases very fast. From the Biot-Savart law one can consider that the decrease for simple loop coil can be roughly approximated as $1/z^3$, z being the distance from the centre of the coil, which may influence the texture features in the direction of the gradient (Figure 7)

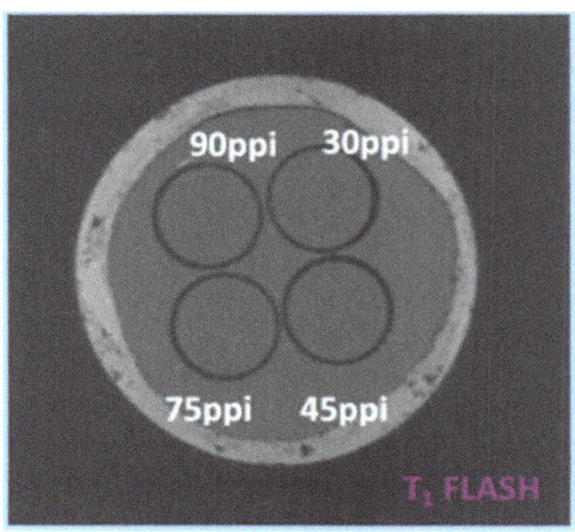

Figure 6 An MR image of the 4 tube texture test object. There is no visual difference between the tube images but image texture analysis can discriminate between them perfectly. This object is extremely useful for studies of how texture results vary with image acquisition changes.

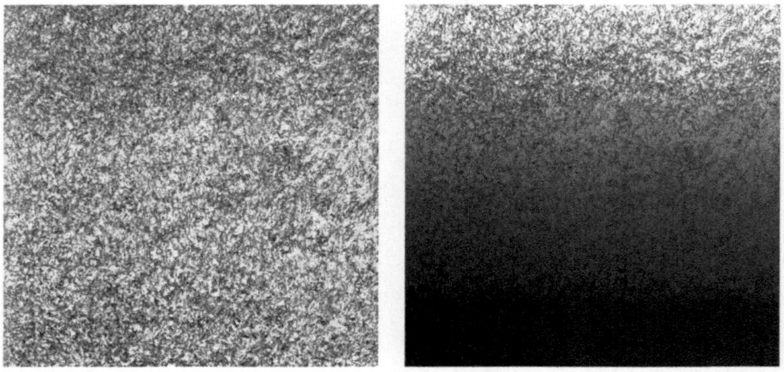

Figure 7 Illustration based on a Brodatz texture (P. Brodatz, "Textures: A Photographic Album for Artists and Designers", Dover Publications, New York, 1966) of the sensitivity coil profile. On the left, the original Brodatz texture. On the right the same texture modified by applying a typical surface coil gradient. Near the coil, the sensitivity is optimal and consequently SNR is increased, while away from the coil centre, the signal decreases.

The extension of this approach is a phased-array coil, which is a combination of multiple coil elements. These coils provide a better sensitivity than a classical RF coil by exploiting the individual high sensitivity of each coil element, but the associated sensitivity map is much more complex and involves many image corrections. These more complex and efficient coils are widely available on clinical systems and are now associated with parallel imaging acquisition strategies that allows shorter acquisition time for a given SNR. Once again, image reconstruction is then much more complex and associated with many (not always editable) options and filters when designing the acquisition protocol on the MRI console. The knowledge and control of all these parameters is not straightforward and can induce bias in the context of quantification and biomarker identification, including texture features.

MRI pulse sequences

The selection of the pulse sequence for MRI acquisition is a key point in the strategy for MRI-TA. Several tissue MR parameters determine the grey level in each voxel: the spin–lattice relaxation time T1, the spin-spin relaxation time T2, the spin density N(H) and the spin displacements (flow and diffusion). Generally for TA the spin echo pulse sequence has been used but fat suppression may also be used (see below). However, the interpretation of voxel grey level is highly complex and must be prudently related to histological changes.

Then the challenge is to develop a pulse sequence potentially able to differentiate histological changes. For instance, to discriminate other modifications from fat infiltration, several fat suppression pulse sequence can be used: short-tau inversion recovery (STIR) [16] has been used to detect muscle edema in DMD though STIR suppresses the signal of all tissues with a short T1 and not selectively the fat. Another pulse sequence, chemical shift selected imaging (CHESS) [17] can provide a selective suppression of fat signal and can be combined with a T2 weighted (T2w) imaging [18]. Chemical shift water–fat separation methods such as multipoint Dixon [19] and iterative decomposition of water and fat with echo asymmetry and least-squares estimation

(IDEAL) [20] separate the water signal from the fat signal using the known phase shift that occurs due to the different precessional frequencies of fat and water. The IDEAL method has also been combined with a Carr–Purcell–Meiboom–Gill (CPMG) imaging sequence to create the IDEAL-CPMG method providing a pair of fat and water decomposed images at each echo time; the water-isolated and fat-isolated signals can then be fitted directly with monoexponential decay curves to measure the T2 of water and the T2 of fat [21]. If fat is totally suppressed, T2 is then a strict water T2; however, fat is not often correctly suppressed minimizing the expected difference between T2 (fat + water) and theoretical water T2. As well, the difference between calculated T2 (the true T2) image, more difficult to obtain and often more noisy, and a T2 weighted (T2w) image must not be underestimated [22].

Different pulse sequences, T1w, T2w, fat-suppressed T1w pre and post contrast agent injection have demonstrated T2w/T1w, T2w and maximal signal enhancement post contrast agent as the more discriminating parameters in dystrophic dog muscle [23]. The images below (Figures 8, 9, 10, 11, 12) illustrate the results obtained with differing pulse sequences in the case of axial slices of the thighs of patients with Pompe disease. Pompe disease is an metabolic disorder which damages muscle and nerve cells throughout the body. It is caused by an accumulation of glycogen in the lysosome and manifests itself through muscle weakness. Muscle texture is affected by this change to tissue structure.

As muscle stiffness is modified in dystrophy, its in-vivo measurement could be relevant. An MR in-vivo method initially proposed in a paper submitted in 1994 and published in 1995 [24], later called Magnetic Resonance Elastography (MRE) [25], has been tested [26]. However, the low resolution which is obtained in MRE make the images without great interest for MRI-TA.

Another MRI method, diffusion tensor imaging (DTI), can provide some geometrical information on fiber size and orientation in muscle dystrophy, as also can MRI-TA [27]. However, DTI in muscle dystrophy is not relevant if the modification caused by the disease is a larger distribution of fibre sizes but with approximately the same mean value, and furthermore the images can be artefacted by fat infiltration.

T1 weighted : TurboSpinEcho sequence (TR270ms TE10ms)

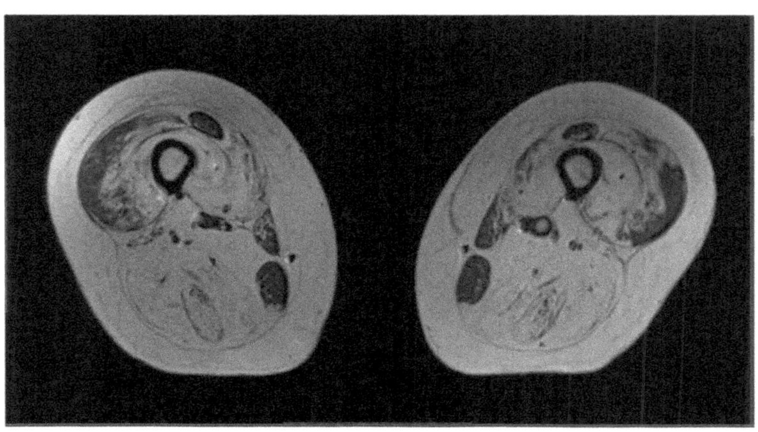

Figure 8 MR image of the thighs of a patient with Pompe disease – T1 weighted.

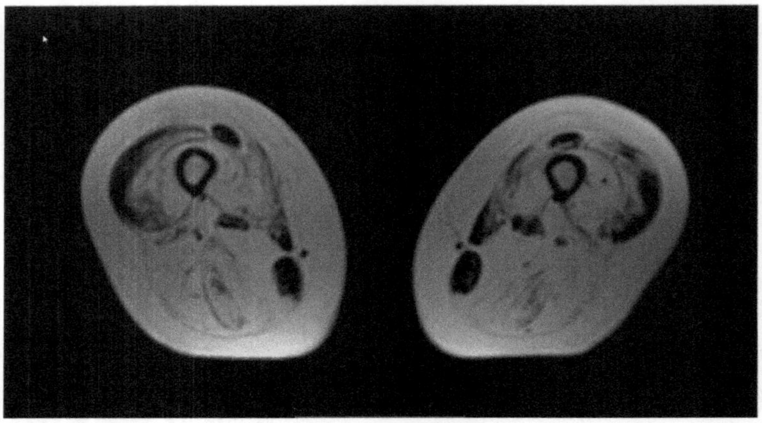

Figure 9 MR image of the thighs of a patient with Pompe disease – T2 weighted.

In summary, the MR pulse sequence selection for MRI-TA has first to take into account the relevance to histological variations, which are not the same during a disease evolution. The total acquisition time is also important, especially with human patients without anesthesia, but is determining for signal-to-noise ratio (SNR) and consequently to spatial resolution: the voxel size (pixel surface x slice thickness) is important; it must be increased to have a better SNR and decreased to have more texture information. Fortunately, as muscle fibers are oriented in a same direction, the slice thickness can be increased if the slice selection is transverse in reference to the main fibers orientation.

Preprocessing (normalization, dynamic range)

In order that TA measurements may be comparable between MR systems (or between different systems) it is necessary to consider whether the image data should be normalised in some manner. Raw image data coming from the scanner will be subject to variations in the electronic gain of the receiver and so the first order amplitude of the image

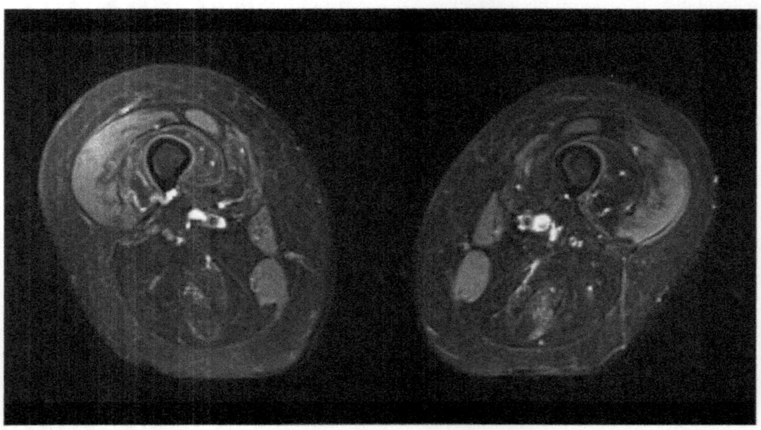

Figure 10 MR image of the thighs of a patient with Pompe disease – T2 weighted with fat saturation used.

Water map : three point Dixon (TR 10ms, TE 2.75/3.95/5.15ms)

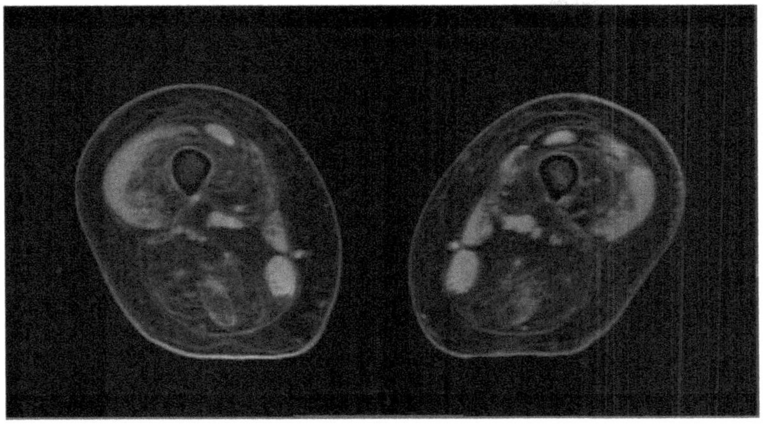

Figure 11 MR image of the thighs of a patient with Pompe disease – water map via 3 point Dixon technique.

(*brightness*) may vary. Further the range of grey shade values (*contrast*) may also vary. Usually the user has no control over the receiver gain, as this setting is automatically set by the scanner software, during set up.

There are many possible approaches to such a normalisation procedure [28]. One common option is to constrain the histogram of the ROI under consideration to a constant image mean and maximum variation of plus and minus 3 times the standard deviation of the pixel values. This is done by adjustment of the software gain and baseline value.

Another matter of great importance in setting up the TA procedure is the number of bits within the image data (dynamic range). It is common in modern MR equipment for the signal data to be digitized to at least 16 if not 24 bits. For several of the common TA procedures eg calculation of the co-occurrence matrix it is undesirable to proceed with this number of bits in the image matrix. This is for two reasons viz. firstly computational complexity is somewhat increased and secondly a resultant sparseness

Fat map : Three point Dixon (TR 10ms, TE 2.75/3.95/5.15ms)

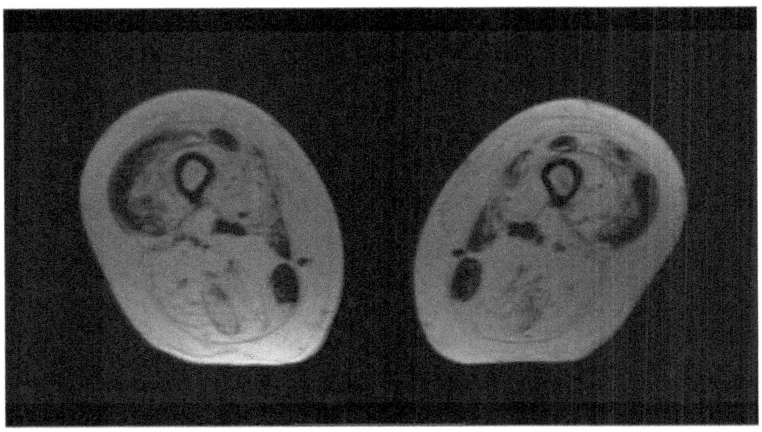

Figure 12 MR image of the thighs of a patient with Pompe disease – fat map via 3 point Dixon technique.

of the resultant co-occurrence matrix, which leads to a sensitivity to noise. A common choice is a reduction of the number of bits in the image to 6 or 8 bits simply by rescaling [29].

Signal-to-noise variations between scanners (different field strengths or receiver characteristics) will also potentially alter texture results. In order to equalize results it may be necessary to consider variation of the NEX (averaging) during acquisition. Increasing NEX obviously increases acquisition time, which in clinical situations, is undesirable.

Window sizing and shaping issues

In the case of TA studies of muscles a choice has to be made between utilising automatic segmentation methods [30] or manual segmentation. As suggested by several authors, the MRI-TA results are often hampered by lack of robust automated muscle segmentation methods [31]. Manual segmentation by the same 'expert' allows better comparisons between texture methods but automated methods are more desirable for clinical use. This issue is still unresolved and requires further work.

There are a number of considerations that determine the choice of Region-of-Interest (ROI) windows for estimating texture features such as the objectives pursued, the nature of the texture and the methods applied for their characterization. The options we have are either geometrical shapes (square or circle windows in 2D, cube or sphere in 3D) or free-form 2D/ 3D shapes (obtained by means of a prior segmentation). These questions are briefly surveyed in this section.

Texture features may be used as a component contributing to automatic object segmentation. In such case, they are combined with edge information in order to improve the extraction results. They are aimed at differentiating the inside from the outside properties of the regions/volumes of interest. Rough estimations of simple features (e.g. first order statistics of image intensities) over geometric windows with small preset sizes are often used. A similar situation is encountered when tracking objects or regions over time like in image sequences. However, if the motion involves also object/region deformations (affine transformation, for instance), invariance of texture features to these deformations should be considered.

Although segmentation, tracking and deformation are of relevance in medical imaging, the main goal remains to precisely characterize the underlying tissue organization and its functional behavior. Such characterization is supposed to be effective in discriminating normal and diseased tissues and needs a large enough number of pixels. Unfortunately, the anatomical structures of interest correspond in general to small-sized objects and have complicated shapes. Geometric windows (as mentioned above) can be interactively placed by the expert. They can also be automatically defined if a prior segmentation has been performed, the size and place of the window being determined by avoiding contour/boundary pixels. However, a manual delineation is often required to extract the target tissue structures in the images. This delineation is not error-free and may involve several experts for cross-validation. This manual interaction becomes highly time-consuming when performed on all the slices of concern. Despite the fact that we have access to the 3D information, e.g. the right dimensional space to work into for tissue texture, in most cases only 2D processing is carried out due to the poor resolution in thickness and between slices. It is also the case that true 3D image acquisitions are extremely time consuming and are unlikely to be practical in a clinical setting. Parallel encoding techniques with gradient echoes do allow 3D acquisitions but this has not, so far, been tested in conjunction with MRI-TA.

In addition, the classical views applied to the nature of textures (structured versus random) may fail to capture the complexity of tissue textures. To get reliable descriptive parameters in the case of structured texture, the window size must be set according to the periodicity of the pattern. Such pattern is not frequently observed in medical imaging. High order statistical parameters computed in the case of random texture require enough pixels to reduce the potential biases. Texture mixtures may also complicate the feature estimations and increase the number of pixels needed. The inherent presence of noise is another critical issue. Noise refers to the intrinsic property of the imaging device and to the interaction between physical waves and tissues. Here too, Gaussian and Laplacian distributions of noise are widely assumed for the definition of samples (e.g. pixels) but are far from being verified in practice.

Summarizing, window sizing and shaping are driven by the data acquired, the nature of the texture, the noise level and type, the texture methods applied and the target we have. Compromises must be carefully examined. Local windows with predefined sizes and shapes, either sliding or not according to the extent of the regions of interest, are most often used. They must have enough pixels in order to get accurate texture feature estimates. 3D TA requires a greater number of voxels not always compatible with the size of the clinical area under examination. For instance, there are nearly 30000 voxel of 1 mm3 in a VOI of 31×31×31 mm. High-order statistics are more sensitive to noise and must be carefully handled. Global windows or bounding boxes of the regions of interest represent another solution but require consideration of proper methods to separate texture from shape.

Overview of the different TA methods

The following describes the various options for methods to investigate the texture properties of digital images. The general categories of the methods are statistical, filter based and model based. Many of the methods can be implemented both in 2D and 3D.

Statistical methods

Grey-level histogram (GLH) - based method

This method allows the calculation of first-order features, based only on the distribution of grey levels (or intensities) of pixels within an analyzed ROI. Such features provide knowledge on the most and the least occurring grey levels, on the concentration of the grey levels around their average, or on the degree of asymmetry in their distribution. Contrariwise, they contain no information on the relationship between neighboring pixels, on the possible direction of the texture, its structure, and other properties resulting from these relationships.

Features

Four features are the most frequently obtained from a grey-level histogram:

- *Average grey level, mean of grey levels*: measure of darkness (or brightness) of the image. It represents the location of the histogram on the grey-level scale.

- *Variance of grey levels*: characterizes the distribution of grey levels around the mean.
- *Skewness*: measures of the asymmetry of the grey-level histogram. It expresses a deviation of the grey-level distribution, compared to a symmetric one. It is equal to zero for the symmetric histograms, negative for right oriented ones (high grey levels are more frequent) and positive for a left oriented histograms (more frequent low grey levels).
- *Kurtosis*: indicates the relative flattening of the grey-level distribution. It is equal to zero for a normal distribution. It takes a negative or a positive value for a distribution whose graph is relatively flat or relatively sharp, respectively.

Autocorrelation (AC) - based method

The correlation between the grey levels of neighboring pixels can be expressed by the normalized autocorrelation coefficients [32]. It is the function of the vertical (Δx) and the horizontal (Δy) distance between the considered pixels in pairs, and can be defined as follows: $\gamma(\Delta x,\Delta y) = A(\Delta x,\Delta y)/A(0,0)$. $A(\Delta x,\Delta y)$ is the average product of pairs of grey levels corresponding to all the pairs of pixels belonging to a ROI, and spaced from one another by a distance $(\Delta x,\Delta y)$. To make the autocorrelation coefficient independent of the image brightness, the grey level mean can be extracted from each grey level before calculating the above formula.

Features

The normalized autocorrelation coefficients can be regarded as textural features. They provide information about the spatial relationship between the texture patterns. For small primitives (texture elements), autocorrelation changes rapidly with change in the distance between the pixels. For the large primitives, it changes slowly. The presence of regular patterns makes regular periodic changes in the autocorrelation.

Gradient (G) - based method

The grey-level gradient at a particular image point depends on the differences between the grey levels of its neighboring pixels, arranged on vertical and horizontal lines intersecting at this point. In most cases, a neighborhood of 3×3 pixels or 5×5 pixels is considered. The gradient matrix contains the values of the absolute gradient at each point of the ROI, except for its boundaries.

Features

On the basis of the gradient matrix, the following features can be calculated [33]: *mean, variance, skewness,* and *kurtosis*. Such features allow us to draw conclusions about the uniformity (homogeneity) or the roughness of the texture. They may also indicate the presence or absence of edges within the ROI.

Grey-level difference matrix (GLDM) - based method

The method consists in studying the absolute values of differences between the grey levels of pixels in pairs belonging to a ROI [34]. Four pixel alignment directions, θ (0°, 45°, 90°, and 135°), and different distances, d, between the pixels in pairs can be considered. The most often, d takes small values. A combination (d,θ) determines the relative position of pixels composing the pairs to analyze. For a specified combination (d,θ), all possible absolute differences in grey levels that can be encoded in the image are taken into account. For each absolute difference, the probability of occurrence of a pair of pixels with just such a difference in the grey levels is calculated. The probabilities sorted in increasing order of corresponding absolute grey level differences form a vector $\mathbf{l}(d,\theta) = [l_0, l_1,\ldots,l_{G-1}]^T$, where G is the number of grey levels possible to occur in the image.

If the distances, d, between pixels in the considered pairs are small, compared to the texture primitives, the grey levels in pairs are quite similar. In this case, the initial components of $\mathbf{l}(d,\theta)$ vector have high values, while the last ones are close to zero. Conversely, if d approaches the texture primitives size, the probabilities of small differences in pixel grey levels are much smaller, and the components of $\mathbf{l}(d,\theta)$ are more spread out. Further, if a texture is directional, the distribution of probabilities in $\mathbf{l}(d,\theta)$ will be different for different θ values. Conversely, if the texture does not seem to have any particular directional elements, the distribution of probabilities in $\mathbf{l}(d,\theta)$ is not significantly affected by the θ.

Features

Five textural features can be derived from the $\mathbf{l}(d,\theta)$ vector. Comparing their values obtained for different pixel distances and/or pixel alignment directions one can draw conclusions about the size, the quantity, and the orientation of texture primitives.

- *Mean* is small for fine textures, when the probabilities of small grey level differences are high, and it is large for diversified textures giving high probabilities of big differences.
- *Energy* or *angular second moment* is small when all the probabilities in the $\mathrm{l}(d,\theta)$ vector are similar (fine textures), and large when some vector components are high, and others low.
- *Contrast* or *inertia* is a measure of intensity contrast between a pixel and its neighbors. It is equal to zero for a constant image, and reaches high values for images with extremely differing grey levels.
- *Inverse difference moment* is the measure of the *local homogeneity.*
- *Entropy* is large for the images that are texturally not uniform, when all the probabilities in the $\mathrm{l}(d,\theta)$ vector are close to each other. The more the probabilities in vector are diversified, the smaller is the entropy.

For non-directed textures, features obtained with the same formula, for the same d, but corresponding to different alignment directions of pixel pairs, θ, can be averaged. Alike, it is possible to average features calculated for different distances.

Co-occurrence matrix (COM) - based method

A co-occurrence matrix $\mathrm{C}(d,\theta)$ is also constructed for a given distance, d, between pixels, still considered in pairs, and a given pixel alignment direction, θ. Each element, c_{ij} $(i,j = 0,\ldots,G\text{-}1)$ represents a probability of occurrence in a ROI of a pair of pixels in which: the first pixel has a grey level i, and the second one – a grey level j [35]. G is the number of grey levels that can be encoded in an image. A matrix dimension is therefore $G \times G$. Like in the GLDM-based method, four pixel alignment directions and several distances between pixels could be considered. In contrast to the previous method, a COM-based one focuses on all the possible combinations of grey levels in pairs.

Features

Several parameters were initially proposed in [35]:

- *Energy* or *angular second moment* is a measure of homogeneity of grey levels characterizing the pixels of a ROI. For relatively homogeneous textures, the grey levels of adjacent pixels are similar. In this case, larger probabilities are associated with elements located on the diagonal (or very close to the diagonal) of the co-occurrence matrix, the other matrix elements are close to zero. The energy for homogeneous textures is greater than for the heterogeneous textures. For the latter ones – c_{ij} probabilities are more uniformly distributed throughout the matrix, and their values are quite small.
- *Contrast* or *inertia*, with a quite large value for heterogeneous regions, characterized by a strong contrast, and quite small value for homogeneous regions.
- *Inverse difference moment* or *local homogeneity.*
- *Entropy*, that quantifies the degree of randomness of the grey levels in a ROI. Its value is the highest when all elements of $C(d,\theta)$ matrix are equal.

Another parameter, *correlation*, can be obtained with the mean and standard deviation of the row and column sums of the $C(d,\theta)$ matrix.

Some other features can be derived from a vector whose components are the co-occurrence probabilities for pixels with a determined sum of the grey levels. All possible sum values are taken into account. The probabilities forming a vector are sorted in increasing order of corresponding sum values. These features are: *sum average, sum variance*, and *sum entropy* [36]. Taking the absolute differences of grey levels, instead of their sums, leads to the features already known from GLDM - based method.

Features calculated for different angles, θ, and/or for different distances, d, can be averaged.

Run-length matrix (RLM) - based method

This method involves counting the number of pixel runs of each grey level, having a given length [37]. All the lengths that can be encountered in the ROI are considered. As in the two preceding methods, four pixels alignment directions, (θ), can be taken into account. The run-length matrix, $R(\theta)$, built for a given θ, has G columns. Each column corresponds to a grey level. In turn, rows of this matrix correspond to all the consecutive run lengths. The number of rows, M, is thus determined by the size of the ROI, and is equal to the maximum length of the pixel run which can exist in a ROI. The element r_{mg} ($r = 1,\ldots,M$, and $g = 0,\ldots,G$) of the matrix $R(\theta)$ is the number of pixel runs of the level g, with the length of m, and oriented in the direction θ.

Homogeneous textures have fairly long pixel runs, while for rough textures, short runs are the most numerous. The features obtained from the RLM method can give the information on the orientation of some texture elements, for example – on the presence of stripes.

Features

The most known features, proposed in the above paper, are: *the short run emphasis*, the *long run emphasis*, the *grey level non-uniformity* (*distribution*), *the run length non-uniformity* (*distribution*), and the *fraction of image in runs.* Two additional parameters were proposed in [38]: *low grey level runs emphasis*, and *high grey level runs emphasis.* Finally, in [39], a *run length entropy was* used as a texture feature.

Filter methods

Laws' texture energy measures

Laws proposed to transform the images using linear filters [40]. An image filtering consists in assigning a new value (grey level) to each pixel of an original ROI, excluding the pixels on its borders. For linear filters such a value depends on a linear combination of the grey levels of the pixels in some neighbourhood of the transformed pixel. The most often two types of neighbourhood are taken into account: 3×3 pixels and 5×5 pixels. The importance of each neighbours contribution to the output pixel value is defined by a mask, called also a convolution matrix. The masks were designed in order to detect different texture elements: ripples, edges, spots.

On the basis of a transformed image, the entropy and several image energy measures can be calculated.

Model-based methods

Fractal model

Several definitions of a fractal object, as well as several methods of calculating a fractal dimension were proposed [41-43]. Benoit Mandelbrot, the creator of fractal geometry, characterized fractals as self-similar objects, whose parts are similar to the whole, and whose topological dimension is not an integer [44]. The fractal dimension of an object reflects the extent to which this object fills the space or the rate of its diversity, the degree of irregularity of the object.

A grey-level image can be considered as a surface in three-dimensional space, where two dimensions are those of the image plane and the third one is the grey level (intensity) of pixels. In this case, we can use the fractal dimension of image surface as a texture descriptor. This feature gives a measure of the irregularity, and of the roughness of the texture. The more a surface is rough, the higher is its fractal dimension.

The fractal dimension of an image (a texture) can be calculated by representing the intensity surface of an image by a fractional Brownian motion model (see also [36]). One of its applications for texture feature extraction can be found in [45].

Image landscapes' fractal dimension (ILF) method

The **ILF** method is based on constructing from the analyzed 2-D grey level images two 1-D sequences that are called *landscapes.*

Stepping through a grey value image length of N pixels and height of M pixels row by row one can calculate the sum of the grey values in each row, G_m $(m = 1, \ldots, M)$, and dividing these numbers by the largest of them, G_R, we obtain the *horizontal landscape, hgs,*

$$NGS_m = G_m/G_R \in [0,\ 1] \qquad m = 1, \ldots, M \tag{1}$$

where $G_m = \sum_{n=1}^{N} g_{mn}$ $G_R = \max(G_m)$ $\qquad m = 1, \ldots, M$

Similarly, stepping through the same image column by column one can calculate the sum of the grey values in each column, G_n $(n = 1, \ldots, N)$, and dividing these numbers by the largest of them, G_C, we obtain the *vertical landscape, vgs*

$$NGS_n = G_n/G_C \in [0,\ 1] \qquad n = 1, \ldots, N \tag{2}$$

where $G_n = \sum_{m=1}^{M} g_{mn};\ G_C = \max(G_n) \qquad n = 1, \ldots, N$

The landscapes obtained in such a way are then analyzed using Higuchi fractal dimension in a sliding window of the length L moved in each step l points (pixels) to the right. Normalization in (1) and (2) is convenient but not really necessary since Higuchi fractal dimension is invariant with respect to scaling of the data.

Features

Using ILF method [46] one can obtain the following characteristics of the analyzed image:

- *Two (or four) graphs of fractal dimension* of the image, along horizontal and vertical axes; if fractal dimension smoothly fluctuates it means that there are non-stationarities/discontinuities in the image; in such a case mean fractal dimensions D_h and D_v are enough to characterize the image.
- *Two (or four) mean fractal dimension values* – if their values are nearly identical then it is enough to characterize the surface (its image) using one number, D_f – the average of D_h and D_v; if D_h and D_v differ considerably it means that the surface (and so its image) shows texture, i.e. has different properties in different directions.

Moment-based methods

These methods belong to the so-called transform methods. They include several families among which complex moments, rotational moments, geometric moments and orthogonal moments are the quantities studied [47-49]. Orthogonal moments, such as Legendre and Zernike moments, present the advantage of being optimal with regard to information redundancy. To provide some cues on these moments, let us take the example of Legendre moments. The two-dimensional (2D) $(n + m)$th order of Legendre moment of an image intensity function $f(x, y)$, L_{nm}, is defined as

$$L_{nm} = \frac{(2n+1)(2m+1)}{4} \int_{-1}^{1} \int_{-1}^{1} P_n(x) P_m(y) f(x, y)\, dx dy \tag{3}$$

where $P_n(x)$ is the nth order of Legendre polynomial given by

$$P_n(x) = \frac{1}{2^n} \sum_{k=0}^{n/2} (-1)^k \frac{(2n-2k)!}{k!(n-k)!(n-2k)!} x^{n-2k} \tag{4}$$

Since the Legendre polynomials are orthogonal over the interval [−1, 1], the image $f(x, y)$ (the ROI in most cases) can be reconstructed from its moments. When an analog original image is digitized to its discrete form, the 2D Legendre moments L_{nm} defined by Eq. (3) is usually approximated by

$$L_{nm} = \frac{(2n+1)(2m+1)}{(N-1)(M-1)} \sum_{i=1}^{N} \sum_{j=1}^{M} P_n(x_i) P_m\left(y_j\right) f(i, j) \tag{5}$$

with $x_i = \frac{2i-N-1}{N-1}$, $y_i = \frac{2j-M-1}{M-1}$.

Features

The texture features are in such case the estimated coefficients $L_{nm.}$ The number of features depends on the order selected. The low order coefficients capture the low frequency components while the high order ones (more sensitive to noise) will provide details of the image or the ROI under consideration.

Data analysis

Once a series of texture features are extracted it is necessary to determine the usefulness of each or combinations of several for a particular classification task. This is often called *pattern recognition* or *data mining* and consists of applying well-developed methods to determine how the features can be used to classify the data. The overall objective of the data mining process is to extract information from a data set and transform it into an understandable structure for subsequent use.

There are a large number of methods which can be applied to these problems eg discriminant analysis, principal component analysis, clustering, neural networks etc. etc. It is not the purpose of this paper to review these in any detail. There are also a significant number of commercial (or freeware) software packages that are available to help with these tasks. Examples of these are Weka, SPSS, The Unscrambler, SAS and STATISTICA.

As an example, Weka is a collection of machine learning algorithms for data mining tasks. The algorithms can either be applied directly to a dataset or called from a users own Java code. Weka contains tools for data pre-processing, classification, regression, clustering, association rules, and visualization. It is also well-suited for developing new machine learning schemes.

Using computers for texture segmentation and analysis, both supervised and unsupervised, is obvious but it raises some questions. The most important question that biomedical researchers have been asked since computer introduction to the labs is '**Do you REALLY understand what your computer calculates?**' Unfortunately, the answer to this question is quite often not positive. It concerns in particular commercial software when the source code is not given to the end users and even IT specialists often have problems with understanding the details of source code written by others. In the case of Biologists and Medical Doctors there is a problem with interpretation of texture parameters, which have little intuitive meaning.

The second question concerns **repeatability of results**. One would like that at least classification of the images from a given series would lead to the same result when repeated. But already the problem of choosing the Region(s) of Interest may lead to differences between researchers. It seems that the whole image is as unbiased as one can get, any remaining bias originated in how the images were acquired. It was exactly the case in the application of a fractal method in the analysis of textures of anatomo-pathological slides for staging in cases of Anal Intraepithelial Neoplasia [46] - some original slides showed white irregular 'holes' but any attempts to analyze ROIs without these holes only worsened the final results.

The third question concerns very, very important but not well-known so called **'The Curse of Dimensionality'**. To create a classifier first one needs descriptors for each object class that can be expressed by numbers. It might seem that to obtain a more accurate classification, one could add more features, maybe even a few hundred of these features. But in fact, after a certain point, increasing the dimensionality of the problem by adding new features would actually degrade the performance of the classifier - as the dimensionality increases, the classifier's performance increases until the optimal number of features is reached; further increasing the dimensionality without increasing the number of training samples results in a decrease in classifier performance; this is called **overfitting** (see [50]). So, with only a small number of images in the database, it is just intractable to check all possible combinations of features in any given method nor to compare different proposed methods of texture analysis to decide which combination or which method is the best one. Even so called feature extraction methods like Principal Component Analysis, that reduce the dimensionality of the classification problem, are not always helpful.

In summary, complicated and very specialized methods are good for research studies (with care over over-interpretation), but if one wants to propose a texture analysis procedure that could become widely applicable in hospitals and clinics, this should be a rather quite simple method, based on only a few texture parameters (preferably just one or two), quick and easily understandable by Medical Doctors. More work is required on this latter point.

Conclusion

The review presented in the sections above leads to the following conclusions. The important points that must be considered when applying texture analysis to in-vivo muscles are the following:

1. Quality assessment of the MR imaging system is essential both at the beginning and throughout any study
2. Careful choice of imaging sequence is required with regard to image resolution, imaging time and signal-to-noise.
3. Many kinds of texture analysis methods exist. To ensure that the best results are obtained, a full range of procedures should be investigated.
4. The Regions of Interest used for the texture analysis should not be too small (>100 pixels in 2D TA)
5. The dynamic range (pixel depth) of the MR images will need to be reduced to allow practical texture analysis to be carried out. Eight or six bits are recommended.
6. Normalisation of the MR images is essential to remove unwanted variations caused by receiver gain changes.
7. Care must be taken to avoid over-interpretation of the texture data due to the large number of texture parameters and the sometimes limited amount of independent image data available.
8. With all these points taken into consideration, texture analysis carried out on MR images is capable of providing useful clinical information. Progression of disease processes can be followed and muscle status assessed.

If attention is paid to all these factors then the published evidence is that texture analysis of MR data may well be useful in a clinical setting. What is required now is an improved understanding of how the texture results relate to the underlying pathological changes in tissue and also a good way of presenting the output of the texture analysis for interpretation by medical specialists.

Abbreviations

2D: Two dimensional imaging; 3D: Three dimensional imaging; B_0: Static magnetic field of the mri device (in Tesla); CFA: Correspondence factorial analysis; DMD: Duchenne muscular dystrophy; GRMD: Golden retriever muscular dystrophy; I Signal: Intensity; MHz: Mega hertz; MRI: Magnetic resonance imaging; NEX: Number of excitations in MRI acquisition; NMR: Nuclear magnetic resonance; N(H): Spin density; PCA: Principal component analysis; ROI: Region of interest; SE: Spin echo pulse sequence; SNR: Signal to noise ratio; TA: Texture analysis; TO: Test object; TR: Repetition time in MRI acquisition; T1: Spin lattice NMR relaxation time; T2: Spin spin NMR relaxation time; VOI: Volume of interest.

Competing interests
The authors declare that they have no competing interests.

Authors' contributions
All authors have contributed to the manuscript, which has been finally edited by RAL. All authors read and approved the final manuscript.

Author details
[1]Medical Research Institute, Ninewells Hospital, Dundee, Scotland, UK. [2]Institute of Myology, University Hospital Pitié-Salpêtrière, Paris, France. [3]Faculty of Computer Science, Bialystok University of Technology, Bialystok, Poland. [4]Nalecz Institute of Biocybernetics and Biomedical Engineering Polish Academy of Sciences, Warsaw, Poland. [5]Laboratory of Image Science and Technology, School of Computer Science and Engineering, Southeast University, Nanjing 210096, China. [6]Centre de Recherche en Information Biomédicale Sino-français (CRIBs), Nanjing, China. [7]INSERM, U1099, Rennes F-35000, France. [8]Université de Rennes 1, LTSI, Rennes F-35042, France. [9]Centre de Recherche en Information Biomédicale Sino-français (CRIBs), Rennes F-35000, France. [10]PRISM-Bio-SCANs, University of Rennes, Rennes, France.

References
1. Haralick R. Statistical and structural approaches to texture. IEEE Proc. 1979;67:786–804.
2. Hajek M, Dezortova M, Materka A, Lerski R. Texture Analysis for Magnetic resonance Imaging. Prague: Med 4 publishing; 2006.
3. Biondetti PR, Ehman RL. Soft-tissue sarcomas: use of textural patterns in skeletal muscle as a diagnostic feature in postoperative MR imaging. Radiology. 1992;183(3):845–8.
4. Skoch A, Jirak D, Vyhnanovska P, Dezortova M, Fendrych P, Rolencova E, et al. Classification of calf muscle MR images by texture analysis. MAGMA. 2004;16:259–67.
5. Wang J, Fan Z, Vandenborne K, Walter G, Shiloh-Malawsky Y,et al. Statistical texture analysis based MRI quantification of Duchenne muscular dystrophy in a canine model. Proc. SPIE 8672, Medical Imaging 2013: Biomedical Applications in Molecular, Structural, and Functional Imaging, 86720 F (March 29, 2013);
6. Nketiah G, Sievanen H, Eskola H. Correlation between hip muscles MRI texture parameters and femoral neck boneareal bone mineral density (aBMD) in different athletes groups. Phys Med. 2014;30(Supplement 1):e38–9.
7. Herlidou S, Rolland Y, Bansard JY, Le Rumeur E, de Certaines JD. Comparison of automated and visual texture analysis in MRI: characterization of normal and diseased skeletal muscle. Magn Reson Imaging. 1999;17(9):1393–7.
8. Mahmoud-Ghoneim D, Cherel Y, Lesoeur J, Lemaire L, Rocher C, de Certaines JD, et al. Texture analysis of magnetic resonance images of rat muscles during atrophy and regeneration. Magn Reson Imaging. 2006;24(2):167–71.
9. Mamhoud-Ghoneim D, Bonny JM, Renou J-P, de Certaines JD. Ex-vivo Magnetic Resonance Imaging Texture Analysis can identify genotypic origin in bovine meat. J Sci Food Agric. 2005;85:629–32.
10. Nguyen F, Eliat PA, Pinot M, Franconi F, Lemaire L, de Certaines JD, et al. Correlations between Magnetic Resonance Imaging histopathology in mdx (X-linked Muscular Dystrophy) murine model of Duchenne Muscular Dystrophy. Edinburgh: 24th congress of the European Society of Veterinary Pathology; 2006.
11. Lerski RA, de Wilde J, Boyce D, Ridgway J. Quality control in magnetic resonance imaging. IPEM Report no 80. 1998. ISBN: 0904181901.
12. European Communities Research Project (COMAC BME II 2.3). Protocols and test objects for the assessment of MRI equipment. Magn Reson Imaging. 1988;6:195–9.
13. Lerski RA, McRobbie DW, Straughan K, Walker PM, de Certaines JD, Bernard AM. Multi-center trial with protocols and prototype test objects for the assessment of MRI equipment. Magn Reson Imaging. 1988;6:201–14.
14. Lerski RA, de Certaines JD. Performance assessment and quality control in MRI by Eurospin test objects and protocols. Magn Reson Imaging. 1993;11:817–33.
15. Jackson EF, Bronskill MJ, Drost DJ, Och J, Pooley RA, Sobel WT, et al. AAPM Report 100: Acceptance Testing and Quality Assurance Procedures for Magnetic Resonance Imaging Facilities. College Park, MD: American Association of Physicists in Medicine; 2010. ISBN 978-1-936366-02-6.
16. Bydder GM, Pennock JM, Steiner RE, Khenia S, Payne JA, Young IR. The short TI inversion recovery sequence—An approach to MR imaging of the abdomen. Magn Reson Med. 1985;3:251–4.
17. Keller PJ, Hunter WW, Schmalbrock P. Multisection fat-water imaging with chemical shift selective presaturation. Radiology. 1987;164:539–41.
18. Kobayashi M, Nakamura A, Hasegawa D, Fujita M, Orima H, Takeda S. Evaluation of dystrophic dog pathology by fat-suppressed T2-weighted imaging. Muscle Nerve. 2009;40:815–26.
19. Glover GH, Schneider E. Three-point Dixon technique for true water/fat decomposition with B0 inhomogeneity correction. Magn Reson Med. 1991;18:371–83.
20. Reeder SB, Pineda AR, Wen Z, Shimakawa A, Yu H, Brittain JH, et al. Iterative decomposition of water and fat with echo asymmetry and least- squares estimation (IDEAL): application with fast spin-echo imaging. Magn Reson Med. 2005;54:636–44.
21. Janiczek RL, Gambarota G, Sinclair CDJ, Yousry TA, Thornton JS, Golay X, et al. Simultaneous T2 and Lipid Quantitation Using IDEAL-CPMG. Magn Reson Med. 2011;66:1293–302.
22. Carlier PG. Global T2 versus water T2 in NMR imaging of fatty infiltrated muscles: different methodology, different information and different implications. Neuromuscul Disord. 2014;24:390–2.
23. Thibaud JL, Monnet A, Bertoldi D, Barthelemy I, Blot S, Carlier PG. Characterization of dystrophic muscle in golden retriever muscular dystrophy dogs by nuclear magnetic resonance imaging. Neuromuscul Disord. 2007;17:575–84.

24. Lewa CJ, de Certaines JD. Viscoelastic property detection by elastic displacement NMR measurements. J Magn Reson Imaging. 1995;5:242–4.
25. Muthupillai R, Lomas DJ, Rossman PJ, Greenleaf JF, Manduca A, Ehman RL. Magnetic resonance elastography by direct visualization of propagating acoustic strain waves. Science. 1995;269(5232):1854.
26. Qin EC, Juge L, Lambert S, Sinkus R, Bilston L. MR-Elastography and diffusion tensor imaging to measure the in-vivo anisotropic elasticity of skeletal muscles of Mdx and healthy mice. Proc Int Soc Mag Reson Med. 2012;20:3269.
27. McMillan A.B., Shi D., Pratt S.J.P., Lovering R.M., Diffusion Tensor MRI to Assess Damage in Healthy and Dystrophic Skeletal Muscle after Lengthening Contractions, Journal of Biomedicine and Biotechnology, vol 2011, Article ID 970726, 10 pages,
28. Collewet G, Strzelecki M, Mariette F. Influence of MRI acquisition protocols and image intensity normalization methods on texture classification. Magn Reson Imaging. 2004;22(1):81–91.
29. Mahmoud-Ghoneim D, Alkaabi MK, de Certaines JD, Goettsche FM. The impact of image dynamic range on texture classification of brain white matter. BMC Med Imaging. 2008;8:18.
30. Wang J, Fan Z, Vandenborne K, Walter G, Shiloh-Malawsky Y, An H, et al. A computerized MRI biomarker quantification scheme for a canine model of Duchenne muscular dystrophy. Int J Comput Assist Radiol Surg. 2013;8(5):763–74.
31. Fan Z, Wang J, Ahn M, Shiloh-Malawsky Y, Chahin N, Elmore S, et al. Characteristics of magnetic resonance imaging biomarkers in a natural history study of golden retriever muscular dystrophy. Neuromuscul Disord. 2014;24(2):178–91.
32. Gonzalez RC, Woods RE. Image Compression. In: Digital Image Processing. 2nd ed. Reading, MA: Addison-Wesley; 2002.
33. Lerski R, Straughan K, Shad L, Boyce D, Bluml S, Zuna I. MR Image Texture Analysis - An Approach to Tissue Characterization. Magn Reson Imaging. 1993;11:873–87.
34. Weszka JS, Dyer CR, Rosenfeld A. A Comparative Study of Texture Measures for Terrain Classification. IEEE Trans Syst Man Cybern. 1976;6:269–85.
35. Haralick RM, Shanmugam K, Dinstein I. Textural features for image classification. IEEE Trans Syst Man Cybern. 1973;3:610–21.
36. Bankman I. N. (Ed.). Handbook of Medical Imaging, Processing and Analysis, Academic Press, 2000.
37. Galloway MM. Texture analysis using grey level run lengths. Comput Graphics Image Process. 1975;4:172–9.
38. Chu A, Sehgal CM, Greenleaf JF. Use of grey value distribution of run lengths for texture analysis. Pattern Recogn Lett. 1990;11:415–20.
39. Albregtsen F, Nielsen B, Danielsen HE. Adaptive grey level run length features from class distance matrices. Proc 15th Int Conf Pattern Recognition. 2000;3:738–41.
40. Laws KI. Textured image segmentation. Los Angeles, California, USA: Unpublished doctoral dissertation. University of Southern California; 1980.
41. Edgar G. A. Measure, Topology and Fractal Geometry, Springer-Verlag, 1990.
42. Falconer K. Fractal Geometry, Mathematical Foundations and Applications, John Wiley & Sons, 1990.
43. Peitgen H. O., Jürgens H., Saupe D. Fractals for the Classroom. Part 1: Introduction to Fractals and Chaos, Springer-Verlag, 1992.
44. Mandelbrot B. The Fractal Geometry of Nature, W. H. Freeman and Co., 1982
45. Chen EL, Chung PC, Chen CL, Tsai HM, Chang CI. An automatic diagnostic system for CT liver image classification. IEEE Trans Biomed Eng. 1998;45(6):783–94.
46. Klonowski W, Pierzchalski M, Stepien P, Stepien R, Sedivy R, Ahammer H. Application of Higuchi's fractal dimension in analysis of images of Anal Intraepithelial Neoplasia. Chaos, Solitons Fractals (Elsevier). 2013;48:54–60.
47. Shu H, Luo L, Coatrieux JL. Moment-Based Approaches in Image, Part 1. Basic Features. IEEE Eng Med Biol Mag. 2007;26(5):70–4.
48. Shu HZ, Luo LM, Coatrieux JL. Moment-based approaches in image, Part2: invariance. IEEE Eng Med Biol Mag. 2008;27(1):81–3.
49. Shu HZ, Luo LM, Coatrieux JL. Moment-based approaches in image, Part 3: computational considerations. IEEE Eng Med Biol Mag. 2008;27(3):89–91.
50. Vincent Spruyt, About the Curse of Dimensionality, June 6, 2014, http://www.datasciencecentral.com/profiles/blogs/about-the-curse-of-dimensionality.

Permissions

All chapters in this book were first published in EPJNBP, by Springer; hereby published with permission under the Creative Commons Attribution License or equivalent. Every chapter published in this book has been scrutinized by our experts. Their significance has been extensively debated. The topics covered herein carry significant findings which will fuel the growth of the discipline. They may even be implemented as practical applications or may be referred to as a beginning point for another development.

The contributors of this book come from diverse backgrounds, making this book a truly international effort. This book will bring forth new frontiers with its revolutionizing research information and detailed analysis of the nascent developments around the world.

We would like to thank all the contributing authors for lending their expertise to make the book truly unique. They have played a crucial role in the development of this book. Without their invaluable contributions this book wouldn't have been possible. They have made vital efforts to compile up to date information on the varied aspects of this subject to make this book a valuable addition to the collection of many professionals and students.

This book was conceptualized with the vision of imparting up-to-date information and advanced data in this field. To ensure the same, a matchless editorial board was set up. Every individual on the board went through rigorous rounds of assessment to prove their worth. After which they invested a large part of their time researching and compiling the most relevant data for our readers.

The editorial board has been involved in producing this book since its inception. They have spent rigorous hours researching and exploring the diverse topics which have resulted in the successful publishing of this book. They have passed on their knowledge of decades through this book. To expedite this challenging task, the publisher supported the team at every step. A small team of assistant editors was also appointed to further simplify the editing procedure and attain best results for the readers.

Apart from the editorial board, the designing team has also invested a significant amount of their time in understanding the subject and creating the most relevant covers. They scrutinized every image to scout for the most suitable representation of the subject and create an appropriate cover for the book.

The publishing team has been an ardent support to the editorial, designing and production team. Their endless efforts to recruit the best for this project, has resulted in the accomplishment of this book. They are a veteran in the field of academics and their pool of knowledge is as vast as their experience in printing. Their expertise and guidance has proved useful at every step. Their uncompromising quality standards have made this book an exceptional effort. Their encouragement from time to time has been an inspiration for everyone.

The publisher and the editorial board hope that this book will prove to be a valuable piece of knowledge for researchers, students, practitioners and scholars across the globe.

List of Contributors

Sergio Alonso
Physikalisch-Technische Bundesanstalt, Abbestrasse 2-12, 10587 Berlin, Germany

Markus Bär
Physikalisch-Technische Bundesanstalt, Abbestrasse 2-12, 10587 Berlin, Germany

Hil GE Meijer
Department of Applied Mathematics, MIRA Institute of Biomedical Technology and Technical Medicine, University of Twente, Postbus 217, 7500 AE, Enschede, The Netherlands

Stephen Coombes
Center for Mathematical Medicine & Biology, School of Mathematical Sciences, University of Nottingham, NG7 2RD, Nottingham, UK

Dimitris A Pinotsis
The Wellcome Trust Centre for Neuroimaging, University College London, Queen Square, London 0WC1N 3BG, UK

Karl J Friston
The Wellcome Trust Centre for Neuroimaging, University College London, Queen Square, London 0WC1N 3BG, UK

Bernd J Kröger
Neurophonetics Group, Department of Phoniatrics, Pedaudiology, and Communication Disorders, Medical School, RWTH Aachen University, Aachen, Germany
Cognitive Computation and Applications Laboratory, School of Computer Science and Technology, Tianjin University, Tianjin, China

Jim Kannampuzha
Neurophonetics Group, Department of Phoniatrics, Pedaudiology, and Communication Disorders, Medical School, RWTH Aachen University, Aachen, Germany

Emily Kaufmann
Education and Rehabilitation of the Deaf and Hard of Hearing, Department of Special Education, Faculty of Human Sciences, University of Cologne, Cologne, Germany

Vlado A Buljan
Brain and Mind Research Institute, Sydney Medical School, The University of Sydney, Sydney, NSW 2050, Australia
Department of Physiology, Sydney Medical School, The University of Sydney, Sydney, NSW 2006, Australia
Australian Nuclear Sciences and Technology Organisation (ANSTO), The Bragg Institute and the Australian Synchrotron, Sydney, Australia

R M Damian Holsinger
Brain and Mind Research Institute, Sydney Medical School, The University of Sydney, Sydney, NSW 2050, Australia
Discipline of Biomedical Science, School of Medical Sciences, Sydney Medical School, University of Sydney, Lidcombe, NSW 1825, Australia

Brett D Hambly
Discipline of Pathology, School of Medical Sciences, Sydney Medical School, Bosch Institute, University of Sydney, Sydney, NSW 2006, Australia

Vangelis Kanellis
Brain and Mind Research Institute, Sydney Medical School, The University of Sydney, Sydney, NSW 2050, Australia

Elie Matar
Brain and Mind Research Institute, Sydney Medical School, The University of Sydney, Sydney, NSW 2050, Australia

Xanthe Larkin
Brain and Mind Research Institute, Sydney Medical School, The University of Sydney, Sydney, NSW 2050, Australia

Guo Jun Liu
Australian Nuclear Sciences and Technology Organisation (ANSTO), The Bragg Institute and the Australian Synchrotron, Sydney, Australia

John J Bohorquez-Florez
Brain and Mind Research Institute, Sydney Medical School, The University of Sydney, Sydney, NSW 2050, Australia

Richard B Banati
Brain and Mind Research Institute, Sydney Medical School, The University of Sydney, Sydney, NSW 2050, Australia
Australian Nuclear Sciences and Technology Organisation (ANSTO), The Bragg Institute and the Australian Synchrotron, Sydney, Australia

Discipline of Medical Radiation Sciences, Faculty of HealthSciences, University of Sydney, Lidcombe, NSW 1825, Australia

Kenneth JE Vos
Department of Physics and Astronomy, University of Lethbridge, 4401 University Drive, T1K 3M4 Lethbridge, Canada

Angela G Martin
Department of Physics and Astronomy, University of Lethbridge, 4401 University Drive, T1K 3M4 Lethbridge, Canada

Maxine G Trimboli
Department of Physics and Astronomy, University of Lethbridge, 4401 University Drive, T1K 3M4 Lethbridge, Canada

Lindsay Forestell
Department of Physics and Astronomy, University of Lethbridge, 4401 University Drive, T1K 3M4 Lethbridge, Canada

Khaled Barakat
Department of Physics, University of Alberta, T6G 2J1 Edmonton, Canada
Department of Engineering, Mathematics and Physics, Fayoum University, Fayoum, Egypt

Jack A Tuszynski
Department of Physics, University of Alberta, T6G 2J1 Edmonton, Canada
Department of Oncology, Faculty of Medicine and Dentistry, University of Alberta, T6G 2J1 Edmonton, Canada

Emeline Fresnel
CORIA-UMR 6614 Normandie Université, CNRS-Université et INSA de Rouen, Campus Universitaire du Madrillet, F-76800 Saint-Etienne du Rouvray, France
ADIR Association, Hôpital de Bois-Guillaume, F-76031 Rouen, France

Jean-François Muir
Service de pneumologie - CHU de Rouen, Hôpital de Bois-Guillaume, F-76230 Bois-Guillaume, France

Christophe Letellier
CORIA-UMR 6614 Normandie Université, CNRS-Université et INSA de Rouen, Campus Universitaire du Madrillet, F-76800 Saint-Etienne du Rouvray, France
ADIR Association, Hôpital de Bois-Guillaume, F-76031 Rouen, France

Sid Visser
School of Mathematical Sciences, University of Nottingham, Nottingham, UK
Department of Applied Mathematics, University of Twente, Enschede, The Netherlands
MIRA—Institute for Biomedical Technology and Technical Medicine, University of Twente, Enschede, The Netherlands

Stephan A Van Gils
Department of Applied Mathematics, University of Twente, Enschede, The Netherlands
MIRA—Institute for Biomedical Technology and Technical Medicine, University of Twente, Enschede, The Netherlands

Pawel Stepien
Nalecz Institute of Biocybernetics and Biomedical Engineering PAS, Warsaw, Poland. Institute of Experimental Medicine NWB RAMS, St. Petersburg, Russia

Wlodzimierz Klonowski
Nalecz Institute of Biocybernetics and Biomedical Engineering PAS, Warsaw, Poland

Suvorov
Institute of Experimental Medicine NWB RAMS, St. Petersburg, Russia

Pawel Stepien
Nalecz Institute of Biocybernetics and Biomedical Engineering PAS, Warsaw, Poland

Peter beim Graben
Dept. of German Studies and Linguistics, Humboldt-Universität zu Berlin, Unter den Linden 6, 10099 Berlin, Germany
Team Neurosys, INRIA Grand Est - Nancy, 615 rue du Jardin Botanique, 54602 Villers-les-Nancy, France

Axel Hutt
Bernstein Center for Computational Neuroscience Berlin, Berlin, Germany

Irena Cosic
RMIT University, Melbourne 3000, VIC, Australia
AMALNA Consulting, 46 Second St, Black Rock 3193, VIC, Australia

Drasko Cosic
AMALNA Consulting, 46 Second St, Black Rock 3193, VIC, Australia

Katarina Lazar
AMALNA Consulting, 46 Second St, Black Rock 3193, VIC, Australia

Richard A Lerski
Medical Research Institute, Ninewells Hospital, Dundee, Scotland, UK

Jacques D de Certaines
Institute of Myology, University Hospital Pitié-Salpêtrière, Paris, France

Dorota Duda
Faculty of Computer Science, Bialystok University of Technology, Bialystok, Poland

Wlodzimierz Klonowski
Nalecz Institute of Biocybernetics and Biomedical Engineering Polish Academy of Sciences, Warsaw, Poland

Guanyu Yang
Laboratory of Image Science and Technology, School of Computer Science and Engineering, Southeast University, Nanjing 210096, China
Centre de Recherche en Information Biomédicale Sino-français (CRIBs), Nanjing, China

Jean Louis Coatrieux
INSERM, U1099, Rennes F-35000, France
Université de Rennes 1, LTSI, Rennes F-35042, France
Centre de Recherche en Information Biomédicale Sino-français (CRIBs), Rennes F-35000, France

Noura Azzabou
Institute of Myology, University Hospital Pitié-Salpêtrière, Paris, France

Pierre-Antoine Eliat
PRISM-Bio-SCANs, University of Rennes, Rennes, France

Jacques D De Certaines
Institute of Myology, NMR laboratory, Paris, France
CEA, I2BM, MIRCen, NMR laboratory, Paris, France

Thibaut Larcher
INRA, UMR 703 PAnTher, Oniris, Nantes, France

Dorota Duda
Faculty of Computer Science, Bialystok University of Technology, Bialystok, Poland

Noura Azzabou
Institute of Myology, NMR laboratory, Paris, France
CEA, I2BM, MIRCen, NMR laboratory, Paris, France

Pierre-Antoine Eliat
PRISM-BioSCANs, University of Rennes, Rennes, France

Luis M Escudero
Instituto de Biomedicina, CSCIC Universidad de Sevilla, Sevilla, Spain

Antonio MG Pinheiro
University da Beira Interior, Covilha, Portugal

Guanyu Yang
LIST, Southeast University, Nanjing, China

Jean-Louis Coatrieux
LTSI, INSERM, University of Rennes, Rennes, France

Eduard Snezkho
Lab. Of Mathematical Cybernetics, UIIP, Belorussian Academy of Science, Minsk Belarus

Alexey Shukelovich
Institute of Myology, NMR laboratory, Paris, France

Manuela Pereira
University da Beira Interior, Covilha, Portugal

Richard A Lerski
Ninewells Hosp. and Medical School, Dundee, Scotland, United Kingdom